The Mathematics of Data

IAS/PARK CITY
MATHEMATICS SERIES
Volume 25

The Mathematics of Data

Michael W. Mahoney
John C. Duchi
Anna C. Gilbert
Editors

American Mathematical Society
Institute for Advanced Study
Society for Industrial and Applied Mathematics

Rafe Mazzeo, Series Editor
Michael W. Mahoney, John C. Duchi, and Anna C. Gilbert, Volume Editors.

IAS/Park City Mathematics Institute runs mathematics education programs that bring together high school mathematics teachers, researchers in mathematics and mathematics education, undergraduate mathematics faculty, graduate students, and undergraduates to participate in distinct but overlapping programs of research and education. This volume contains the lecture notes from the Graduate Summer School program

2010 *Mathematics Subject Classification.* Primary 15-02, 52-02, 60-02, 62-02, 65-02, 68-02, 90-02.

Library of Congress Cataloging-in-Publication Data

Names: Mahoney, Michael W., editor. | Duchi, John, editor. | Gilbert, Anna C. (Anna Catherine), 1972– editor. | Institute for Advanced Study (Princeton, N.J.) | Society for Industrial and Applied Mathematics. | Park City Mathematics Institute.

Title: The mathematics of data / Michael W. Mahoney, John C. Duchi, Anna C. Gilbert, editors.

Description: Providence : American Mathematical Society, 2018. | Series: IAS/Park City mathematics series ; Volume 25 | "Institute for Advanced Study." | "Society for Industrial and Applied Mathematics." | Based on a series of lectures held July 2016, at the Park City Mathematics Institute. | Includes bibliographical references.

Identifiers: LCCN 2018024239 | ISBN 9781470435752 (alk. paper)

Subjects: LCSH: Mathematics teachers–Training of–Congresses. | Mathematics–Study and teaching–Congresses | Big data–Congresses. | AMS: Linear and multilinear algebra; matrix theory – Research exposition (monographs, survey articles). msc | Convex and discrete geometry – Research exposition (monographs, survey articles). msc | Probability theory and stochastic processes – Research exposition (monographs, survey articles). msc | Statistics – Research exposition (monographs, survey articles). msc | Numerical analysis – Research exposition (monographs, survey articles). msc | Computer science – Research exposition (monographs, survey articles). msc | Operations research, mathematical programming – Research exposition (monographs, survey articles). msc

Classification: LCC QA11.A1 M345 2018 | DDC 510–dc23

LC record available at https://lccn.loc.gov/2018024239

∞ The paper used in this book is acid-free and falls within the guidelines
established to ensure permanence and durability.
Visit the AMS home page at `https://www.ams.org/`

10 9 8 7 6 5 4 3 2 1   23 22 21 20 19 18

Contents

IAS/Park City Mathematics Series
Volume 25, Pages -7—8
https://doi.org/10.1090/pcms/025/00827

Preface

The IAS/Park City Mathematics Institute (PCMI) was founded in 1991 as part of the Regional Geometry Institute initiative of the National Science Foundation. In mid-1993 the program found an institutional home at the Institute for Advanced Study (IAS) in Princeton, New Jersey.

The IAS/Park City Mathematics Institute encourages both research and education in mathematics and fosters interaction between the two. The three-week summer institute offers programs for researchers and postdoctoral scholars, graduate students, undergraduate students, high school students, undergraduate faculty, K-12 teachers, and international teachers and education researchers. The Teacher Leadership Program also includes weekend workshops and other activities during the academic year.

One of PCMI's main goals is to make all of the participants aware of the full range of activities that occur in research, mathematics training and mathematics education: the intention is to involve professional mathematicians in education and to bring current concepts in mathematics to the attention of educators. To that end, late afternoons during the summer institute are devoted to seminars and discussions of common interest to all participants, meant to encourage interaction among the various groups. Many deal with current issues in education: others treat mathematical topics at a level which encourages broad participation.

Each year the Research Program and Graduate Summer School focuses on a different mathematical area, chosen to represent some major thread of current mathematical interest. Activities in the Undergraduate Summer School and Undergraduate Faculty Program are also linked to this topic, the better to encourage interaction between participants at all levels. Lecture notes from the Graduate Summer School are published each year in this series. The prior volumes are:

- Volume 1: *Geometry and Quantum Field Theory* (1991)
- Volume 2: *Nonlinear Partial Differential Equations in Differential Geometry* (1992)
- Volume 3: *Complex Algebraic Geometry* (1993)
- Volume 4: *Gauge Theory and the Topology of Four-Manifolds* (1994)
- Volume 5: *Hyperbolic Equations and Frequency Interactions* (1995)
- Volume 6: *Probability Theory and Applications* (1996)
- Volume 7: *Symplectic Geometry and Topology* (1997)
- Volume 8: *Representation Theory of Lie Groups* (1998)
- Volume 9: *Arithmetic Algebraic Geometry* (1999)

- Volume 10: *Computational Complexity Theory* (2000)
- Volume 11: *Quantum Field Theory, Supersymmetry, and Enumerative Geometry* (2001)
- Volume 12: *Automorphic Forms and their Applications* (2002)
- Volume 13: *Geometric Combinatorics* (2004)
- Volume 14: *Mathematical Biology* (2005)
- Volume 15: *Low Dimensional Topology* (2006)
- Volume 16: *Statistical Mechanics* (2007)
- Volume 17: *Analytical and Algebraic Geometry* (2008)
- Volume 18: *Arthimetic of L-functions* (2009)
- Volume 19: *Mathematics in Image Processing* (2010)
- Volume 20: *Moduli Spaces of Riemann Surfaces* (2011)
- Volume 21: *Geometric Group Theory* (2012)
- Volume 22: *Geometric Analysis* (2013)
- Volume 23: *Mathematics and Materials* (2014)
- Volume 24: *Geometry of Moduli Spaces and Representation Theory* (2015)

The American Mathematical Society publishes material from the Undergraduate Summer School in their Student Mathematical Library and from the Teacher Leadership Program in the series IAS/PCMI—The Teacher Program.

After more than 25 years, PCMI retains its intellectual vitality and continues to draw a remarkable group of participants each year from across the entire spectrum of mathematics, from Fields Medalists to elementary school teachers.

Rafe Mazzeo
PCMI Director
March 2017

IAS/Park City Mathematics Series
Volume 25, Pages -9—12
https://doi.org/10.1090/pcms/025/00828

Introduction

Michael W. Mahoney, John C. Duchi, and Anna C. Gilbert

"The Mathematics of Data" was the topic for the 26th annual Park City Mathematics Institute (PCMI) summer session, held in July 2016. To those more familiar with very abstract areas of mathematics or more applied areas of data—the latter going these days by names such as "big data" or "data science"—it may come as a surprise that such an area even exists. A moment's thought, however, should dispel such a misconception. After all, data must be modeled, e.g., by a matrix or a graph or a flat table, and if one performs similar operations on very different types of data, then there is an expectation that there must be some sort of common mathematical structure, e.g., from linear algebra or graph theory or logic. So too, ignorance or errors or noise in the data can be modeled, and it should be plausible that how well operations perform on data depend not just on how well data are modeled but also on how well ignorance or noise or errors are modeled. So too, the operations themselves can be modeled, e.g., to make statements such as whether the operations answer a precise question, exactly or approximately, or whether they will return a solution in a reasonable amount of time.

As such, "The Mathematics of Data" fits squarely in applied mathematics—when that term is broadly, not narrowly, defined. Technically, it represents some combination of what is traditionally the domain of linear algebra and probability and optimization and other related areas. Moreover, while some of the work in this area takes place in mathematics departments, much of the work in the area takes place in computer science, statistics, and other related departments. This was the challenge and opportunity we faced, both in designing the graduate summer school portion of the PCMI summer session, as well as in designing this volume. With respect to the latter, while the area is not sufficiently mature to say the final word, we have tried to capture the major trends in the mathematics of data sufficiently broadly and at a sufficiently introductory level that this volume could be used as a teaching resource for students with backgrounds in any of the wide range of areas related to the mathematics of data.

The first chapter, "Lectures on Randomized Numerical Linear Algebra," provides an overview of linear algebra, probability, and ways in which they interact fruitfully in many large-scale data applications. Matrices are a common way to model data, e.g., an $m \times n$ matrix provides a natural way to describe m objects, each of which is described by n features, and thus linear algebra, as well as more sophisticated variants such as functional analysis and linear operator theory, are central to the mathematics of data. An interesting twist is that, while work in numerical linear algebra and scientific computing typically focuses on deterministic algorithms that return answers to machine precision, randomness can be used in novel algorithmic and statistical ways in matrix algorithms for data. While randomness is often assumed to be a property of the data (e.g., think of noise being modeled by random variables drawn from a Gaussian distribution), it can also be a powerful algorithmic resource to speed up algorithms (e.g., think of Monte Carlo and Markov Chain Monte Carlo methods), and many of the most interesting and exciting developments in the mathematics of data explore this algorithmic-statistical interface. This chapter, in particular, describes the use of these methods for the development of improved algorithms for fundamental and ubiquitous matrix problems such as matrix multiplication, least-squares approximation, and low-rank matrix approximation.

The second chapter, "Optimization Algorithms for Data Analysis," goes one step beyond basic linear algebra problems, which themselves are special cases of optimization problems, to consider more general optimization problems. Optimization problems are ubiquitous throughout data science, and a wide class of problems can be formulated as optimizing smooth functions, possibly with simple constraints or structured nonsmooth regularizers. This chapter describes some canonical problems in data analysis and their formulation as optimization problems. It also describes iterative algorithms (i.e., those that generate a sequence of points) that, for convex objective functions, converge to the set of solutions of such problems. Algorithms covered include first-order methods that depend on gradients, so-called accelerated gradient methods, and Newton's second-order method that can guarantee convergence to points that approximately satisfy second-order conditions for a local minimizer of a smooth nonconvex function.

The third chapter, "Introductory Lectures on Stochastic Optimization," covers the basic analytical tools and algorithms necessary for stochastic optimization. Stochastic optimization problems are problems whose definition involves randomness, e.g., minimizing the expectation of some function; and stochastic optimization algorithms are algorithms that generate and use random variables to find the solution of a (perhaps deterministic) problem. As with the use of randomness in Randomized Numerical Linear Algebra, there is an interesting synergy between the two ways in which stochasticity appears. This chapter builds the necessary convex analytic and other background, and it describes gradient and subgradient first-order methods for the solution of these types of problems. These methods tend to be simple methods that are slower to converge than more advanced methods—such as Newton's or other second-order methods—for deterministic problems, but they have the advantage that they can be robust to noise in the optimization problem itself. Also covered are mirror descent and adaptive methods, as well as methods for proving upper and lower bounds on such stochastic algorithms.

The fourth chapter, "Randomized Methods for Matrix Computations," goes into more detail on randomized methods for computing efficiently a low-rank approximation to a given matrix. One often wants to decompose a large $m \times n$ matrix A, where m and n are both large, into two lower-rank more-rectangular matrices E and F such that $A \approx EF$. Examples include low-rank approximations to the eigenvalue decomposition or the singular value decomposition. While low-rank approximation problems of this type form a cornerstone of traditional applied mathematics and scientific computing, they also arise in a broad range of data science applications. Importantly, though, the questions one asks of these matrix decompositions (e.g., whether one is interested in numerical precision or statistical inference objectives) and even how one accesses these matrices (e.g., within the RAM model idealization or in a single-pass streaming setting where the data can't even be stored) are very different. Randomness can be useful in many ways here. This chapter describes randomized algorithms that obtain better worst-case running time, both in the RAM model and a streaming model, how randomness can be used to obtain improved communication properties for algorithms, and also several data-driven decompositions such as the Nyström method, the Interpolative Decomposition, and the CUR decomposition.

The fifth chapter, "Four Lectures on Probabilistic Methods for Data Science," describes modern methods of high dimensional probability and illustrates how these methods can be used in data science. Methods of high-dimensional probability play a central role in applications for statistics, signal processing, theoretical computer science, and related fields. For example, they can be used within a randomized algorithm to obtain improved running time properties, and/or they can be used as random models for data, in which case they are needed to obtain inferential guarantees. Indeed, they are used (explicitly or implicitly) in all of the previous chapters. This chapter presents a sample of particularly useful tools of high-dimensional probability, focusing on the classical and matrix Bernstein's inequality and the uniform matrix deviation inequality, and it illustrates these tools with applications for dimension reduction, network analysis, covariance estimation, matrix completion, and sparse signal recovery.

The sixth and final chapter, "Homological Algebra and Data," provides an example of how methods from more pure mathematics, in this case topology, might be used fruitfully in data science and the mathematics of data, as outlined in the previous chapters. Topology is—informally—the study of shape, and topological data analysis provides a framework to analyze data in a manner that should be insensitive to the particular metric chosen, e.g., to measure the similarity between data points. It involves replacing a set of data points with a family of simplicial complexes, and then using ideas from persistent homology to try to determine the large scale structure of the set. This chapter approaches topological data analysis from the perspective of homological algebra, where homology is an algebraic compression scheme that excises all but the essential topological features from a class of data structures. An important point is that linear algebra can be enriched to cover not merely linear transformations—the 99.9% use case—but also sequences of linear transformations that form complexes, thus opening the possibility of further mathematical developments.

Overall, the 2016 PCMI summer program included minicourses by Petros Drineas, John Duchi, Cynthia Dwork and Kunal Talwar, Robert Ghrist, Piotr Indyk, Mauro Maggioni, Gunnar Martinsson, Roman Vershynin, and Stephen Wright. This volume consists of contributions, summarized above, by Petros Drineas (with Michael Mahoney), Stephen Wright, John Duchi, Gunnar Martinsson, Roman Vershynin, and Robert Ghrist. Each chapter in this volume was written by a different author, and so each chapter has it's own unique style, including notational differences, but we have taken some effort to ensure that they can fruitfully be read together.

Putting together such an effort—both the entire summer session as well as this volume—is not a minor undertaking, but for us it was not difficult, due to the large amount of support we received. We would first like to thank Richard Hain, the former PCMI Program Director, who first invited us to organize the summer school, as well as Rafe Mazzeo, the current PCMI Program Director, who provided seamless guidance throughout the entire process. In terms of running the summer session, a special thank you goes out to the entire PCMI staff, and in particular to Beth Brainard and Dena Vigil as well as Bryna Kra and Michelle Wachs. We received a lot of feedback from participants who enjoyed the event, and Beth and Dena deserve much of the credit for making it run smoothly; and Bryna and Michelle's role with the graduate steering committee helped us throughout the entire process. In terms of this volume, in addition to thanking the authors for their efforts and (usually) getting back to us in a timely manner, we would like to thank Ian Morrison, who is the PCMI Publisher. Putting together a volume such as this can be a tedious task, but for us it was not, and this is in large part due to Ian's help and guidance.

IAS/Park City Mathematics Series
Volume 25, Pages 1–48
https://doi.org/10.1090/pcms/025/00829

Lectures on Randomized Numerical Linear Algebra

Petros Drineas and Michael W. Mahoney

Contents

2010 *Mathematics Subject Classification.* Primary 68W20; Secondary 65Fxx, 62Jxx.
Key words and phrases. Random sampling, random projection, matrix multiplication, least-squares approximation, low-rank matrix approximation, Park City Mathematics Institute.

1. Introduction

Matrices are ubiquitous in computer science, statistics, and applied mathematics. An $m \times n$ matrix can encode information about m objects (each described by n features), or the behavior of a discretized differential operator on a finite element mesh; an $n \times n$ positive-definite matrix can encode the correlations between all pairs of n objects, or the edge-connectivity between all pairs of n nodes in a social network; and so on. Motivated largely by technological developments that generate extremely large scientific and Internet data sets, recent years have witnessed exciting developments in the theory and practice of matrix algorithms. Particularly remarkable is the use of *randomization*—typically assumed to be a property of the input data due to, e.g., noise in the data generation mechanisms—as an algorithmic or computational resource for the development of improved algorithms for fundamental matrix problems such as matrix multiplication, least-squares (LS) approximation, low-rank matrix approximation, etc.

Randomized Numerical Linear Algebra (RandNLA) is an interdisciplinary research area that exploits randomization as a computational resource to develop improved algorithms for large-scale linear algebra problems. From a foundational perspective, RandNLA has its roots in theoretical computer science (TCS), with deep connections to mathematics (convex analysis, probability theory, metric embedding theory) and applied mathematics (scientific computing, signal processing, numerical linear algebra). From an applied perspective, RandNLA is a vital new tool for machine learning, statistics, and data analysis. Well-engineered implementations have already outperformed highly-optimized software libraries for ubiquitous problems such as least-squares regression, with good scalability in parallel and distributed environments. Moreover, RandNLA promises a sound algorithmic and statistical foundation for modern large-scale data analysis.

This chapter serves as a self-contained, gentle introduction to three fundamental RandNLA algorithms: randomized matrix multiplication, randomized least-squares solvers, and a randomized algorithm to compute a low-rank approximation to a matrix. As such, this chapter has strong connections with many areas of applied mathematics, and in particular it has strong connections with several other chapters in this volume. Most notably, this includes that of G. Martinsson, who uses these methods to develop improved low-rank matrix approximation solvers [19]; R. Vershynin, who develops probabilistic tools that are used in the analysis of RandNLA algorithms [28]; and J. Duchi, who uses stochastic and randomized methods in a complementary manner for more general optimization problems [12].

We start this chapter with a review of basic linear algebraic facts in Section 2; we review basic facts from discrete probability in Section 3; we present a randomized algorithm for matrix multiplication in Section 4; we present a randomized algorithm for least-squares regression problems in Section 5; and finally we present a randomized algorithm for low-rank approximation in Section 6. We conclude this introduction by noting that [10, 17] might also be of interest to a reader who wants to go through other introductory texts on RandNLA.

2. Linear Algebra

In this section, we present a brief overview of basic linear algebraic facts and notation that will be useful in this chapter. We assume basic familiarity with linear algebra (e.g., inner/outer products of vectors, basic matrix operations such as addition, scalar multiplication, transposition, upper/lower triangular matrices, matrix-vector products, matrix multiplication, matrix trace, etc.).

2.1. Basics. We will entirely focus on matrices and vectors over the *reals*. We will use the notation $\mathbf{x} \in \mathbb{R}^n$ to denote an n-dimensional vector: notice the use of bold latin *lowercase* letters for vectors. Vectors will always be assumed to be column vectors, unless explicitly noted otherwise. The vector of all zeros will be denoted as $\mathbf{0}$, while the vector of all ones will be denoted as $\mathbf{1}$; dimensions will be implied from context or explicitly included as a subscript.

We will use bold latin *uppercase* letters for matrices, e.g., $\mathbf{A} \in \mathbb{R}^{m \times n}$ denotes an $m \times n$ matrix $\mathbf{A}$. We will use the notation $\mathbf{A}_{i*}$ to denote the i-th row of $\mathbf{A}$ as a row vector and $\mathbf{A}_{*i}$ to denote the i-th column of $\mathbf{A}$ as a column vector. The (square) identity matrix will be denoted as $\mathbf{I}_n$ where n denotes the number of rows and columns. Finally, we use $\mathbf{e}_i$ to denote the i-th column of $\mathbf{I}_n$, i.e., the i-th *canonical* vector.

Matrix Inverse. A matrix $\mathbf{A} \in \mathbb{R}^{n \times n}$ is nonsingular or invertible if there exists a matrix $\mathbf{A}^{-1} \in \mathbb{R}^{n \times n}$ such that

$$\mathbf{A}\mathbf{A}^{-1} = \mathbf{I}_{n \times n} = \mathbf{A}^{-1}\mathbf{A}.$$

The inverse exists when all the columns (or all the rows) of $\mathbf{A}$ are linearly independent. In other words, there does not exist a non-zero vector $\mathbf{x} \in \mathbb{R}^n$ such that $\mathbf{A}\mathbf{x} = \mathbf{0}$. Standard properties of the inverse include: $(\mathbf{A}^{-1})^\top = (\mathbf{A}^\top)^{-1} = \mathbf{A}^{-\top}$ and $(\mathbf{A}\mathbf{B})^{-1} = \mathbf{B}^{-1}\mathbf{A}^{-1}$.

Orthogonal matrix. A matrix $\mathbf{A} \in \mathbb{R}^{n \times n}$ is orthogonal if $\mathbf{A}^\top = \mathbf{A}^{-1}$. Equivalently, for all i and j between one and n,

$$\mathbf{A}_{*i}^\top \mathbf{A}_{*j} = \begin{cases} 0, & \text{if } i \neq j \\ 1, & \text{if } i = j \end{cases}.$$

The same property holds for the rows of $\mathbf{A}$. In words, the columns (rows) of $\mathbf{A}$ are pairwise orthogonal and normal vectors.

QR Decomposition. Any matrix $\mathbf{A} \in \mathbb{R}^{n \times n}$ can be decomposed into the product of an orthogonal matrix and an upper triangular matrix as:

$$\mathbf{A} = \mathbf{Q}\mathbf{R},$$

where $\mathbf{Q} \in \mathbb{R}^{n \times n}$ is an orthogonal matrix and $\mathbf{R} \in \mathbb{R}^{n \times n}$ is an upper triangular matrix. The QR decomposition is useful in solving systems of linear equations, has computational complexity $O(n^3)$, and is numerically stable. To solve the linear system $\mathbf{A}\mathbf{x} = \mathbf{b}$ using the QR decomposition we first premultiply both sides by $\mathbf{Q}^\top$, thus getting $\mathbf{Q}^\top\mathbf{Q}\mathbf{R}\mathbf{x} = \mathbf{R}\mathbf{x} = \mathbf{Q}^\top\mathbf{b}$. Then, we solve $\mathbf{R}\mathbf{x} = \mathbf{Q}^\top\mathbf{b}$ using backward substitution [13].

2.2. Norms. Norms are used to measure the size or mass of a matrix or, relatedly, the length of a vector. They are functions that map an object from $\mathbb{R}^{m \times n}$ (or $\mathbb{R}^n$) to $\mathbb{R}$. Formally:

Definition 2.2.1. Any function, $\|\cdot\|$: $\mathbb{R}^{m \times n} \to \mathbb{R}$ that satisfies the following properties is called a **norm**:

(1) Non-negativity: $\|\mathbf{A}\| \geqslant 0$; $\|\mathbf{A}\| = 0$ if and only if $\mathbf{A} = \mathbf{0}$.
(2) Triangle inequality: $\|\mathbf{A} + \mathbf{B}\| \leqslant \|\mathbf{A}\| + \|\mathbf{B}\|$.
(3) Scalar multiplication: $\|\alpha\mathbf{A}\| = |\alpha|\|\mathbf{A}\|$, for all $\alpha \in \mathbb{R}$.

The following properties are easy to prove for any norm: $\|-\mathbf{A}\| = \|\mathbf{A}\|$ and

$$|\|\mathbf{A}\| - \|\mathbf{B}\|| \leqslant \|\mathbf{A} - \mathbf{B}\|.$$

The latter property is known as the reverse triangle inequality.

2.3. Vector norms. Given $\mathbf{x} \in \mathbb{R}^n$ and an integer $p \geqslant 1$, we define the vector p-norm as:

$$\|\mathbf{x}\|_p = \left(\sum_{i=1}^{n} |x_i|^p\right)^{1/p}.$$

The most common vector p-norms are:

- One norm: $\|\mathbf{x}\|_1 = \sum_{i=1}^{n} |x_i|$.
- Euclidean (two) norm: $\|\mathbf{x}\|_2 = \sqrt{\sum_{i=1}^{n} |x_i|^2} = \sqrt{\mathbf{x}^\top \mathbf{x}}$.
- Infinity (max) norm: $\|\mathbf{x}\|_\infty = \max_{1 \leqslant i \leqslant n} |x_i|$.

Given $\mathbf{x}, \mathbf{y} \in \mathbb{R}^n$ we can bound the inner product $\mathbf{x}^\top \mathbf{y} = \sum_{i=1}^{n} x_i y_i$ using p-norms. The Cauchy-Schwartz inequality states that:

$$|\mathbf{x}^\top \mathbf{y}| \leqslant \|\mathbf{x}\|_2 \|\mathbf{y}\|_2.$$

In words, it gives an upper bound for the inner product of two vectors in terms of the Euclidean norm of the two vectors. Hölder's inequality states that

$$|\mathbf{x}^\top \mathbf{y}| \leqslant \|\mathbf{x}\|_1 \|\mathbf{y}\|_\infty \quad \text{and} \quad |\mathbf{x}^\top \mathbf{y}| \leqslant \|\mathbf{x}\|_\infty \|\mathbf{y}\|_1.$$

The following inequalities between common vector p-norms are easy to prove:

$$\begin{aligned} \|\mathbf{x}\|_\infty &\leqslant \|\mathbf{x}\|_1 \leqslant n\|\mathbf{x}\|_\infty, \\ \|\mathbf{x}\|_2 &\leqslant \|\mathbf{x}\|_1 \leqslant \sqrt{n}\|\mathbf{x}\|_2, \\ \|\mathbf{x}\|_\infty &\leqslant \|\mathbf{x}\|_2 \leqslant \sqrt{n}\|\mathbf{x}\|_\infty. \end{aligned}$$

Also, $\|\mathbf{x}\|_2^2 = \mathbf{x}^T\mathbf{x}$. We can now define the notion of orthogonality for a pair of vectors and state the Pythagorean theorem.

Theorem 2.3.1. Two vectors $\mathbf{x}, \mathbf{y} \in \mathbb{R}^n$ are orthogonal, i.e., $\mathbf{x}^\top \mathbf{y} = 0$, if and only if

$$\|\mathbf{x} \pm \mathbf{y}\|_2^2 = \|\mathbf{x}\|_2^2 + \|\mathbf{y}\|_2^2.$$

Theorem 2.3.1 is also known as the Pythagorean Theorem. Another interesting property of the Euclidean norm is that it does not change after pre(post)-multiplication by a matrix with orthonormal columns (rows).

Theorem 2.3.2. Given a vector $\mathbf{x} \in \mathbb{R}^n$ and a matrix $\mathbf{V} \in \mathbb{R}^{m \times n}$ with $m \geqslant n$ and $\mathbf{V}^\top \mathbf{V} = \mathbf{I}_n$:

$$\|\mathbf{V}\mathbf{x}\|_2 = \|\mathbf{x}\|_2 \quad \text{and} \quad \|\mathbf{x}^T \mathbf{V}^T\|_2 = \|\mathbf{x}\|_2.$$

2.4. Induced matrix norms. Given a matrix $\mathbf{A} \in \mathbb{R}^{m \times n}$ and an integer $p \geqslant 1$ we define the matrix p-norm as:

$$\|\mathbf{A}\|_p = \max_{\mathbf{x} \neq 0} \frac{\|\mathbf{A}\mathbf{x}\|_p}{\|\mathbf{x}\|_p} = \max_{\|\mathbf{x}\|_p = 1} \|\mathbf{A}\mathbf{x}\|_p.$$

The most frequent matrix p-norms are:

- One norm: the maximum absolute column sum,

$$\|\mathbf{A}\|_1 = \max_{1 \leqslant j \leqslant n} \sum_{i=1}^{m} |\mathbf{A}_{ij}| = \max_{1 \leqslant j \leqslant n} \|\mathbf{A}\mathbf{e}_j\|_1.$$

- Infinity norm: the maximum absolute row sum,

$$\|\mathbf{A}\|_\infty = \max_{1 \leqslant i \leqslant m} \sum_{j=1}^{n} |\mathbf{A}_{ij}| = \max_{1 \leqslant j \leqslant m} \|\mathbf{A}^\top \mathbf{e}_i\|_1.$$

- Two (or spectral) norm :

$$\|\mathbf{A}\|_2 = \max_{\|\mathbf{x}\|_2=1} \|\mathbf{A}\mathbf{x}\|_2 = \max_{\|\mathbf{x}\|_2=1} \sqrt{\mathbf{x}^\top \mathbf{A}^\top \mathbf{A}\mathbf{x}}.$$

This family of norms is named "induced" because they are realized by a non-zero vector $\mathbf{x}$ that varies depending on $\mathbf{A}$ and p. Thus, there exists a unit norm vector (unit norm in the p-norm) $\mathbf{x}$ such that $\|\mathbf{A}\|_p = \|\mathbf{A}\mathbf{x}\|_p$. The induced matrix p-norms follow the submultiplicativity laws:

$$\|\mathbf{A}\mathbf{x}\|_p \leqslant \|\mathbf{A}\|_p \|\mathbf{x}\|_p \qquad \text{and} \qquad \|\mathbf{A}\mathbf{B}\|_p \leqslant \|\mathbf{A}\|_p \|\mathbf{B}\|_p.$$

Furthermore, matrix p-norms are invariant to permutations: $\|\mathbf{P}\mathbf{A}\mathbf{Q}\|_p = \|\mathbf{A}\|_p$, where $\mathbf{P}$ and $\mathbf{Q}$ are permutation matrices of appropriate dimensions. Also, if we consider the matrix with permuted rows and columns

$$\mathbf{P}\mathbf{A}\mathbf{Q} = \begin{pmatrix} \mathbf{B} & \mathbf{A}_{12} \\ \mathbf{A}_{21} & \mathbf{A}_{22} \end{pmatrix},$$

then the norm of the submatrix is related to the norm of the full unpermuted matrix as follows: $\|\mathbf{B}\|_p \leqslant \|\mathbf{A}\|_p$. The following relationships between matrix p-norms are relatively easy to prove. Given a matrix $\mathbf{A} \in \mathbb{R}^{m \times n}$,

$$\frac{1}{\sqrt{n}}\|\mathbf{A}\|_\infty \leqslant \|\mathbf{A}\|_2 \leqslant \sqrt{m}\|\mathbf{A}\|_\infty,$$

$$\frac{1}{\sqrt{m}}\|\mathbf{A}\|_1 \leqslant \|\mathbf{A}\|_2 \leqslant \sqrt{n}\|\mathbf{A}\|_1.$$

It is also the case that $\|\mathbf{A}^\top\|_1 = \|\mathbf{A}\|_\infty$ and $\|\mathbf{A}^\top\|_\infty = \|\mathbf{A}\|_1$. While transposition affects the infinity and one norm of a matrix, it does not affect the two norm, i.e., $\|\mathbf{A}^\top\|_2 = \|\mathbf{A}\|_2$. Also, the matrix two-norm is not affected by pre-(or post-) multiplication with matrices whose columns (or rows) are orthonormal vectors: $\|\mathbf{U}\mathbf{A}\mathbf{V}^\top\|_2 = \|\mathbf{A}\|_2$, where $\mathbf{U}$ and $\mathbf{V}$ are orthonormal matrices ($\mathbf{U}^\mathsf{T}\mathbf{U} = \mathbf{I}$ and $\mathbf{V}^\mathsf{T}\mathbf{V} = \mathbf{I}$) of appropriate dimensions.

2.5. The Frobenius norm. The Frobenius norm is not an induced norm, as it belongs to the family of Schatten norms (to be discussed in Section 2.8).

Definition 2.5.1. Given a matrix $\mathbf{A} \in \mathbb{R}^{m \times n}$, we define the Frobenius norm as:

$$\|\mathbf{A}\|_F = \sqrt{\sum_{j=1}^{n}\sum_{i=1}^{m} \mathbf{A}_{ij}^2} = \sqrt{\mathbf{Tr}\left(\mathbf{A}^\top \mathbf{A}\right)},$$

where $\mathbf{Tr}(\cdot)$ denotes the matrix trace (where, recall, the trace of a square matrix is defined to be the sum of the elements on the main diagonal).

Informally, the Frobenius norm measures the variance or variability (which can be given an interpretation of size or mass) of a matrix. Given a vector $\mathbf{x} \in \mathbb{R}^n$, its Frobenius norm is equal to its Euclidean norm, i.e., $\|\mathbf{x}\|_F = \|\mathbf{x}\|_2$. Transposition of a matrix $\mathbf{A} \in \mathbb{R}^{m \times n}$ does not affect its Frobenius norm, i.e., $\|\mathbf{A}\|_F = \|\mathbf{A}^\top\|_F$. Similar to the two norm, the Frobenius norm does not change under permutations or under pre(post)- multiplication with any matrix with orthonormal columns (rows):

$$\|\mathbf{U}\mathbf{A}\mathbf{V}^\top\|_F = \|\mathbf{A}\|_F,$$

where $\mathbf{U}$ and $\mathbf{V}$ are orthonormal matrices ($\mathbf{U}^\top\mathbf{U} = \mathbf{I}$ and $\mathbf{V}^\top\mathbf{V} = \mathbf{I}$) of appropriate dimensions. The two and the Frobenius norm can be related by:

$$\|\mathbf{A}\|_2 \leqslant \|\mathbf{A}\|_F \leqslant \sqrt{\mathrm{rank}(\mathbf{A})}\|\mathbf{A}\|_2 \leqslant \sqrt{\min\{m,n\}}\|\mathbf{A}\|_2.$$

The Frobenius norm satisfies the so-called strong sub-multiplicativity property, namely:

$$\|\mathbf{A}\mathbf{B}\|_F \leqslant \|\mathbf{A}\|_2\|\mathbf{B}\|_F \quad \text{and} \quad \|\mathbf{A}\mathbf{B}\|_F \leqslant \|\mathbf{A}\|_F\|\mathbf{B}\|_2.$$

For any vectors $\mathbf{x} \in \mathbb{R}^m$ and $\mathbf{y} \in \mathbb{R}^n$, the Frobenius norm of their outer product is equal to the product of the Euclidean norms of the two vectors forming the outer product:

$$\|\mathbf{x}\mathbf{y}^\top\|_F = \|\mathbf{x}\|_2\|\mathbf{y}\|_2.$$

Finally, we state a matrix version of the Pythagorean theorem.

Lemma 2.5.2 (Matrix Pythagoras). Let $\mathbf{A}, \mathbf{B} \in \mathbb{R}^{m\times n}$. If $\mathbf{A}^\top\mathbf{B} = \mathbf{0}$ then

$$\|\mathbf{A}+\mathbf{B}\|_F^2 = \|\mathbf{A}\|_F^2 + \|\mathbf{B}\|_F^2.$$

2.6. The Singular Value Decomposition. The Singular Value Decomposition (SVD) is the most important matrix decomposition and exists for every matrix.

Definition 2.6.1. Given a matrix $\mathbf{A} \in \mathbb{R}^{m\times n}$, we define its full SVD as:

$$\mathbf{A} = \mathbf{U}\mathbf{\Sigma}\mathbf{V}^\top = \sum_{i=1}^{\min\{m,n\}} \sigma_i \mathbf{u}_i \mathbf{v}_i^\top,$$

where $\mathbf{U} \in \mathbb{R}^{m\times m}$ and $\mathbf{V} \in \mathbb{R}^{n\times n}$ are orthogonal matrices that contain the left and right singular vectors of $\mathbf{A}$, respectively, and $\mathbf{\Sigma} \in \mathbb{R}^{m\times n}$ is a diagonal matrix, with the singular values of $\mathbf{A}$ in decreasing order on the diagonal.

We will often use $\mathbf{u}_i$ (respectively, $\mathbf{v}_j$), $i = 1, \ldots, m$ (respectively, $j = 1, \ldots, n$) to denote the columns of the matrix $\mathbf{U}$ (respectively, $\mathbf{V}$). Similarly, we will use σ_i, $i = 1, \ldots, \min\{m, n\}$ to denote the singular values:

$$\sigma_1 \geqslant \sigma_2 \geqslant \cdots \geqslant \sigma_{\min\{m,n\}} \geqslant 0.$$

The singular values of $\mathbf{A}$ are non-negative and their number is equal to $\min\{m, n\}$. The number of non-zero singular values of $\mathbf{A}$ is equal to the rank of $\mathbf{A}$. Due to orthonormal invariance, we get:

$$\mathbf{\Sigma}_{\mathbf{P}\mathbf{A}\mathbf{Q}^\top} = \mathbf{\Sigma}_{\mathbf{A}},$$

where $\mathbf{P}$ and $\mathbf{Q}$ are orthonormal matrices ($\mathbf{P}^\top\mathbf{P} = \mathbf{I}$ and $\mathbf{Q}^\top\mathbf{Q} = \mathbf{I}$) of appropriate dimensions. In words, the singular values of $\mathbf{PAQ}$ are the same as the singular values of $\mathbf{A}$. The following inequalities involving the singular values of the matrices $\mathbf{A}$ and $\mathbf{B}$ are important. First, if both $\mathbf{A}$ and $\mathbf{B}$ are in $\mathbb{R}^{m\times n}$, for all

$i = 1, \ldots, \min\{m, n\}$,

$$|\sigma_i(\mathbf{A}) - \sigma_i(\mathbf{B})| \leqslant \|\mathbf{A} - \mathbf{B}\|_2. \tag{2.6.2}$$

Second, if $\mathbf{A} \in \mathbb{R}^{p \times m}$ and $\mathbf{B} \in \mathbb{R}^{m \times n}$, for all $i = 1, \ldots, \min\{m, n\}$,

$$\sigma_i(\mathbf{AB}) \leqslant \sigma_1(\mathbf{A})\sigma_i(\mathbf{B}), \tag{2.6.3}$$

where, recall, $\sigma_1(\mathbf{A}) = \|\mathbf{A}\|_2$. We are often interested in keeping only the non-zero singular values and the corresponding left and right singular vectors of a matrix $\mathbf{A}$. Given a matrix $\mathbf{A} \in \mathbb{R}^{m \times n}$ with $\text{rank}(\mathbf{A}) = \rho$, its thin SVD can be defined as follows.

Definition 2.6.4. Given a matrix $\mathbf{A} \in \mathbb{R}^{m \times n}$ of rank $\rho \leqslant \min\{m, n\}$, we define its thin SVD as:

$$\mathbf{A} = \underbrace{\mathbf{U}}_{m \times \rho} \underbrace{\mathbf{\Sigma}}_{\rho \times \rho} \underbrace{\mathbf{V}^\top}_{\rho \times n} = \sum_{i=1}^{\rho} \sigma_i \mathbf{u}_i \mathbf{v}_i^\top,$$

where $\mathbf{U} \in \mathbb{R}^{m \times \rho}$ and $\mathbf{V} \in \mathbb{R}^{n \times \rho}$ are matrices with pairwise orthonormal columns (i.e., $\mathbf{U}^\top\mathbf{U} = \mathbf{I}$ and $\mathbf{V}^\top\mathbf{V} = \mathbf{I}$) that contain the left and right singular vectors of $\mathbf{A}$ corresponding to the non-zero singular values; $\mathbf{\Sigma} \in \mathbb{R}^{\rho \times \rho}$ is a diagonal matrix with the non-zero singular values of $\mathbf{A}$ in decreasing order on the diagonal.

If $\mathbf{A}$ is a nonsingular matrix, we can compute its inverse using the SVD:

$$\mathbf{A}^{-1} = (\mathbf{U\Sigma V}^\top)^{-1} = \mathbf{V\Sigma}^{-1}\mathbf{U}^\top.$$

(If $\mathbf{A}$ is nonsingular, then it is square and full rank, in which case the thin SVD is the same as the full SVD.) The SVD is so important since, as is well-known, the best rank-k approximation to any matrix can be computed via the SVD.

Theorem 2.6.5. Let $\mathbf{A} = \mathbf{U\Sigma V}^\top \in \mathbb{R}^{m \times n}$ be the thin SVD of $\mathbf{A}$; let k be an integer less than $\rho = \text{rank}(\mathbf{A})$; and let $\mathbf{A}_k = \sum_{i=1}^{k} \sigma_i \mathbf{u}_i \mathbf{v}_i^\top = \mathbf{U}_k \mathbf{\Sigma}_k \mathbf{V}_k^\top$. Then,

$$\sigma_{k+1} = \min_{\mathbf{B} \in \mathbb{R}^{m \times n},\ \text{rank}(\mathbf{B}) = k} \|\mathbf{A} - \mathbf{B}\|_2 = \|\mathbf{A} - \mathbf{A}_k\|_2$$

and

$$\sum_{j=k+1}^{\rho} \sigma_j^2 = \min_{\mathbf{B} \in \mathbb{R}^{m \times n},\ \text{rank}(\mathbf{B}) = k} \|\mathbf{A} - \mathbf{B}\|_F^2 = \|\mathbf{A} - \mathbf{A}_k\|_F^2.$$

In words, the above theorem states that if we seek a rank k approximation to a matrix $\mathbf{A}$ that minimizes the two or the Frobenius norm of the "error" matrix, i.e., of the difference between $\mathbf{A}$ and its approximation, then it suffices to keep the top k singular values of $\mathbf{A}$ and the corresponding left and right singular vectors.

We will often use the following notation: let $\mathbf{U}_k \in \mathbb{R}^{m \times k}$ and $\mathbf{V}_k \in \mathbb{R}^{n \times k}$ denote the matrices of the top k left and top k right singular vectors of $\mathbf{A}$; and let $\mathbf{\Sigma}_k \in \mathbb{R}^{k \times k}$ denote the diagonal matrix containing the top k singular values of $\mathbf{A}$. Similarly, let $\mathbf{U}_{k,\perp} \in \mathbb{R}^{m \times (\rho - k)}$ (respectively, $\mathbf{V}_{k,\perp} \in \mathbb{R}^{n \times (\rho - k)}$) denote the matrix of the bottom $\rho - k$ nonzero left (respectively, right) singular vectors of $\mathbf{A}$; and let $\mathbf{\Sigma}_{k,\perp} \in \mathbb{R}^{(\rho-k)\times(\rho-k)}$ denote the diagonal matrix containing the bottom $\rho - k$ singular values of $\mathbf{A}$. Then,

$$\mathbf{A}_k = \mathbf{U}_k \mathbf{\Sigma}_k \mathbf{V}_k^T \quad \text{and} \quad \mathbf{A}_{k,\perp} = \mathbf{A} - \mathbf{A}_k = \mathbf{U}_{k,\perp}\mathbf{\Sigma}_{k,\perp}\mathbf{V}_{k,\perp}^T. \tag{2.6.6}$$

2.7. SVD and Fundamental Matrix Spaces. Any matrix $\mathbf{A} \in \mathbb{R}^{m\times n}$ defines four fundamental spaces:

The Column Space of A: This space is spanned by the columns of $\mathbf{A}$:

$$\text{range}(\mathbf{A}) = \{\mathbf{b} : \mathbf{A}\mathbf{x} = \mathbf{b}, \quad \mathbf{x} \in \mathbb{R}^n\} \subset \mathbb{R}^m.$$

The Null Space of A: This space is spanned by all vectors $\mathbf{x} \in \mathbb{R}^n$ such that $\mathbf{A}\mathbf{x} = \mathbf{0}$:

$$\text{null}(\mathbf{A}) = \{\mathbf{x} : \mathbf{A}\mathbf{x} = \mathbf{0}\} \subset \mathbb{R}^n.$$

The Row Space of A: This space is spanned by the rows of $\mathbf{A}$:

$$\text{range}(\mathbf{A}^\top) = \{\mathbf{d} : \mathbf{A}^\top\mathbf{y} = \mathbf{d}, \quad \mathbf{y} \in \mathbb{R}^m\} \subset \mathbb{R}^n.$$

The Left Null Space of A: This space is spanned by all vectors $\mathbf{y} \in \mathbb{R}^m$ such that $\mathbf{A}^\top\mathbf{y} = \mathbf{0}$:

$$\text{null}(\mathbf{A}^\top) = \{\mathbf{y} : \mathbf{A}^\top\mathbf{y} = \mathbf{0}\} \subset \mathbb{R}^m.$$

The SVD reveals orthogonal bases for all these spaces. Given a matrix $\mathbf{A} \in \mathbb{R}^{m\times n}$, with $\text{rank}(\mathbf{A}) = \rho$, its SVD can be written as:

$$\mathbf{A} = \begin{pmatrix} \mathbf{U}_\rho & \mathbf{U}_{\rho,\perp} \end{pmatrix} \begin{pmatrix} \mathbf{\Sigma}_\rho & \mathbf{0} \\ \mathbf{0} & \mathbf{0} \end{pmatrix} \begin{pmatrix} \mathbf{V}_\rho^\top \\ \mathbf{V}_{\rho,\perp}^\top \end{pmatrix}.$$

It is easy to prove that:

$$\begin{aligned} \text{range}(\mathbf{A}) &= \text{range}(\mathbf{U}_\rho), \\ \text{null}(\mathbf{A}) &= \text{range}(\mathbf{V}_{\rho,\perp}), \\ \text{range}(\mathbf{A}^\top) &= \text{range}(\mathbf{V}_\rho), \\ \text{null}(\mathbf{A}^\top) &= \text{range}(\mathbf{U}_{\rho,\perp}). \end{aligned}$$

Theorem 2.7.1 (Basic Theorem of Linear Algebra.)**.** The column space of $\mathbf{A}$ is orthogonal to the null space of $\mathbf{A}^\top$ and their union is $\mathbb{R}^m$. The column space of $\mathbf{A}^\top$ is orthogonal to the null space of $\mathbf{A}$ and their union is $\mathbb{R}^n$.

2.8. Matrix Schatten norms. The matrix Schatten norms are a special family of norms that are defined on the vector containing the singular values of a matrix. Given a matrix $\mathbf{A} \in \mathbb{R}^{m\times n}$ with singular values $\sigma_1 \geqslant \cdots \geqslant \sigma_\rho > 0$, we define the Schatten p-norm as:

$$\|\mathbf{A}\|_p = \left(\sum_{i=1}^{\rho} \sigma_i^p\right)^{\frac{1}{p}}.$$

Common Schatten norms of a matrix $\mathbf{A} \in \mathbb{R}^{m\times n}$ are:

Schatten one-norm: The nuclear norm, i.e., the sum of the singular values.
Schatten two-norm: The Frobenius norm, i.e., the square root of the sum of the squares of the singular values.
Schatten infinity-norm: The spectral norm, defined as the limit as $p \to \infty$ of the Schatten p-norm, i.e., the largest singular value.

Schatten norms are orthogonally invariant and submultiplicative, and they satisfy Hölder's inequality.

2.9. The Moore-Penrose pseudoinverse. A generalization of the well-known notion of matrix inverse is the Moore-Penrose pseudoinverse. Formally, given a matrix $\mathbf{A} \in \mathbb{R}^{m \times n}$, a matrix $\mathbf{A}^\dagger$ is the Moore Penrose pseudoinverse of $\mathbf{A}$ if it satisfies the following properties:

(1) $\mathbf{A}\mathbf{A}^\dagger\mathbf{A} = \mathbf{A}$.
(2) $\mathbf{A}^\dagger\mathbf{A}\mathbf{A}^\dagger = \mathbf{A}^\dagger$.
(3) $(\mathbf{A}\mathbf{A}^\dagger)^\top = \mathbf{A}\mathbf{A}^\dagger$.
(4) $(\mathbf{A}^\dagger\mathbf{A})^\top = \mathbf{A}^\dagger\mathbf{A}$.

Given a matrix $\mathbf{A} \in \mathbb{R}^{m \times n}$ of rank ρ and its thin SVD

$$\mathbf{A} = \sum_{i=1}^{\rho} \sigma_i \mathbf{u}_i \mathbf{v}_i^\top,$$

its Moore-Penrose pseudoinverse $\mathbf{A}^\dagger$ is

$$\mathbf{A}^\dagger = \sum_{i=1}^{\rho} \frac{1}{\sigma_i} \mathbf{v}_i \mathbf{u}_i^\top.$$

If a matrix $\mathbf{A} \in \mathbb{R}^{n \times n}$ has full rank, then $\mathbf{A}^\dagger = \mathbf{A}^{-1}$. If a matrix $\mathbf{A} \in \mathbb{R}^{m \times n}$ has full column rank, then $\mathbf{A}^\dagger\mathbf{A} = \mathbf{I}_n$, and $\mathbf{A}\mathbf{A}^\dagger$ is a projection matrix onto the column span of $\mathbf{A}$; while if it has full row rank, then $\mathbf{A}\mathbf{A}^\dagger = \mathbf{I}_m$, and $\mathbf{A}^\dagger\mathbf{A}$ is a projection matrix onto the row span of $\mathbf{A}$.

A particularly important property regarding the pseudoinverse of the product of two matrices is the following: for matrices $\mathbf{Y}_1 \in \mathbb{R}^{m \times p}$ and $\mathbf{Y}_2 \in \mathbb{R}^{p \times n}$, satisfying $\text{rank}(\mathbf{Y}_1) = \text{rank}(\mathbf{Y}_2)$, [5, Theorem 2.2.3] states that

$$(\mathbf{Y}_1\mathbf{Y}_2)^\dagger = \mathbf{Y}_2^\dagger\mathbf{Y}_1^\dagger. \tag{2.9.1}$$

(We emphasize that the condition on the ranks is crucial: while the inverse of the product of two matrices always equals the product of the inverses of those matrices, the analogous statement is not true in full generality for the Moore-Penrose pseudoinverse [5].)

The fundamental spaces of the Moore-Penrose pseudoinverse are connected with those of the actual matrix. Given a matrix $\mathbf{A}$ and its Moore-Penrose pseudoinverse $\mathbf{A}^\dagger$, the column space of $\mathbf{A}^\dagger$ can be defined as:

$$\text{range}(\mathbf{A}^\dagger) = \text{range}(\mathbf{A}^\top\mathbf{A}) = \text{range}(\mathbf{A}^\top),$$

and it is orthogonal to the null space of $\mathbf{A}$. The null space of $\mathbf{A}^\dagger$ can be defined as:

$$\text{null}(\mathbf{A}^\dagger) = \text{null}(\mathbf{A}\mathbf{A}^\top) = \text{null}(\mathbf{A}^\top),$$

and it is orthogonal to the column space of $\mathbf{A}$.

2.10. References. We refer the interested reader to [5,13,26,27] for additional background on linear algebra and matrix computations, as well as to [4,25] for additional background on matrix perturbation theory.

3. Discrete Probability

In this section, we present a brief overview of discrete probability. More advanced results (in particular, Bernstein-type inequalities for real-valued and matrix-valued random variables) will be introduced in the appropriate context later in the chapter. It is worth noting that most of RandNLA builds upon simple, fundamental principles of discrete (instead of continuous) probability.

3.1. Random experiments: basics. A random experiment is any procedure that can be infinitely repeated and has a well-defined set of possible outcomes. Typical examples are the roll of a dice or the toss of a coin. The sample space Ω of a random experiment is the set of all possible outcomes of the random experiment. If the random experiment only has two possible outcomes (e.g., success and failure) then it is often called a Bernoulli trial. In discrete probability, the sample space Ω is finite. (We will *not* cover countably or uncountably infinite sample spaces in this chapter.)

An event is any subset of the sample space Ω. Clearly, the set of all possible events is the powerset (the set of all possible subsets) of Ω, often denoted as 2^Ω. As an example, consider the following random experiment: toss a coin three times. Then, the sample space Ω is

$$\Omega = \{\text{HHH}, \text{HHT}, \text{HTH}, \text{HTT}, \text{THH}, \text{THT}, \text{TTH}, \text{TTT}\}$$

and an event $\mathcal{E}$ could be described in words as "the output of the random experiment was either all heads or all tails". Then, $\mathcal{E} = \{\text{HHH}, \text{TTT}\}$. The *probability measure* or *probability function* maps the (finite) sample space Ω to the interval $[0,1]$. Formally, let the function $\mathbf{Pr}[\omega]$ for all $\omega \in \Omega$ be a function whose domain is Ω and whose range is the interval $[0,1]$. This function has the so-called normalization property, namely

$$\sum_{\omega \in \Omega} \mathbf{Pr}[\omega] = 1.$$

If $\mathcal{E}$ is an event, then

$$\mathbf{Pr}[\mathcal{E}] = \sum_{\omega \in \mathcal{E}} \mathbf{Pr}[\omega], \tag{3.1.1}$$

namely the probability of an event is the sum of the probabilities of its elements. It follows that the probability of the empty event (the event $\mathcal{E}$ that corresponds to the empty set) is equal to zero, whereas the probability of the event Ω (clearly Ω itself is an event) is equal to one. Finally, the uniform probability function is defined as $\mathbf{Pr}[\omega] = 1/|\Omega|$, for all $\omega \in \Omega$.

3.2. Properties of events. Recall that events are sets and thus set operations (union, intersection, complementation) are applicable. Assuming finite sample spaces and using Eqn. (3.1.1), it is easy to prove the following property for the union of two events $\mathcal{E}_1$ and $\mathcal{E}_2$:

$$\mathbf{Pr}[\mathcal{E}_1 \cup \mathcal{E}_2] = \mathbf{Pr}[\mathcal{E}_1] + \mathbf{Pr}[\mathcal{E}_2] - \mathbf{Pr}[\mathcal{E}_1 \cap \mathcal{E}_2].$$

This property follows from the well-known inclusion-exclusion principle for set union and can be generalized to more than two sets and thus to more than two events. Similarly, one can prove that $\mathbf{Pr}[\bar{\mathcal{E}}] = 1 - \mathbf{Pr}[\mathcal{E}]$. In the above, $\bar{\mathcal{E}}$ denotes the complement of the event $\mathcal{E}$. Finally, it is trivial to see that if $\mathcal{E}_1$ is a subset of $\mathcal{E}_2$ then $\mathbf{Pr}[\mathcal{E}_1] \leqslant \mathbf{Pr}[\mathcal{E}_2]$.

3.3. The union bound. The union bound is a fundamental result in discrete probability and can be used to bound the probability of a union of events without any special assumptions on the relationships between the events. Indeed, let $\mathcal{E}_i$ for all $i = 1, \ldots, n$ be events defined over a finite sample space Ω. Then, the union bound states that

$$\mathbf{Pr}\left[\bigcup_{i=1}^{n} \mathcal{E}_i\right] \leqslant \sum_{i=1}^{n} \mathbf{Pr}[\mathcal{E}_i].$$

The proof of the union bound is quite simple and can be done by induction, using the inclusion-exclusion principle for two sets that was discussed in the previous section.

3.4. Disjoint events and independent events. Two events $\mathcal{E}_1$ and $\mathcal{E}_2$ are called *disjoint* or *mutually exclusive* if their intersection is the empty set, i.e., if

$$\mathcal{E}_1 \cap \mathcal{E}_2 = \emptyset.$$

This can be generalized to any number of events by necessitating that the events are all pairwise disjoint. Two events $\mathcal{E}_1$ and $\mathcal{E}_2$ are called *independent* if the occurrence of one does not affect the probability of the other. Formally, they must satisfy

$$\mathbf{Pr}[\mathcal{E}_1 \cap \mathcal{E}_2] = \mathbf{Pr}[\mathcal{E}_1] \cdot \mathbf{Pr}[\mathcal{E}_2].$$

Again, this can be generalized to more than two events by necessitating that the events are all pairwise independent.

3.5. Conditional probability. For any two events $\mathcal{E}_1$ and $\mathcal{E}_2$, the conditional probability $\mathbf{Pr}[\mathcal{E}_1|\mathcal{E}_2]$ is the probability that $\mathcal{E}_1$ occurs given that $\mathcal{E}_2$ occurs. Formally,

$$\mathbf{Pr}[\mathcal{E}_1|\mathcal{E}_2] = \frac{\mathbf{Pr}[\mathcal{E}_1 \cap \mathcal{E}_2]}{\mathbf{Pr}[\mathcal{E}_2]}.$$

Obviously, the probability of $\mathcal{E}_2$ in the denominator must be non-zero for this to be well-defined. The well-known Bayes rule states that for any two events $\mathcal{E}_1$ and $\mathcal{E}_2$ such that $\mathbf{Pr}[\mathcal{E}_1] > 0$ and $\mathbf{Pr}[\mathcal{E}_2] > 0$,

$$\mathbf{Pr}[\mathcal{E}_2|\mathcal{E}_1] = \frac{\mathbf{Pr}[\mathcal{E}_1|\mathcal{E}_2]\,\mathbf{Pr}[\mathcal{E}_2]}{\mathbf{Pr}[\mathcal{E}_1]}.$$

Using the Bayes rule and the fact that the sample space Ω can be partitioned as $\Omega = \mathcal{E}_2 \cup \overline{\mathcal{E}_2}$, it follows that

$$\mathbf{Pr}[\mathcal{E}_1] = \mathbf{Pr}[\mathcal{E}_1|\mathcal{E}_2]\,\mathbf{Pr}[\mathcal{E}_2] + \mathbf{Pr}\left[\mathcal{E}_1|\overline{\mathcal{E}_2}\right]\mathbf{Pr}\left[\overline{\mathcal{E}_2}\right].$$

We note that the probabilities of both events $\mathcal{E}_1$ and $\mathcal{E}_2$ must be in the open interval $(0,1)$. We can now revisit the notion of independent events. Indeed, for any two events $\mathcal{E}_1$ and $\mathcal{E}_2$ such that $\mathbf{Pr}[\mathcal{E}_1] > 0$ and $\mathbf{Pr}[\mathcal{E}_2] > 0$ the following statements are equivalent:

(1) $\mathbf{Pr}[\mathcal{E}_1|\mathcal{E}_2] = \mathbf{Pr}[\mathcal{E}_1]$,
(2) $\mathbf{Pr}[\mathcal{E}_2|\mathcal{E}_1] = \mathbf{Pr}[\mathcal{E}_2]$, and
(3) $\mathbf{Pr}[\mathcal{E}_1 \cap \mathcal{E}_2] = \mathbf{Pr}[\mathcal{E}_1]\,\mathbf{Pr}[\mathcal{E}_2]$.

Recall that the last statement was the definition of independence in the previous section.

3.6. Random variables. Random variables are *functions* mapping the sample space Ω to the real numbers $\mathbb{R}$. Note that even though they are called variables, in reality they are functions. Let Ω be the sample space of a random experiment. A formal definition for the random variable X would be as follows: let $\alpha \in \mathbb{R}$ be a real number (not necessarily positive) and note that the function

$$X^{-1}(\alpha) = \{\omega \in \Omega \;:\; X(\omega) = \alpha\}$$

returns a subset of Ω and thus is an event. Therefore, the function $X^{-1}(\alpha)$ has a probability. We will abuse notation and write:

$$\mathbf{Pr}[X = \alpha]$$

instead of the more proper notation $\mathbf{Pr}\left[X^{-1}(\alpha)\right]$. This function of α is of great interest and it is easy to generalize as follows:

$$\mathbf{Pr}[X \leqslant \alpha] = \mathbf{Pr}\left[X^{-1}(\alpha') : \alpha' \in (-\infty, \alpha]\right] = \mathbf{Pr}[\omega \in \Omega \;:\; X(\omega) \leqslant \alpha].$$

3.7. Probability mass function and cumulative distribution function. Two common functions associated with random variables are the probability mass function (PMF) and the cumulative distribution function (CDF). The first measures the probability that a random variable takes a particular value $\alpha \in \mathbb{R}$, and the second measures the probability that a random variable takes any value below $\alpha \in \mathbb{R}$.

Definition 3.7.1 (Probability Mass Function (PMF))**.** For a random variable X and a real number α, the function $f(\alpha) = \mathbf{Pr}[X = \alpha]$ is called the *probability mass function (PMF).*

Definition 3.7.2 (Cumulative Distribution Function (CDF)). For a random variable X and a real number α, the the function $F(\alpha) = \mathbf{Pr}[X \leqslant \alpha]$ is called the *cumulative distribution function (CDF).*

It is obvious from the above definitions that $F(\alpha) = \sum_{x \leqslant \alpha} f(x)$.

3.8. Independent random variables. Following the notion of independence for events, we can now define the notion of independence for random variables. Indeed, two random variables X and Y are independent if for all reals a and b,

$$\mathbf{Pr}[X = a \text{ and } Y = b] = \mathbf{Pr}[X = a] \cdot \mathbf{Pr}[Y = b].$$

3.9. Expectation of a random variable. Given a random variable X, its expectation $\mathbf{E}[X]$ is defined as

$$\mathbf{E}[X] = \sum_{x \in X(\Omega)} x \cdot \mathbf{Pr}[X = x].$$

In the above, $X(\Omega)$ is the image of the random variable X over the sample space Ω; recall that X is a function. That is, the sum is over the range of the random variable X. Alternatively, $\mathbf{E}[X]$ can be expressed in terms of a sum over the domain of X, i.e., over Ω. For finite sample spaces Ω, such as those that arise in discrete probability, we get

$$\mathbf{E}[X] = \sum_{\omega \in \Omega} X(\omega)\mathbf{Pr}[\omega].$$

We now discuss fundamental properties of the expectation. The most important property is linearity of expectation: for any random variables X and Y and real number λ,

$$\mathbf{E}[X + Y] = \mathbf{E}[X] + \mathbf{E}[Y], \text{ and}$$
$$\mathbf{E}[\lambda X] = \lambda \mathbf{E}[X].$$

The first property generalizes to any finite sum of random variables and does not need any assumptions on the random variables involved in the summation. If two random variables X and Y are independent then we can manipulate the expectation of their product as follows:

$$\mathbf{E}[XY] = \mathbf{E}[X] \cdot \mathbf{E}[Y].$$

3.10. Variance of a random variable. Given a random variable X, its variance $\mathbf{Var}[X]$ is defined as

$$\mathbf{Var}[X] = \mathbf{E}[X - \mathbf{E}[X]]^2.$$

In words, the variance measures the average of the (square) of the difference $X - \mathbf{E}[X]$. The standard deviation is the square root of the variance and is often denoted by σ. It is easy to prove that

$$\mathbf{Var}[X] = \mathbf{E}\left[X^2\right] - \mathbf{E}[X]^2.$$

This obviously implies

$$\mathbf{Var}[X] \leqslant \mathbf{E}\left[X^2\right],$$

which is often all we need in order to get an upper bound for the variance. Unlike the expectation, the variance does not have a linearity property, unless the random variables involved are independent. Indeed, if the random variables X and Y are independent, then

$$\mathbf{Var}[X+Y] = \mathbf{Var}[X] + \mathbf{Var}[Y].$$

The above property generalizes to sums of more than two random variables, assuming that all involved random variables are pairwise independent. Also, for any real λ,

$$\mathbf{Var}[\lambda X] = \lambda^2 \mathbf{Var}[X].$$

3.11. Markov's inequality. Let X be a non-negative random variable; for any $\alpha > 0$,

$$\mathbf{Pr}[X \geqslant \alpha] \leqslant \frac{\mathbf{E}[X]}{\alpha}.$$

This is a very simple inequality to apply and only needs an upper bound for the expectation of X. An equivalent formulation is the following: let X be a non-negative random variable; for any $k > 1$,

$$\mathbf{Pr}[X \geqslant k \cdot \mathbf{E}[X]] \leqslant \frac{1}{k},$$

or, equivalently,

$$\mathbf{Pr}[X \leqslant k \cdot \mathbf{E}[X]] \geqslant 1 - \frac{1}{k}.$$

In words, the probability that a random variable exceeds k times its expectation is at most $1/k$. In order to prove Markov's inequality, we will show,

$$\mathbf{Pr}[X \geqslant t] \leqslant \frac{\mathbf{E}[X]}{t}$$

assuming

$$k = \frac{t}{\mathbf{E}[X]},$$

for any $t > 0$. In order to prove the above inequality, we define the following function

$$f(X) = \begin{cases} 1, & \text{if } X \geqslant t \\ 0, & \text{otherwise} \end{cases}$$

with expectation:

$$\mathbf{E}[f(X)] = 1 \cdot \mathbf{Pr}[X \geqslant t] + 0 \cdot \mathbf{Pr}[X < t] = \mathbf{Pr}[X \geqslant t].$$

Clearly, from the function definition, $f(X) \leqslant \frac{X}{t}$. Taking expectation on both sides:

$$\mathbf{E}[f(X)] \leqslant \mathbf{E}\left[\frac{X}{t}\right] = \frac{\mathbf{E}[X]}{t}.$$

Thus,

$$\mathbf{Pr}[X \geqslant t] \leqslant \frac{\mathbf{E}[X]}{t}.$$

Hence, we conclude the proof of Markov's inequality.

3.12. The Coupon Collector Problem. Suppose there are m types of coupons and we seek to collect them in independent trials, where in each trial the probability of obtaining any one coupon is $1/m$ (uniform). Let X denote the number of trials that we need in order to collect at least one coupon of each type. Then, one can prove that [20, Section 3.6]:

$$\mathbf{E}[X] = m \ln m + \Theta(m), \text{ and}$$

$$\mathbf{Var}[X] = \frac{\pi^2}{6} m^2 + \Theta(m \ln m).$$

The occurrence of the additional $\ln m$ factor in the expectation is common in sampling-based approaches that attempt to recover m different types of objects using sampling in independent trials. Such factors will appear in many RandNLA sampling-based algorithms.

3.13. References. There are numerous texts covering discrete probability; most of the material in this chapter was adapted from [20].

4. Randomized Matrix Multiplication

Our first randomized algorithm for a numerical linear algebra problem is a simple, sampling-based approach to approximate the product of two matrices $\mathbf{A} \in \mathbb{R}^{m \times n}$ and $\mathbf{B} \in \mathbb{R}^{n \times p}$. This randomized matrix multiplication algorithm is at the heart of all of the RandNLA algorithms that we will discuss in this chapter, and indeed all of RandNLA more generally. It is of interest both pedagogically and in and of itself, and it is also used in an essential way in the analysis of the least squares approximation and low-rank approximation algorithms discussed below.

We start by noting that the product $\mathbf{AB}$ may be written as the sum of n rank one matrices:

$$\mathbf{AB} = \sum_{t=1}^{n} \underbrace{\mathbf{A}_{*t}\mathbf{B}_{t*}}_{\in \mathbb{R}^{m \times n}}, \tag{4.0.1}$$

where each of the summands is the *outer product* of a column of $\mathbf{A}$ and the corresponding row of $\mathbf{B}$. Recall that the standard definition of matrix multiplication states that the (i,j)-th entry of the matrix product $\mathbf{AB}$ is equal to the *inner product* of the i-th row of $\mathbf{A}$ and the j-th column of $\mathbf{B}$, namely

$$(\mathbf{AB})_{ij} = \mathbf{A}_{i*}\mathbf{B}_{*j} \in \mathbb{R}.$$

It is easy to see that the two definitions are equivalent. However, when matrix multiplication is formulated as in Eqn. (4.0.1), a simple randomized algorithm to approximate the product **AB** suggests itself: in independent identically distributed (i.i.d.) trials, randomly sample (and appropriately rescale) a few rank-one matrices from the n terms in the summation of Eqn. (4.0.1); and then output the sum of the (rescaled) terms as an estimator for **AB**.

Input: $\mathbf{A} \in \mathbb{R}^{m \times n}$, $\mathbf{B} \in \mathbb{R}^{n \times p}$, integer c ($1 \leqslant c \leqslant n$), and $\{p_k\}_{k=1}^{n}$ s.t. $p_k \geqslant 0$ and $\sum_{k=1}^{n} p_k = 1$.
Output: $\mathbf{C} \in \mathbb{R}^{m \times c}$ and $\mathbf{R} \in \mathbb{R}^{c \times p}$.

(1) For $t = 1$ to c,
- Pick $i_t \in \{1, \ldots, n\}$ with $\mathbf{Pr}[i_t = k] = p_k$, independently and with replacement.
- Set $\mathbf{C}_{*t} = \frac{1}{\sqrt{c p_{i_t}}} \mathbf{A}_{*i_t}$ and $\mathbf{R}_{t*} = \frac{1}{\sqrt{c p_{i_t}}} \mathbf{B}_{i_t *}$.

(2) Return $\mathbf{CR} = \sum_{t=1}^{c} \frac{1}{c p_{i_t}} \mathbf{A}_{*i_t} \mathbf{B}_{i_t *}$.

Algorithm 4.0.2: The RANDMATRIXMULTIPLY algorithm

Consider the RANDMATRIXMULTIPLY algorithm (Algorithm 4.0.2), which makes this simple idea precise. When this algorithm is given as input two matrices **A** and **B**, a probability distribution $\{p_k\}_{k=1}^{n}$, and a number c of column-row pairs to choose, it returns as output an estimator for the product **AB** of the form

$$\sum_{t=1}^{c} \frac{1}{c p_{i_t}} \mathbf{A}_{*i_t} \mathbf{B}_{i_t *}.$$

Equivalently, the above estimator can be thought of as the product of the two matrices **C** and **R** formed by the RANDMATRIXMULTIPLY algorithm, where **C** consists of c (rescaled) columns of **A** and **R** consists of the corresponding (rescaled) rows of **B**.

Observe that

$$\mathbf{CR} = \sum_{t=1}^{c} \mathbf{C}_{*t} \mathbf{R}_{t*} = \sum_{t=1}^{c} \left(\sqrt{\frac{1}{c p_{i_t}}} \mathbf{A}_{*i_t} \right) \left(\sqrt{\frac{1}{c p_{i_t}}} \mathbf{B}_{i_t *} \right) = \frac{1}{c} \sum_{t=1}^{c} \frac{1}{p_{i_t}} \mathbf{A}_{*i_t} \mathbf{B}_{i_t *}.$$

Therefore, the procedure used for sampling and scaling column-row pairs in the RANDMATRIXMULTIPLY algorithm corresponds to sampling and rescaling terms in Eqn. (4.0.1).

Remark 4.0.3. The analysis of RandNLA algorithms has benefited enormously from formulating algorithms using the so-called *sampling-and-rescaling matrix formalism*. Let's define the sampling-and-rescaling matrix $\mathbf{S} \in \mathbb{R}^{n \times c}$ to be a matrix with $\mathbf{S}_{i_t t} = 1/\sqrt{c p_{i_t}}$ if the i_t-th column of **A** is chosen in the t-th trial (all other entries of **S** are set to zero). Then

$$\mathbf{C} = \mathbf{AS} \text{ and } \mathbf{R} = \mathbf{S}^{\mathsf{T}} \mathbf{B},$$

so that $\mathbf{CR} = \mathbf{ASS}^T\mathbf{B} \approx \mathbf{AB}$. Obviously, the matrix $\mathbf{S}$ is very sparse, having a single non-zero entry per column, for a total of c non-zero entries, and so it is not explicitly constructed and stored by the algorithm.

Remark 4.0.4. The choice of the sampling probabilities $\{p_k\}_{k=1}^n$ in the RANDMATRIXMULTIPLY algorithm is very important. As we will prove in Lemma 4.1.1, the estimator returned by the RANDMATRIXMULTIPLY algorithm is (in an element-wise sense) unbiased, regardless of our choice of the sampling probabilities. However, a natural notion of the *variance* of our estimator (see Theorem 4.1.2 for a precise definition) is minimized when the sampling probabilities are set to

$$p_k = \frac{\|\mathbf{A}_{*k}\mathbf{B}_{k*}\|_F}{\sum_{k'=1}^n \|\mathbf{A}_{*k'}\mathbf{B}_{k'*}\|_F} = \frac{\|\mathbf{A}_{*k}\|_2\|\mathbf{B}_{k*}\|_2}{\sum_{k'=1}^n \|\mathbf{A}_{*k'}\|_2\|\mathbf{B}_{k'*}\|_2}.$$

In words, the best choice when sampling rank-one matrices from the summation of Eqn. (4.0.1) is to select rank-one matrices that have larger Frobenius norms with higher probabilities. This is equivalent to selecting column-row pairs that have larger (products of) Euclidean norms with higher probability.

Remark 4.0.5. This approach for approximating matrix multiplication has several advantages. First, it is conceptually simple. Second, since the heart of the algorithm involves matrix multiplication of smaller matrices, it can use any algorithms that exist in the literature for performing the desired matrix multiplication. Third, this approach does not tamper with the sparsity of the input matrices. Finally, the algorithm can be easily implemented in one pass over the input matrices $\mathbf{A}$ and $\mathbf{B}$, given the sampling probabilities $\{p_k\}_{k=1}^n$. See [9, Section 4.2] for a detailed discussion regarding the implementation of the RANDMATRIXMULTIPLY algorithm in the pass-efficient and streaming models of computation.

4.1. Analysis of the RANDMATRIXMULTIPLY algorithm. This section provides upper bounds for the error matrix $\|\mathbf{AB} - \mathbf{CR}\|_F^2$, where $\mathbf{C}$ and $\mathbf{R}$ are the outputs of the RANDMATRIXMULTIPLY algorithm.

Our first lemma proves that the expectation of the (i,j)-th element of the estimator $\mathbf{CR}$ is equal to the (i,j)-th element of the exact product $\mathbf{AB}$, regardless of the choice of the sampling probabilities. It further bounds the variance of the (i,j)-th element of the estimator, although this bound does depend on our choice of the sampling probabilities.

Lemma 4.1.1. Let $\mathbf{C}$ and $\mathbf{R}$ be constructed as described in the RANDMATRIXMULTIPLY algorithm. Then,

$$\mathbf{E}\left[(\mathbf{CR})_{ij}\right] = (\mathbf{AB})_{ij}$$

and

$$\mathbf{Var}\left[(\mathbf{CR})_{ij}\right] \leqslant \frac{1}{c}\sum_{k=1}^n \frac{\mathbf{A}_{ik}^2\mathbf{B}_{kj}^2}{p_k}.$$

Proof. Fix some pair i, j. For $t = 1, \dots, c$, define

$$X_t = \left(\frac{\mathbf{A}_{*i_t}\mathbf{B}_{i_t*}}{cp_{i_t}}\right)_{ij} = \frac{\mathbf{A}_{ii_t}\mathbf{B}_{i_tj}}{cp_{i_t}}.$$

Thus, for any t,

$$\mathbf{E}[X_t] = \sum_{k=1}^{n} p_k \frac{\mathbf{A}_{ik}\mathbf{B}_{kj}}{cp_k} = \frac{1}{c}\sum_{k=1}^{n} \mathbf{A}_{ik}\mathbf{B}_{kj} = \frac{1}{c}(\mathbf{AB})_{ij}.$$

Since we have $(\mathbf{CR})_{ij} = \sum_{t=1}^{c} X_t$, it follows that

$$\mathbf{E}\left[(\mathbf{CR})_{ij}\right] = \mathbf{E}\left[\sum_{t=1}^{c} X_t\right] = \sum_{t=1}^{c} \mathbf{E}[X_t] = (\mathbf{AB})_{ij}.$$

Hence, $\mathbf{CR}$ is an unbiased estimator of $\mathbf{AB}$, regardless of the choice of the sampling probabilities. Using the fact that $(\mathbf{CR})_{ij}$ is the sum of c independent random variables, we get

$$\mathbf{Var}\left[(\mathbf{CR})_{ij}\right] = \mathbf{Var}\left[\sum_{t=1}^{c} X_t\right] = \sum_{t=1}^{c} \mathbf{Var}[X_t].$$

Using $\mathbf{Var}[X_t] \leqslant \mathbf{E}\left[X_t^2\right] = \sum_{k=1}^{n} \frac{\mathbf{A}_{ik}^2\mathbf{B}_{kj}^2}{c^2p_k}$, we get

$$\mathbf{Var}\left[(\mathbf{CR})_{ij}\right] = \sum_{t=1}^{c} \mathbf{Var}[X_t] \leqslant c\sum_{k=1}^{n} \frac{\mathbf{A}_{ik}^2\mathbf{B}_{kj}^2}{c^2p_k} = \frac{1}{c}\sum_{k=1}^{n} \frac{\mathbf{A}_{ik}^2\mathbf{B}_{kj}^2}{p_k},$$

which concludes the proof of the lemma. □

Our next result bounds the expectation of the Frobenius norm of the error matrix $\mathbf{AB} - \mathbf{CR}$. Notice that this error metric depends on our choice of the sampling probabilities $\{p_k\}_{k=1}^{n}$.

Theorem 4.1.2. Construct $\mathbf{C}$ and $\mathbf{R}$ using the RANDMATRIXMULTIPLY algorithm and let $\mathbf{CR}$ be an approximation to $\mathbf{AB}$. Then,

$$\mathbf{E}\left[\|\mathbf{AB} - \mathbf{CR}\|_F^2\right] \leqslant \sum_{k=1}^{n} \frac{\|\mathbf{A}_{*k}\|_2^2\|\mathbf{B}_{k*}\|_2^2}{cp_k}. \tag{4.1.3}$$

Furthermore, if

$$p_k = \frac{\|\mathbf{A}_{*k}\|_2\|\mathbf{B}_{k*}\|_2}{\sum_{k'=1}^{n}\|\mathbf{A}_{*k'}\|_2\|\mathbf{B}_{k'*}\|_2}, \tag{4.1.4}$$

for all $k = 1, \dots, n$, then

$$\mathbf{E}\left[\|\mathbf{AB} - \mathbf{CR}\|_F^2\right] \leqslant \frac{1}{c}\left(\sum_{k=1}^{n}\|\mathbf{A}_{*k}\|_2\|\mathbf{B}_{k*}\|_2\right)^2. \tag{4.1.5}$$

This choice for $\{p_k\}_{k=1}^{n}$ minimizes $\mathbf{E}\left[\|\mathbf{AB} - \mathbf{CR}\|_F^2\right]$, among possible choices for the sampling probabilities.

Proof. First of all, since $\mathbf{CR}$ is an unbiased estimator of $\mathbf{AB}$, $\mathbf{E}\left[(\mathbf{AB}-\mathbf{CR})_{ij}\right]=0$. Thus,

$$\mathbf{E}\left[\|\mathbf{AB}-\mathbf{CR}\|_F^2\right]=\sum_{i=1}^{m}\sum_{j=1}^{p}\mathbf{E}\left[(\mathbf{AB}-\mathbf{CR})_{ij}^2\right]=\sum_{i=1}^{m}\sum_{j=1}^{p}\mathbf{Var}\left[(\mathbf{CR})_{ij}\right].$$

Using Lemma 4.1.1, we get

$$\begin{aligned}\mathbf{E}\left[\|\mathbf{AB}-\mathbf{CR}\|_F^2\right]&\leqslant\frac{1}{c}\sum_{k=1}^{n}\frac{1}{p_k}\left(\sum_i \mathbf{A}_{ik}^2\right)\left(\sum_j \mathbf{B}_{kj}^2\right)\\&=\frac{1}{c}\sum_{k=1}^{n}\frac{1}{p_k}\|\mathbf{A}_{*k}\|_2^2\|\mathbf{B}_{k*}\|_2^2.\end{aligned}$$

Let p_k be as in Eqn. (4.1.4); then

$$\mathbf{E}\left[\|\mathbf{AB}-\mathbf{CR}\|_F^2\right]\leqslant\frac{1}{c}\left(\sum_{k=1}^{n}\|\mathbf{A}_{*k}\|_2\|\mathbf{B}_{k*}\|_2\right)^2.$$

Finally, to prove that the aforementioned choice for the $\{p_k\}_{k=1}^n$ minimizes the quantity $\mathbf{E}\left[\|\mathbf{AB}-\mathbf{CR}\|_F^2\right]$, define the function

$$f(p_1,\dots p_n)=\sum_{k=1}^{n}\frac{1}{p_k}\|\mathbf{A}_{*k}\|_2^2\|\mathbf{B}_{k*}\|_2^2,$$

which characterizes the dependence of $\mathbf{E}\left[\|\mathbf{AB}-\mathbf{CR}\|_F^2\right]$ on the p_k's. In order to minimize f subject to $\sum_{k=1}^n p_k=1$, we can introduce the Lagrange multiplier λ and define the function

$$g(p_1,\dots p_n)=f(p_1,\dots p_n)+\lambda\left(\sum_{k=1}^{n}p_k-1\right).$$

We then have the minimum at

$$0=\frac{\partial g}{\partial p_k}=\frac{-1}{p_k^2}\|\mathbf{A}_{*k}\|_2^2\|\mathbf{B}_{k*}\|_2^2+\lambda.$$

Thus,

$$p_k=\frac{\|\mathbf{A}_{*k}\|_2\|\mathbf{B}_{k*}\|_2}{\sqrt{\lambda}}=\frac{\|\mathbf{A}_{*k}\|_2\|\mathbf{B}_{k*}\|_2}{\sum_{k'=1}^{n}\|\mathbf{A}_{*k'}\|_2\|\mathbf{B}_{k'*}\|_2},$$

where the second equality comes from solving for $\sqrt{\lambda}$ in $\sum_{k=1}^n p_k=1$. These probabilities are minimizers of f because $\frac{\partial^2 g}{\partial {p_k}^2}>0$ for all k. □

We conclude this section by pointing out that we can apply Markov's inequality on the expectation bound of Theorem 4.1.2 in order to get bounds for the Frobenius norm of the error matrix $\mathbf{AB}-\mathbf{CR}$ that hold with constant probability. We refer the reader to [9, Section 4.4] for a tighter analysis, arguing for a better (in the sense of better dependence on the failure probability than provided by Markov's inequality) concentration of the Frobenius norm of the error matrix around its mean using a martingale argument.

4.2. Analysis of the algorithm for nearly optimal probabilities. We now discuss three different choices for the sampling probabilities that are easy to analyze and will be useful in this chapter. We summarize these results in the following list; all three bounds can be easily proven following the proof of Theorem 4.1.2.

Nearly optimal probabilities, depending on both A and B: If the $\{p_k\}_{k=1}^n$ satisfy

$$\sum_{k=1}^{n} p_k = 1 \quad \text{and} \quad p_k \geqslant \frac{\beta \|\mathbf{A}_{*k}\|_2 \|\mathbf{B}_{k*}\|_2}{\sum_{k'=1}^{n} \|\mathbf{A}_{*k'}\|_2 \|\mathbf{B}_{k'*}\|_2}, \tag{4.2.1}$$

for some positive constant $\beta \leqslant 1$, then,

$$\mathbf{E}\left[\|\mathbf{AB} - \mathbf{CR}\|_F^2\right] \leqslant \frac{1}{\beta c}\left(\sum_{k=1}^{n} \|\mathbf{A}_{*k}\|_2 \|\mathbf{B}_{k*}\|_2\right)^2. \tag{4.2.2}$$

Nearly optimal probabilities, depending only on A: If the $\{p_k\}_{k=1}^n$ satisfy

$$\sum_{k=1}^{n} p_k = 1 \quad \text{and} \quad p_k \geqslant \frac{\beta \|\mathbf{A}_{*k}\|_2^2}{\|\mathbf{A}\|_F^2}, \tag{4.2.3}$$

for some positive constant $\beta \leqslant 1$, then,

$$\mathbf{E}\left[\|\mathbf{AB} - \mathbf{CR}\|_F^2\right] \leqslant \frac{1}{\beta c}\|\mathbf{A}\|_F^2 \|\mathbf{B}\|_F^2. \tag{4.2.4}$$

Nearly optimal probabilities, depending only on B: If the $\{p_k\}_{k=1}^n$ satisfy

$$\sum_{k=1}^{n} p_k = 1 \quad \text{and} \quad p_k \geqslant \frac{\beta \|\mathbf{B}_{k*}\|_2^2}{\|\mathbf{B}\|_F^2}, \tag{4.2.5}$$

for some positive constant $\beta \leqslant 1$, then,

$$\mathbf{E}\left[\|\mathbf{AB} - \mathbf{CR}\|_F^2\right] \leqslant \frac{1}{\beta c}\|\mathbf{A}\|_F^2 \|\mathbf{B}\|_F^2. \tag{4.2.6}$$

We note that, from the Cauchy-Schwartz inequality,

$$\left(\sum_{k=1}^{n} \|\mathbf{A}_{*k}\|_2 \|\mathbf{B}_{k*}\|_2\right)^2 \leqslant \|\mathbf{A}\|_F^2 \|\mathbf{B}\|_F^2,$$

and thus the bound of Eqn. (4.2.2) is generally better than those of Eqns. (4.2.4) and (4.2.6). See [9, Section 4.3, Table 1] for other sampling probabilities and respective error bounds that might be of interest.

4.3. Bounding the two norm. In both applications of the RandMatrixMultiply algorithm that we will discuss in this chapter (see least-squares approximation and low-rank matrix approximation in Sections 5 and 6, respectively), we will be particularly interested in approximating the product $\mathbf{U}^T\mathbf{U}$, where $\mathbf{U}$ is a tall-and-thin matrix, by sampling (and rescaling) a few rows of $\mathbf{U}$. (The matrix $\mathbf{U}$ will be a matrix spanning the column space or the "important" part of the column space of some other matrix of interest.) It turns out that, without loss of generality, we can focus on the special case where $\mathbf{U} \in \mathbb{R}^{n \times d}$ ($n \gg d$) is a matrix with orthonormal columns (i.e., $\mathbf{U}^T\mathbf{U} = \mathbf{I}_d$). Then, if we let $\mathbf{R} \in \mathbb{R}^{c \times d}$ be a sample of c (rescaled) rows of $\mathbf{U}$ constructed using the RANDMATRIXMULTIPLY algorithm, and note that the corresponding c (rescaled) columns of $\mathbf{U}^T$ form the matrix $\mathbf{R}^T$, then Theorem 4.1.2 implies that

$$\mathbf{E}\left[\|\mathbf{U}^T\mathbf{U}-\mathbf{R}^T\mathbf{R}\|_F^2\right]=\mathbf{E}\left[\|\mathbf{I}_d-\mathbf{R}^T\mathbf{R}\|_F^2\right]\leqslant\frac{d^2}{\beta c}. \tag{4.3.1}$$

In the above, we used the fact that $\|\mathbf{U}\|_F^2=d$. For the above bound to hold, it suffices to use sampling probabilities p_k $(k=1,\ldots,n)$ that satisfy

$$\sum_{k=1}^{n}p_k=1 \quad \text{and} \quad p_k\geqslant\frac{\beta\|\mathbf{U}_{k*}\|_2^2}{d}. \tag{4.3.2}$$

(The quantities $\|\mathbf{U}_{k*}\|_2^2$ are known as leverage scores [17]; and the probabilities given by Eqn. (4.3.2) are nearly-optimal, in the sense of Eqn. (4.2.1), i.e., in the sense that they approximate the optimal probabilities for approximating the matrix product shown in Eqn (4.3.1), up to a β factor.) Applying Markov's inequality to the bound of Eqn. (4.3.1) and setting

$$c=\frac{10d^2}{\beta\epsilon^2}, \tag{4.3.3}$$

we get that, with probability at least 9/10,

$$\|\mathbf{U}^T\mathbf{U}-\mathbf{R}^T\mathbf{R}\|_F=\|\mathbf{I}_d-\mathbf{R}^T\mathbf{R}\|_F\leqslant\epsilon. \tag{4.3.4}$$

Clearly, the above equation also implies a two-norm bound. Indeed, with probability at least 9/10,

$$\|\mathbf{U}^T\mathbf{U}-\mathbf{R}^T\mathbf{R}\|_2=\|\mathbf{I}_d-\mathbf{R}^T\mathbf{R}\|_2\leqslant\epsilon$$

by setting c to the value of Eqn. (4.3.3).

In the remainder of this section, we will state and prove a theorem that also guarantees $\|\mathbf{U}^T\mathbf{U}-\mathbf{R}^T\mathbf{R}\|_2\leqslant\epsilon$, while setting c to a value that is *smaller* than the one in Eqn. (4.3.3). For related concentration techniques, see the chapter by Vershynin in this volume [28].

Theorem 4.3.5. Let $\mathbf{U}\in\mathbb{R}^{n\times d}$ $(n\gg d)$ satisfy $\mathbf{U}^T\mathbf{U}=\mathbf{I}_d$. Construct $\mathbf{R}$ using the RANDMATRIXMULTIPLY algorithm and let the sampling probabilities $\{p_k\}_{k=1}^n$ satisfy the conditions of Eqn. (4.3.2), for all $k=1,\ldots,n$ and for some constant $\beta\in(0,1]$.

Let $\epsilon\in(0,1)$ be an accuracy parameter and let

$$c\geqslant\frac{96d}{\beta\epsilon^2}\ln\left(\frac{96d}{\beta\epsilon^2\sqrt{\delta}}\right). \tag{4.3.6}$$

Then, with probability at least $1-\delta$,

$$\|\mathbf{U}^T\mathbf{U}-\mathbf{R}^T\mathbf{R}\|_2=\|\mathbf{I}_d-\mathbf{R}^T\mathbf{R}\|_2\leqslant\epsilon.$$

Prior to proving the above theorem, we state a matrix-Bernstein inequality that is due to Oliveira [22, Lemma 1].

Lemma 4.3.7. Let $\mathbf{x}^1, \mathbf{x}^2, \ldots, \mathbf{x}^c$ be independent identically distributed copies of a d-dimensional random vector $\mathbf{x}$ with

$$\|\mathbf{x}\|_2 \leqslant M \qquad \text{and} \qquad \|\mathbf{E}\left[\mathbf{x}\mathbf{x}^T\right]\|_2 \leqslant 1.$$

Then, for any $\alpha > 0$,

$$\|\frac{1}{c}\sum_{i=1}^{c} \mathbf{x}^i {\mathbf{x}^i}^T - \mathbf{E}\left[\mathbf{x}\mathbf{x}^T\right]\|_2 \leqslant \alpha$$

holds with probability at least

$$1 - \left(2c^2\right)\exp\left(-\frac{c\alpha^2}{16M^2 + 8M^2\alpha}\right).$$

This inequality essentially gives a bound for the probability that the matrix $\frac{1}{c}\sum_{i=1}^{c}\mathbf{x}^i{\mathbf{x}^i}^T$ deviates significantly from its expectation where the deviation is measured with respect to the two norm (namely the largest singular value) of the error matrix.

Proof. (of Theorem 4.3.5) Define the random *row* vector $\mathbf{y} \in \mathbb{R}^d$ as

$$\mathbf{Pr}\left[\mathbf{y} = \frac{1}{\sqrt{p_k}}\mathbf{U}_{k*}\right] = p_k \geqslant \frac{\beta\|\mathbf{U}_{k*}\|_2^2}{d},$$

for $k = 1, \ldots, n$. In words, $\mathbf{y}$ is set to be the (rescaled) k-th row of $\mathbf{U}$ with probability p_k. Thus, the matrix $\mathbf{R}$ has rows $\frac{1}{\sqrt{c}}\mathbf{y}^1, \frac{1}{\sqrt{c}}\mathbf{y}^2, \ldots, \frac{1}{\sqrt{c}}\mathbf{y}^c$, where $\mathbf{y}^1, \mathbf{y}^2, \ldots, \mathbf{y}^c$ are c independent copies of $\mathbf{y}$. Using this notation, it follows that

$$\mathbf{E}\left[\mathbf{y}^T\mathbf{y}\right] = \sum_{k=1}^{n} p_k\left(\frac{1}{\sqrt{p_k}}\mathbf{U}_{k*}^T\right)\left(\frac{1}{\sqrt{p_k}}\mathbf{U}_{k*}\right) = \mathbf{U}^T\mathbf{U} = \mathbf{I}_d. \tag{4.3.8}$$

Also,

$$\mathbf{R}^T\mathbf{R} = \frac{1}{c}\sum_{t=1}^{c}\underbrace{{\mathbf{y}^t}^T\mathbf{y}^t}_{\mathbb{R}^{d\times d}}.$$

For this vector $\mathbf{y}$, let

$$M \geqslant \|\mathbf{y}\|_2 = \frac{1}{\sqrt{p_k}}\|\mathbf{U}_{k*}\|_2. \tag{4.3.9}$$

Notice that from Eqn. (4.3.8) we immediately get $\|\mathbf{E}\left[\mathbf{y}^T\mathbf{y}\right]\|_2 = \|\mathbf{I}_d\|_2 = 1$. Applying Lemma 4.3.7 (with $\mathbf{x} = \mathbf{y}^T$), we get

$$\|\mathbf{R}^T\mathbf{R} - \mathbf{U}^T\mathbf{U}\|_2 < \epsilon, \tag{4.3.10}$$

with probability at least $1 - (2c)^2\exp\left(-\frac{c\epsilon^2}{16M^2+8M^2\epsilon}\right)$.

Let δ be the failure probability of Theorem 4.3.5; we seek an appropriate value of c in order to guarantee $(2c)^2\exp\left(-\frac{c\epsilon^2}{16M^2+8M^2\epsilon}\right) \leqslant \delta$. Equivalently, we need to satisfy

$$\frac{c}{\ln\left(2c/\sqrt{\delta}\right)} \geqslant \frac{2}{\epsilon^2}\left(16M^2 + 8M^2\epsilon\right).$$

Combine Eqns. (4.3.9) and (4.3.2) to get $M^2 \leqslant \|\mathbf{U}\|_F^2/\beta = d/\beta$. Recall that $\epsilon < 1$ to conclude that it suffices to choose a value of c such that

$$\frac{c}{\ln\left(2c/\sqrt{\delta}\right)} \geqslant \frac{48d}{\beta\epsilon^2},$$

or, equivalently,

$$\frac{2c/\sqrt{\delta}}{\ln\left(2c/\sqrt{\delta}\right)} \geqslant \frac{96d}{\beta\epsilon^2\sqrt{\delta}}.$$

We now use the fact that for any $\eta \geqslant 4$, if $x \geqslant 2\eta \ln \eta$, then $x/\ln x \geqslant \eta$. Let $x = 2c/\sqrt{\delta}$, let $\eta = 96d/\left(\beta\epsilon^2\sqrt{\delta}\right)$, and note that $\eta \geqslant 4$ since $d \geqslant 1$ and β, ϵ, and δ are at most one. Thus, it suffices to set

$$\frac{2c}{\sqrt{\delta}} \geqslant 2\frac{96d}{\beta\epsilon^2\sqrt{\delta}} \ln\left(\frac{96d}{\beta\epsilon^2\sqrt{\delta}}\right),$$

which concludes the proof of the theorem. □

Remark 4.3.11. Let $\delta = 1/10$ and let ϵ and β be constants. Then, we can compare the bound of Eqn. (4.3.3) with the bound of Eqn. (4.3.6) of Theorem 4.3.5: both values of c guarantee the same accuracy ϵ and the same success probability (say 9/10). However, asymptotically, the bound of Theorem 4.3.5 holds by setting $c = O(d \ln d)$, while the bound of Eqn. (4.3.3) holds by setting $c = O(d^2)$. Thus, the bound of Theorem 4.3.5 is much better. By the Coupon Collector Problem (see Section 3.12), sampling-based approaches necessitate at least $\Omega(d \ln d)$ samples, thus making our algorithm asymptotically optimal. We should note, however, that deterministic methods exist (see, for example, [24]) that achieve the same bound with $c = O(d/\epsilon^2)$ samples.

Remark 4.3.12. We made no effort to optimize the constants in the expression for c in Eqn. (4.3.6). Better constants are known, by using tighter matrix-Bernstein inequalities. For a state-of-the-art bound see, for example, [16, Theorem 5.1].

4.4. References. Our presentation in this chapter follows closely the derivations in [9]; see [9] for a detailed discussion of prior work on this topic. We also refer the interested reader to [16] and references therein for more recent work on randomized matrix multiplication.

5. RandNLA Approaches for Regression Problems

In this section, we will present a simple randomized algorithm for least-squares regression. In many applications in mathematics and statistical data analysis, it is of interest to find an approximate solution to a system of linear equations that has no exact solution. For example, let a matrix $\mathbf{A} \in \mathbb{R}^{n\times d}$ and a vector $\mathbf{b} \in \mathbb{R}^n$ be given. If $n \gg d$, there will not in general exist a vector $\mathbf{x} \in \mathbb{R}^d$ such that $\mathbf{Ax} = \mathbf{b}$, and yet it is often of interest to find a vector $\mathbf{x}$ such that $\mathbf{Ax} \approx \mathbf{b}$ in some precise sense. The method of least squares, whose original formulation is often credited to Gauss and Legendre, accomplishes this by minimizing the sum of squares of the elements of the residual vector, i.e., by solving the optimization problem

$$\mathcal{Z} = \min_{x \in \mathbb{R}^d} \|\mathbf{A}x - \mathbf{b}\|_2. \tag{5.0.1}$$

The minimum ℓ_2-norm vector among those satisfying Eqn. (5.0.1) is

$$x_{\text{opt}} = \mathbf{A}^{\dagger}\mathbf{b}, \tag{5.0.2}$$

where $\mathbf{A}^{\dagger}$ denotes the Moore-Penrose generalized inverse of the matrix $\mathbf{A}$. This solution vector has a very natural statistical interpretation as providing an optimal estimator among all linear unbiased estimators, and it has a very natural geometric interpretation as providing an orthogonal projection of the vector $\mathbf{b}$ onto the span of the columns of the matrix $\mathbf{A}$.

Recall that to minimize the quantity in Eqn. (5.0.1), we can set the derivative of $\|\mathbf{A}x - \mathbf{b}\|_2^2 = (\mathbf{A}x - \mathbf{b})^T(\mathbf{A}x - \mathbf{b})$ with respect to x equal to zero, from which it follows that the minimizing vector x_{opt} is a solution of the so-called normal equations

$$\mathbf{A}^T\mathbf{A}x_{\text{opt}} = \mathbf{A}^T\mathbf{b}. \tag{5.0.3}$$

Computing $\mathbf{A}^T\mathbf{A}$, and thus computing x_{opt} in this way, takes $O(nd^2)$ time, assuming $n \geqslant d$. Geometrically, Eqn. (5.0.3) requires the residual vector $\mathbf{b}^{\perp} = \mathbf{b} - \mathbf{A}x_{\text{opt}}$ to be orthogonal to the column space of $\mathbf{A}$, i.e., ${\mathbf{b}^{\perp}}^T\mathbf{A} = 0$. While solving the normal equations squares the condition number of the input matrix (and thus is typically not recommended in practice), direct methods (such as the QR decomposition, see Section 2.1) also solve the problem of Eqn. (5.0.1) in $O(nd^2)$ time, assuming that $n \geqslant d$. Finally, an alternative expression for the vector x_{opt} of Eqn. (5.0.2) emerges by leveraging the SVD of $\mathbf{A}$. If $\mathbf{A} = \mathbf{U}_A\mathbf{\Sigma}_A\mathbf{V}_A^T$ denotes the SVD of $\mathbf{A}$, then

$$x_{\text{opt}} = \mathbf{V}_A\mathbf{\Sigma}_A^{-1}\mathbf{U}_A^T\mathbf{b} = \mathbf{A}^{\dagger}\mathbf{b}.$$

Computing x_{opt} in this way also takes $O(nd^2)$ time, again assuming $n \geqslant d$. In this section, we will describe a randomized algorithm that will provide accurate relative-error approximations to the minimal ℓ_2-norm solution vector x_{opt} of Eqn. (5.0.2) faster than these "exact" algorithms for a large class of over-constrained least-squares problems.

5.1. The Randomized Hadamard Transform. The Randomized Hadamard Transform was introduced in [1] as one step in the development of a fast version of the Johnson-Lindenstrauss lemma. Recall that the $n \times n$ Hadamard matrix (assuming n is a power of two) $\tilde{\mathbf{H}}_n$, may be defined recursively as follows:

$$\tilde{\mathbf{H}}_n = \begin{bmatrix} \tilde{\mathbf{H}}_{n/2} & \tilde{\mathbf{H}}_{n/2} \\ \tilde{\mathbf{H}}_{n/2} & -\tilde{\mathbf{H}}_{n/2} \end{bmatrix}, \quad \text{with} \quad \tilde{\mathbf{H}}_2 = \begin{bmatrix} +1 & +1 \\ +1 & -1 \end{bmatrix}.$$

We can now define the *normalized* Hadamard transform $\mathbf{H}_n$ as $(1/\sqrt{n})\tilde{\mathbf{H}}_n$; it is easy to see that $\mathbf{H}_n\mathbf{H}_n^T = \mathbf{H}_n^T\mathbf{H}_n = \mathbf{I}_n$. Now consider a diagonal matrix

$\mathbf{D} \in \mathbb{R}^{n\times n}$ such that $\mathbf{D}_{ii}$ is set to +1 with probability 1/2 and to -1 with probability 1/2. The product $\mathbf{HD}$ is the *Randomized Hadamard Transform* and has three useful properties. First, when applied to a vector, it "spreads out" the mass/energy of that vector, in the sense of providing a bound for the largest element, or infinity norm, of the transformed vector. Second, computing the product $\mathbf{HDx}$ for any vector $\mathbf{x} \in \mathbb{R}^n$ takes $O(n\log_2 n)$ time. Even better, if we only need to access, say, r elements in the transformed vector, then those r elements can be computed in $O(n\log_2 r)$ time. We will expand on the latter observation in Section 5.5, where we will discuss the running time of the proposed algorithm. Third, the Randomized Hadamard Transform is an orthogonal transformation, since $\mathbf{HDD}^T\mathbf{H}^T = \mathbf{H}^T\mathbf{D}^T\mathbf{DH} = \mathbf{I}_n$.

5.2. The main algorithm and main theorem. We are now ready to provide an overview of the RANDLEASTSQUARES algorithm (Algorithm 5.2.1). Let the matrix product $\mathbf{HD}$ denote the $n\times n$ Randomized Hadamard Transform discussed in the previous section. (For simplicity, we restrict our discussion to the case that n is a power of two, although this restriction can easily be removed by using variants of the Randomized Hadamard Transform [17].) Our algorithm is a *preconditioned random sampling algorithm*: after premultiplying $\mathbf{A}$ and $\mathbf{b}$ by $\mathbf{HD}$, our algorithm samples uniformly at random r constraints from the preprocessed problem. (See Eqn. (5.2.3), as well as the remarks after Theorem 5.2.2 for the precise value of r.) Then, this algorithm solves the least squares problem on just those sampled constraints to obtain a vector $\tilde{\mathbf{x}}_{opt} \in \mathbb{R}^d$ such that Theorem 5.2.2 is satisfied.

Input: $\mathbf{A} \in \mathbb{R}^{n\times d}$, $\mathbf{b} \in \mathbb{R}^n$, and an error parameter $\epsilon \in (0,1)$.

Output: $\tilde{\mathbf{x}}_{opt} \in \mathbb{R}^d$.

(1) Let r assume the value of Eqn. (5.2.3).

(2) Let $\mathbf{S}$ be an empty matrix.

(3) **For** $t = 1,\dots,r$ (i.i.d. trials with replacement) **select uniformly at random** an integer from $\{1,2,\dots,n\}$.

- **If** i is selected, **then** append the column vector $\left(\sqrt{n/r}\right)\mathbf{e}_i$ to $\mathbf{S}$, where $\mathbf{e}_i \in \mathbb{R}^n$ is the i-th canonical vector.

(4) Let $\mathbf{H} \in \mathbb{R}^{n\times n}$ be the normalized Hadamard transform matrix.

(5) Let $\mathbf{D} \in \mathbb{R}^{n\times n}$ be a diagonal matrix with

$$\mathbf{D}_{ii} = \begin{cases} +1 & \text{, with probability } 1/2 \\ -1 & \text{, with probability } 1/2 \end{cases}$$

(6) Compute and return $\tilde{\mathbf{x}}_{opt} = \left(\mathbf{S}^T\mathbf{HDA}\right)^{\dagger}\mathbf{S}^T\mathbf{HDb}$.

Algorithm 5.2.1: The RANDLEASTSQUARES algorithm

Formally, we will let $\mathbf{S} \in \mathbb{R}^{n \times r}$ denote a sampling-and-rescaling matrix specifying which of the n (preprocessed) constraints are to be sampled and how they are to be rescaled. This matrix is initially empty and is constructed as described in the RANDLEASTSQUARES algorithm. (We are describing this algorithm in terms of the matrix $\mathbf{S}$, but as with the RANDMATRIXMULTIPLY algorithm, we do not need to construct it explicitly in an actual implementation [3].) Then, we can consider the problem

$$\tilde{\mathcal{Z}} = \min_{\mathbf{x} \in \mathbb{R}^d} \|\mathbf{S}^T\mathbf{HDAx} - \mathbf{S}^T\mathbf{HDb}\|_2,$$

which is a least squares approximation problem involving only the r constraints, where the r constraints are uniformly sampled from the matrix $\mathbf{A}$ after the preprocessing with the Randomized Hadamard Transform. The minimum ℓ_2-norm vector $\tilde{\mathbf{x}}_{opt} \in \mathbb{R}^d$ among those that achieve the minimum value $\tilde{\mathcal{Z}}$ in this problem is

$$\tilde{\mathbf{x}}_{opt} = \left(\mathbf{S}^T\mathbf{HDA}\right)^\dagger \mathbf{S}^T\mathbf{HDb},$$

which is the output of the RANDLEASTSQUARES algorithm. One can prove (and the proof is provided below) the following theorem about this algorithm.

Theorem 5.2.2. Suppose $\mathbf{A} \in \mathbb{R}^{n \times d}$ is a matrix of rank d, with n being a power of two. Let $\mathbf{b} \in \mathbb{R}^n$ and let $\epsilon \in (0,1)$. Run the RANDLEASTSQUARES algorithm with

$$r = \max\left\{48^2 d \ln(40nd) \ln\left(100^2 d \ln(40nd)\right), 40d\ln(40nd)/\epsilon\right\} \tag{5.2.3}$$

and return $\tilde{\mathbf{x}}_{opt}$. Then, with probability at least .8, the following two claims hold: first, $\tilde{\mathbf{x}}_{opt}$ satisfies

$$\|\mathbf{A}\tilde{\mathbf{x}}_{opt} - \mathbf{b}\|_2 \leqslant (1+\epsilon)\mathcal{Z},$$

where, recall, that $\mathcal{Z}$ is given in Eqn. (5.0.1); and, second, if we assume that $\|\mathbf{U}_A\mathbf{U}_A^T\mathbf{b}\|_2 \geqslant \gamma\|\mathbf{b}\|_2$ for some $\gamma \in (0,1]$, then $\tilde{\mathbf{x}}_{opt}$ satisfies

$$\|\mathbf{x}_{opt} - \tilde{\mathbf{x}}_{opt}\|_2 \leqslant \sqrt{\epsilon}\left(\kappa(\mathbf{A})\sqrt{\gamma^{-2}-1}\right)\|\mathbf{x}_{opt}\|_2.$$

Finally,

$$n(d+1) + 2n(d+1)\log_2(r+1) + O(rd^2)$$

time suffices to compute the solution $\tilde{\mathbf{x}}_{opt}$.

It is worth noting that the claims of Theorem 5.2.2 can be made to hold with probability $1-\delta$, for any $\delta > 0$, by repeating the algorithm $\lceil \ln(1/\delta)/\ln(5) \rceil$ times. Also, we note that if n is not a power of two we can pad $\mathbf{A}$ and $\mathbf{b}$ with all-zero rows in order to satisfy the assumption; this process at most doubles the size of the input matrix.

Remark 5.2.4. Assuming that $d \leqslant n \leqslant e^d$, and using $\max\{a_1, a_2\} \leqslant a_1 + a_2$, we get that

$$r = \mathcal{O}\left(d(\ln d)(\ln n) + \frac{d \ln n}{\epsilon}\right).$$

Thus, the running time of the RANDLEASTSQUARES algorithm becomes

$$\mathcal{O}\left(nd\ln\frac{d}{\epsilon}+d^3(\ln d)(\ln n)+\frac{d^3\ln n}{\epsilon}\right).$$

Assuming that $n/\ln n=\Omega(d^2)$, the above running time reduces to

$$\mathcal{O}\left(nd\ln\frac{d}{\epsilon}+\frac{nd\ln d}{\epsilon}\right).$$

For fixed ϵ, these improve the standard $O(nd^2)$ running time of traditional deterministic algorithms. It is worth noting that improvements over the standard $O(nd^2)$ time could be derived with weaker assumptions on n and d. However, for the sake of clarity of presentation, we only focus on the above setting.

Remark 5.2.5. The matrix $\mathbf{S}^\mathsf{T}\mathbf{HD}$ can be viewed in one of two equivalent ways: as a random preprocessing or random preconditioning, which "uniformizes" the leverage scores of the input matrix $\mathbf{A}$ (see Lemma 5.4.1 for a precise statement), followed by a uniform sampling operation; or as a Johnson-Lindenstrauss style random projection, which preserves the geometry of the entire span of $\mathbf{A}$, rather than just a discrete set of points (see Lemma 5.4.5 for a precise statement).

5.3. RandNLA algorithms as preconditioners. Stepping back, recall that the RANDLEASTSQUARES algorithm may be viewed as preconditioning the input matrix $\mathbf{A}$ and the target vector $\mathbf{b}$ with a carefully-constructed data-independent random matrix $\mathbf{X}$. (Since the analysis of the RANDLOWRANK algorithm, our main algorithm for low-rank matrix approximation, in Section 6 below, boils down to very similar ideas as the analysis of the RANDLEASTSQUARES algorithm, the ideas underlying the following discussion also apply to the RANDLOWRANK algorithm.) For our random sampling algorithm, we let $\mathbf{X}=\mathbf{S}^\mathsf{T}\mathbf{HD}$, where $\mathbf{S}$ is a matrix that represents the sampling operation and $\mathbf{HD}$ is the Randomized Hadamard Transform. Thus, we replace the least squares approximation problem of Eqn. (5.0.1) with the least squares approximation problem

$$\tilde{\mathcal{Z}}=\min_{\mathbf{x}\in\mathbb{R}^d}\|\mathbf{X}(\mathbf{Ax}-\mathbf{b})\|_2. \tag{5.3.1}$$

We explicitly compute the solution to the above problem using a traditional deterministic algorithm, e.g., by computing the vector

$$\tilde{\mathbf{x}}_{\text{opt}}=(\mathbf{XA})^\dagger\mathbf{Xb}. \tag{5.3.2}$$

Alternatively, one could use standard iterative methods such as the the Conjugate Gradient Normal Residual method, which can produce an ϵ-approximation to the optimal solution of Eqn. (5.3.1) in $O(\kappa(\mathbf{XA})rd\ln(1/\epsilon))$ time, where $\kappa(\mathbf{XA})$ is the condition number of $\mathbf{XA}$ and r is the number of rows of $\mathbf{XA}$. This was indeed the strategy implemented in the popular Blendenpik/LSRN approach [3].

We now state and prove a lemma that establishes sufficient conditions on *any* matrix $\mathbf{X}$ such that the solution vector $\tilde{\mathbf{x}}_{\text{opt}}$ to the least squares problem of Eqn. (5.3.1) will satisfy the relative-error bounds of Theorem 5.2.2. Recall that the SVD of $\mathbf{A}$ is $\mathbf{A} = \mathbf{U}_A \mathbf{\Sigma}_A \mathbf{V}_A^T$. In addition, for notational simplicity, we let $\mathbf{b}^{\perp} = \mathbf{U}_A^{\perp} {\mathbf{U}_A^{\perp}}^T \mathbf{b}$ denote the part of the right hand side vector $\mathbf{b}$ lying outside of the column space of $\mathbf{A}$.

The two conditions that we will require for the matrix $\mathbf{X}$ are:

$$\sigma_{min}^2(\mathbf{X}\mathbf{U}_A) \geqslant 1/\sqrt{2}; \text{ and} \tag{5.3.3}$$

$$\|\mathbf{U}_A^T \mathbf{X}^T \mathbf{X} \mathbf{b}^{\perp}\|_2^2 \leqslant \epsilon \mathcal{Z}^2/2, \tag{5.3.4}$$

for some $\epsilon \in (0, 1)$. Several things should be noted about these conditions.

- First, although Condition (5.3.3) only states that $\sigma_i^2(\mathbf{X}\mathbf{U}_A) \geqslant 1/\sqrt{2}$, for all $i = 1, \ldots, d$, our randomized algorithm satisfies the two-sided inequality $\left|1 - \sigma_i^2(\mathbf{X}\mathbf{U}_A)\right| \leqslant 1 - 1/\sqrt{2}$, for all $i = 1, \ldots, d$. This is equivalent to
$$\|I - \mathbf{U}_A^T \mathbf{X}^T \mathbf{X} \mathbf{U}_A\|_2 \leqslant 1 - 1/\sqrt{2}.$$
Thus, one should think of $\mathbf{X}\mathbf{U}_A$ as an approximate isometry.
- Second, the lemma is a deterministic statement, since it makes no explicit reference to a particular randomized algorithm and since $\mathbf{X}$ is not assumed to be constructed from a randomized process. Failure probabilities will enter later when we show that our randomized algorithm constructs an $\mathbf{X}$ that satisfies Conditions (5.3.3) and (5.3.4) with some probability.
- Third, Conditions (5.3.3) and (5.3.4) define what has come to be known as a *subspace embedding*, since it is an embedding that preserves the geometry of the entire subspace of the matrix $\mathbf{A}$. Such a subspace embedding can be *oblivious* (meaning that it is constructed without knowledge of the input matrix, as with random projection algorithms) or *non-oblivious* (meaning that it is constructed from information in the input matrix, as with data-dependent nonuniform sampling algorithms). This style of analysis represented a major advance in RandNLA algorithms, since it premitted much stronger bounds to be obtained than had been possible with previous methods. See [11] for the journal version (which was a combination and extension of two previous conference papers) of the first paper to use this style of analysis.
- Fourth, Condition (5.3.4) simply states that $\mathbf{X}\mathbf{b}^{\perp} = \mathbf{X}\mathbf{U}_A^{\perp} {\mathbf{U}_A^{\perp}}^T \mathbf{b}$ remains approximately orthogonal to $\mathbf{X}\mathbf{U}_A$. Clearly, before applying $\mathbf{X}$, it holds that $\mathbf{U}_A^T \mathbf{b}^{\perp} = 0$.
- Fifth, although Condition (5.3.4) depends on the right hand side vector $\mathbf{b}$, the RANDLEASTSQUARES algorithm will satisfy it without using any information from $\mathbf{b}$. (See Lemma 5.4.9 below.)

Given Conditions (5.3.3) and (5.3.4), we can establish the following lemma.

Lemma 5.3.5. Consider the overconstrained least squares approximation problem of Eqn. (5.0.1) and let the matrix $\mathbf{U}_A \in \mathbb{R}^{n \times d}$ contain the top d left singular vectors of $\mathbf{A}$. Assume that the matrix $\mathbf{X}$ satisfies Conditions (5.3.3) and (5.3.4) above, for some $\epsilon \in (0, 1)$. Then, the solution vector $\tilde{x}_{opt}$ to the least squares approximation problem (5.3.1) satisfies:

$$\|\mathbf{A}\tilde{\mathbf{x}}_{\text{opt}} - \mathbf{b}\|_2 \leqslant (1+\epsilon)\mathcal{Z}, \text{ and} \tag{5.3.6}$$

$$\|\mathbf{x}_{\text{opt}} - \tilde{\mathbf{x}}_{\text{opt}}\|_2 \leqslant \frac{1}{\sigma_{min}(\mathbf{A})}\sqrt{\epsilon}\mathcal{Z}. \tag{5.3.7}$$

Proof. Let us first rewrite the down-scaled regression problem induced by $\mathbf{X}$ as

$$\begin{aligned}
\min_{x\in\mathbb{R}^d} \|\mathbf{X}\mathbf{b} - \mathbf{X}\mathbf{A}\mathbf{x}\|_2^2 &= \min_{x\in\mathbb{R}^d} \|\mathbf{X}\mathbf{A}x - \mathbf{X}\mathbf{b}\|_2^2 \\
&= \min_{y\in\mathbb{R}^d} \|\mathbf{X}\mathbf{A}(\mathbf{x}_{\text{opt}} + y) - \mathbf{X}(\mathbf{A}x_{\text{opt}} + \mathbf{b}^\perp)\|_2^2 \qquad (5.3.8)\\
&= \min_{y\in\mathbb{R}^d} \|\mathbf{X}\mathbf{A}y - \mathbf{X}\mathbf{b}^\perp\|_2^2 \\
&= \min_{z\in\mathbb{R}^d} \|\mathbf{X}\mathbf{U}_A z - \mathbf{X}\mathbf{b}^\perp\|_2^2. \qquad (5.3.9)
\end{aligned}$$

Eqn. (5.3.8) follows since $\mathbf{b} = \mathbf{A}\mathbf{x}_{\text{opt}} + \mathbf{b}^\perp$ and Eqn. (5.3.9) follows since the columns of the matrix $\mathbf{A}$ span the same subspace as the columns of $\mathbf{U}_A$. Now, let $z_{\text{opt}} \in \mathbb{R}^d$ be such that $\mathbf{U}_A z_{\text{opt}} = \mathbf{A}(\tilde{\mathbf{x}}_{\text{opt}} - \mathbf{x}_{\text{opt}})$. Using this value for z_{opt}, we will prove that z_{opt} is minimizer of the above optimization problem, as follows:

$$\begin{aligned}
\|\mathbf{X}\mathbf{U}_A z_{\text{opt}} - \mathbf{X}\mathbf{b}^\perp\|_2^2 &= \|\mathbf{X}\mathbf{A}(\tilde{\mathbf{x}}_{\text{opt}} - \mathbf{x}_{\text{opt}}) - \mathbf{X}\mathbf{b}^\perp\|_2^2 \\
&= \|\mathbf{X}\mathbf{A}\tilde{\mathbf{x}}_{\text{opt}} - \mathbf{X}\mathbf{A}\mathbf{x}_{\text{opt}} - \mathbf{X}\mathbf{b}^\perp\|_2^2 \\
&= \|\mathbf{X}\mathbf{A}\tilde{\mathbf{x}}_{\text{opt}} - \mathbf{X}\mathbf{b}\|_2^2 \qquad (5.3.10)\\
&= \min_{x\in\mathbb{R}^d} \|\mathbf{X}\mathbf{A}\mathbf{x} - \mathbf{X}\mathbf{b}\|_2^2 \\
&= \min_{z\in\mathbb{R}^d} \|\mathbf{X}\mathbf{U}_A z - \mathbf{X}\mathbf{b}^\perp\|_2^2.
\end{aligned}$$

Eqn. (5.3.10) follows since $\mathbf{b} = \mathbf{A}\mathbf{x}_{\text{opt}} + \mathbf{b}^\perp$ and the last equality follows from Eqn. (5.3.9). Thus, by the normal equations (5.0.3), we have that

$$(\mathbf{X}\mathbf{U}_A)^T\mathbf{X}\mathbf{U}_A z_{\text{opt}} = (\mathbf{X}\mathbf{U}_A)^T\mathbf{X}\mathbf{b}^\perp.$$

Taking the norm of both sides and observing that under Condition (5.3.3) we have $\sigma_i((\mathbf{X}\mathbf{U}_A)^T\mathbf{X}\mathbf{U}_A) = \sigma_i^2(\mathbf{X}\mathbf{U}_A) \geqslant 1/\sqrt{2}$, for all i, it follows that

$$\|z_{\text{opt}}\|_2^2/2 \leqslant \|(\mathbf{X}\mathbf{U}_A)^T\mathbf{X}\mathbf{U}_A z_{\text{opt}}\|_2^2 = \|(\mathbf{X}\mathbf{U}_A)^T\mathbf{X}\mathbf{b}^\perp\|_2^2. \tag{5.3.11}$$

Using Condition (5.3.4) we observe that

$$\|z_{\text{opt}}\|_2^2 \leqslant \epsilon\mathcal{Z}^2. \tag{5.3.12}$$

To establish the first claim of the lemma, let us rewrite the norm of the residual vector as

$$\begin{aligned}
\|\mathbf{b} - \mathbf{A}\tilde{\mathbf{x}}_{\text{opt}}\|_2^2 &= \|\mathbf{b} - \mathbf{A}\mathbf{x}_{\text{opt}} + \mathbf{A}\mathbf{x}_{\text{opt}} - \mathbf{A}\tilde{\mathbf{x}}_{\text{opt}}\|_2^2 \\
&= \|\mathbf{b} - \mathbf{A}\mathbf{x}_{\text{opt}}\|_2^2 + \|\mathbf{A}\mathbf{x}_{\text{opt}} - \mathbf{A}\tilde{\mathbf{x}}_{\text{opt}}\|_2^2 \qquad (5.3.13)\\
&= \mathcal{Z}^2 + \|-\mathbf{U}_A z_{\text{opt}}\|_2^2 \qquad (5.3.14)\\
&\leqslant \mathcal{Z}^2 + \epsilon\mathcal{Z}^2, \qquad (5.3.15)
\end{aligned}$$

where Eqn. (5.3.13) follows by the Pythagorean theorem, since $\mathbf{b} - \mathbf{A}\mathbf{x}_{\mathrm{opt}} = \mathbf{b}^{\perp}$ is orthogonal to $\mathbf{A}$ and consequently to $\mathbf{A}(\mathbf{x}_{\mathrm{opt}} - \tilde{\mathbf{x}}_{\mathrm{opt}})$; Eqn. (5.3.14) follows by the definition of $\mathbf{z}_{\mathrm{opt}}$ and $\mathcal{Z}$; and Eqn. (5.3.15) follows by (5.3.12) and fact that $\mathbf{U}_A$ has orthonormal columns. The first claim of the lemma follows since $\sqrt{1+\epsilon} \leqslant 1+\epsilon$.

To establish the second claim of the lemma, recall that $\mathbf{A}(\mathbf{x}_{\mathrm{opt}} - \tilde{\mathbf{x}}_{\mathrm{opt}}) = \mathbf{U}_A \mathbf{z}_{\mathrm{opt}}$. If we take the norm of both sides of this expression, we have that

$$\|\mathbf{x}_{\mathrm{opt}} - \tilde{\mathbf{x}}_{\mathrm{opt}}\|_2^2 \leqslant \frac{\|\mathbf{U}_A \mathbf{z}_{\mathrm{opt}}\|_2^2}{\sigma_{\min}^2(\mathbf{A})} \tag{5.3.16}$$

$$\leqslant \frac{\epsilon \mathcal{Z}^2}{\sigma_{\min}^2(\mathbf{A})}, \tag{5.3.17}$$

where Eqn. (5.3.16) follows since $\sigma_{\min}(\mathbf{A})$ is the smallest singular value of $\mathbf{A}$ and since the rank of $\mathbf{A}$ is d; and Eqn. (5.3.17) follows by Eqn. (5.3.12) and the orthonormality of the columns of $\mathbf{U}_A$. Taking the square root, the second claim of the lemma follows. □

If we make no assumption on $\mathbf{b}$, then Eqn. (5.3.7) from Lemma 5.3.5 may provide a weak bound in terms of $\|\mathbf{x}_{\mathrm{opt}}\|_2$. If, on the other hand, we make the additional assumption that a constant fraction of the norm of $\mathbf{b}$ lies in the subspace spanned by the columns of $\mathbf{A}$, then Eqn. (5.3.7) can be strengthened. Such an assumption is reasonable, since most least-squares problems are practically interesting if at least some part of $\mathbf{b}$ lies in the subspace spanned by the columns of $\mathbf{A}$.

Lemma 5.3.18. Using the notation of Lemma 5.3.5, and additionally assuming that $\|\mathbf{U}_A \mathbf{U}_A^T \mathbf{b}\|_2 \geqslant \gamma \|\mathbf{b}\|_2$, for some fixed $\gamma \in (0,1]$, it follows that

$$\|\mathbf{x}_{\mathrm{opt}} - \tilde{\mathbf{x}}_{\mathrm{opt}}\|_2 \leqslant \sqrt{\epsilon}\left(\kappa(\mathbf{A})\sqrt{\gamma^{-2}-1}\right)\|\mathbf{x}_{\mathrm{opt}}\|_2. \tag{5.3.19}$$

Proof. Since $\|\mathbf{U}_A \mathbf{U}_A^T \mathbf{b}\|_2 \geqslant \gamma \|\mathbf{b}\|_2$, it follows that

$$\begin{aligned}
\mathcal{Z}^2 &= \|\mathbf{b}\|_2^2 - \|\mathbf{U}_A \mathbf{U}_A^T \mathbf{b}\|_2^2 \\
&\leqslant (\gamma^{-2}-1)\|\mathbf{U}_A \mathbf{U}_A^T \mathbf{b}\|_2^2 \\
&\leqslant \sigma_{\max}^2(\mathbf{A})(\gamma^{-2}-1)\|\mathbf{x}_{\mathrm{opt}}\|_2^2.
\end{aligned}$$

This last inequality follows from $\mathbf{U}_A \mathbf{U}_A^T \mathbf{b} = \mathbf{A}\mathbf{x}_{\mathrm{opt}}$, which implies

$$\|\mathbf{U}_A \mathbf{U}_A^T \mathbf{b}\|_2 = \|\mathbf{A}\mathbf{x}_{\mathrm{opt}}\|_2 \leqslant \|\mathbf{A}\|_2 \|\mathbf{x}_{\mathrm{opt}}\|_2 = \sigma_{\max}(\mathbf{A})\, \|\mathbf{x}_{\mathrm{opt}}\|_2.$$

By combining this with Eqn. (5.3.7) of Lemma 5.3.5, the lemma follows. □

5.4. The proof of Theorem 5.2.2. To prove Theorem 5.2.2, we adopt the following approach: we first show that the Randomized Hadamard Transform has the effect preprocessing or preconditioning the input matrix to make the leverage scores approximately uniform; and we then show that Condition (5.3.3) and (5.3.4) can be satisfied by sampling uniformly on the preconditioned input. The theorem will then follow from Lemma 5.3.5.

The effect of the Randomized Hadamard Transform. We start by stating a lemma that quantifies the manner in which $\mathbf{HD}$ approximately "uniformizes" information in the left singular subspace of the matrix $\mathbf{A}$; this will allow us to sample uniformly and apply our randomized matrix multiplication results from Section 4 in order to analyze the proposed algorithm. We state the lemma for a general $n \times d$ orthogonal matrix $\mathbf{U}$ such that $\mathbf{U}^T\mathbf{U} = \mathbf{I}_d$.

Lemma 5.4.1. Let $\mathbf{U}$ be an $n \times d$ orthogonal matrix and let the product $\mathbf{HD}$ be the $n \times n$ Randomized Hadamard Transform of Section 5.1. Then, with probability at least .95,

$$\|(\mathbf{HDU})_{i*}\|_2^2 \leqslant \frac{2d\ln(40nd)}{n}, \qquad \text{for all } i = 1, \ldots, n. \tag{5.4.2}$$

The following well-known inequality [15, Theorem 2] will be useful in the proof. (See also the chapter by Vershynin in this volume [28] for related results.)

Lemma 5.4.3. Let X_i, $i = 1, \ldots, n$ be independent random variables with finite first and second moments such that, for all i, $a_i \leqslant X_i \leqslant b_i$. Then, for any $t > 0$,

$$\mathbf{Pr}\left[\left|\sum_{i=1}^{n} X_i - \sum_{i=1}^{n} \mathbf{E}[X_i]\right| \geqslant nt\right] \leqslant 2\exp\left(-\frac{2n^2t^2}{\sum_{i=1}^{n}(a_i - b_i)^2}\right).$$

Given this lemma, we now provide the proof of Lemma 5.4.1.

Proof. (of Lemma 5.4.1) Consider $(\mathbf{HDU})_{ij}$ for some i, j (recalling that $i = 1, \ldots, n$ and $j = 1, \ldots, d$). Recall that $\mathbf{D}$ is a diagonal matrix; then,

$$(\mathbf{HDU})_{ij} = \sum_{\ell=1}^{n} \mathbf{H}_{i\ell}\mathbf{D}_{\ell\ell}\mathbf{U}_{\ell j} = \sum_{\ell=1}^{n} \mathbf{D}_{\ell\ell}\left(\mathbf{H}_{i\ell}\mathbf{U}_{\ell j}\right) = \sum_{\ell=1}^{n} X_\ell.$$

Let $X_\ell = \mathbf{D}_{\ell\ell}\left(\mathbf{H}_{i\ell}\mathbf{U}_{\ell j}\right)$ be our set of n (independent) random variables. By the construction of $\mathbf{D}$ and $\mathbf{H}$, it is easy to see that $\mathbf{E}[X_\ell] = 0$; also,

$$|X_\ell| = \left|\mathbf{D}_{\ell\ell}\left(\mathbf{H}_{i\ell}\mathbf{U}_{\ell j}\right)\right| \leqslant \frac{1}{\sqrt{n}}\left|\mathbf{U}_{\ell j}\right|.$$

Applying Lemma 5.4.3, we get

$$\mathbf{Pr}\left[\left|(\mathbf{HDU})_{ij}\right| \geqslant nt\right] \leqslant 2\exp\left(-\frac{2n^3t^2}{4\sum_{\ell=1}^{n}\mathbf{U}_{\ell j}^2}\right) = 2\exp\left(-n^3t^2/2\right).$$

In the last equality we used the fact that $\sum_{\ell=1}^{n}\mathbf{U}_{\ell j}^2 = 1$, i.e., that the columns of $\mathbf{U}$ are unit-length. Let the right-hand side of the above inequality be equal to δ and solve for t to get

$$\mathbf{Pr}\left[\left|(\mathbf{HDU})_{ij}\right| \geqslant \sqrt{\frac{2\ln(2/\delta)}{n}}\right] \leqslant \delta.$$

Let $\delta = 1/(20nd)$ and apply the union bound over all nd possible index pairs (i, j) to get that, with probability at least 1-1/20=0.95, for all i, j,

$$\left|(\mathbf{HDU})_{ij}\right| \leqslant \sqrt{\frac{2\ln(40nd)}{n}}.$$

Thus,

$$\| (\mathbf{HDU})_{i*} \|_2^2 = \sum_{j=1}^{d} (\mathbf{HDU})_{ij}^2 \leqslant \frac{2d\ln(40nd)}{n} \tag{5.4.4}$$

for all $i = 1, \ldots, n$, which concludes the proof of the lemma. □

Satisfying Condition (5.3.3). We next prove the following lemma, which states that all the singular values of $\mathbf{S}^T\mathbf{HDU}_A$ are close to one, and in particular that Condition (5.3.3) is satisfied by the RandLeastSquares algorithm. The proof of this Lemma 5.4.5 essentially follows from our results in Theorem 4.3.5 for the RandMatrixMultiply algorithm (for approximating the product of a matrix and its transpose).

Lemma 5.4.5. Assume that Eqn. (5.4.2) holds. If

$$r \geqslant 48^2 d\ln(40nd)\ln\left(100^2 d\ln(40nd)\right), \tag{5.4.6}$$

then, with probability at least .95,

$$\left|1 - \sigma_i^2\left(\mathbf{S}^T\mathbf{HDU}_A\right)\right| \leqslant 1 - \frac{1}{\sqrt{2}}$$

holds for all $i = 1, \ldots, d$.

Proof. (of Lemma 5.4.5) Note that for all $i = 1, \ldots, d$,

$$\begin{aligned} &\left|1 - \sigma_i^2\left(\mathbf{S}^T\mathbf{HDU}_A\right)\right| \\ = &\left|\sigma_i\left(\mathbf{U}_A^T\mathbf{DH}^T\mathbf{HDU}_A\right) - \sigma_i\left(\mathbf{U}_A^T\mathbf{DH}^T\mathbf{SS}^T\mathbf{HDU}_A\right)\right| \\ \leqslant &\ \|\mathbf{U}_A^T\mathbf{DH}^T\mathbf{HDU}_A - \mathbf{U}_A^T\mathbf{DH}^T\mathbf{SS}^T\mathbf{HDU}_A\|_2. \end{aligned} \tag{5.4.7}$$

In the above, we used the fact that $\mathbf{U}_A^T\mathbf{DH}^T\mathbf{HDU}_A = \mathbf{I}_d$ and inequality (2.6.2) that was discussed in our Linear Algebra review in Section 2.6. We now view $\mathbf{U}_A^T\mathbf{DH}^T\mathbf{SS}^T\mathbf{HDU}_A$ as an approximation to the product of two matrices, namely $\mathbf{U}_A^T\mathbf{DH}^T = (\mathbf{HDU}_A)^T$ and $\mathbf{HDU}_A$, constructed by randomly sampling and rescaling columns of $(\mathbf{HDU}_A)^T$. Thus, we can leverage Theorem 4.3.5.

More specifically, consider the matrix $(\mathbf{HDU}_A)^T$. Obviously, since $\mathbf{H}$, $\mathbf{D}$, and $\mathbf{U}_A$ are orthogonal matrices, $\|\mathbf{HDU}_A\|_2 = 1$ and $\|\mathbf{HDU}_A\|_F = \|\mathbf{U}_A\|_F = \sqrt{d}$. Let $\beta = (2\ln(40nd))^{-1}$; since we assumed that Eqn. (5.4.2) holds, we note that the columns of $(\mathbf{HDU}_A)^T$, which correspond to the rows of $\mathbf{HDU}_A$, satisfy

$$\frac{1}{n} \geqslant \beta\frac{\| (\mathbf{HDU}_A)_{i*} \|_2^2}{\|\mathbf{HDU}_A\|_F^2}, \qquad \text{for all } i = 1, \ldots, n. \tag{5.4.8}$$

Thus, applying Theorem 4.3.5 taking $\beta = (2\ln(40nd))^{-1}$, $\epsilon = 1 - 1/\sqrt{2}$, and $\delta = 1/20$ implies that

$$\|\mathbf{U}_A^T\mathbf{DH}^T\mathbf{HU}_A - \mathbf{U}_A^T\mathbf{DH}^T\mathbf{SS}^T\mathbf{HDU}_A\|_2 \leqslant 1 - \frac{1}{\sqrt{2}}$$

holds with probability at least $1 - 1/20 = .95$. For the above bound to hold, we need r to assume the value of Eqn. (5.4.6). Finally, we note that since we have $\|\mathbf{HDU}_A\|_F^2 = d \geqslant 1$, the assumption of Theorem 4.3.5 on the Frobenius norm of the input matrix is always satisfied. Combining the above with inequality (5.4.8) concludes the proof of the lemma. □

Satisfying Condition (5.3.4). We next prove the following lemma, which states that Condition (5.3.4) is satisfied by the RANDLEASTSQUARES algorithm. The proof of this Lemma 5.4.9 again essentially follows from our bounds for the RANDMATRIXMULTIPLY algorithm from Section 4 (except here it is used for approximating the product of a matrix and a vector).

Lemma 5.4.9. Assume that Eqn. (5.4.2) holds. If $r \geqslant 40d\ln(40nd)/\epsilon$, then, with probability at least .9,

$$\| \left(S^T H D U_A\right)^T S^T H D b^\perp \|_2^2 \leqslant \epsilon \mathcal{Z}^2/2.$$

Proof. (of Lemma 5.4.9) Recall that $b^\perp = U_A^\perp {U_A^\perp}^T b$ and that $\mathcal{Z} = \|b^\perp\|_2$. We start by noting that since $\|U_A^T D H^T H D b^\perp\|_2^2 = \|U_A^T b^\perp\|_2^2 = 0$ it follows that

$$\| \left(S^T H D U_A\right)^T S^T H D b^\perp \|_2^2 = \|U_A^T D H^T S S^T H D b^\perp - U_A^T D H^T H D b^\perp\|_2^2.$$

Thus, $\left(S^T H D U_A\right)^T S^T H D b^\perp$ can be viewed as approximating the product of the two matrices, $(HDU_A)^T$ and $HDb^\perp$, by randomly sampling columns from $(HDU_A)^T$ and rows (elements) from $HDb^\perp$. Note that the sampling probabilities are uniform and do not depend on the norms of the columns of $(HDU_A)^T$ or the rows of $Hb^\perp$. We will apply the bounds of Eqn. (4.2.4), after arguing that the assumptions of Eqn. (4.2.3) are satisfied. Indeed, since we condition on Eqn. (5.4.2) holding, the rows of HDU_A (which of course correspond to columns of $(HDU_A)^T$) satisfy

$$\frac{1}{n} \geqslant \beta \frac{\|(HDU_A)_{i*}\|_2^2}{\|HDU_A\|_F^2}, \qquad \text{for all } i = 1, \ldots, n, \tag{5.4.10}$$

for $\beta = (2\ln(40nd))^{-1}$. Thus, Eqn. (4.2.4) implies

$$\mathbf{E}\left[\| \left(S^T H D U_A\right)^T S^T H D b^\perp \|_2^2\right] \leqslant \frac{1}{\beta r}\|HDU_A\|_F^2 \|HDb^\perp\|_2^2 = \frac{d\mathcal{Z}^2}{\beta r}.$$

In the above we used $\|HDU_A\|_F^2 = d$. Markov's inequality now implies that with probability at least .9,

$$\| \left(S^T H D U_A\right)^T S^T H D b^\perp \|_2^2 \leqslant \frac{10 d\mathcal{Z}^2}{\beta r}.$$

Setting $r \geqslant 20d/(\beta\epsilon)$ and using the value of β specified above concludes the proof of the lemma. □

Completing the proof of Theorem 5.2.2. The theorem follows since Lemma 5.4.5 and Lemma 5.4.9 establish that the sufficient conditions of Lemma 5.3.5 hold. In more detail, we now complete the proof of Theorem 5.2.2. First, let $\mathcal{E}_{(5.4.2)}$ denote the event that Eqn. (5.4.2) holds; clearly, $\mathbf{Pr}\left[\mathcal{E}_{(5.4.2)}\right] \geqslant .95$. Second, let

$\mathcal{E}_{5.4.5,5.4.9|(5.4.2)}$ denote the event that both Lemmas 5.4.5 and 5.4.9 hold conditioned on $\mathcal{E}_{(5.4.2)}$ holding. Then,

$$\begin{aligned}
\mathcal{E}_{5.4.5,5.4.9|(5.4.2)} &= 1 - \overline{\mathcal{E}_{5.4.5,5.4.9|(5.4.2)}} \\
&= 1 - \mathbf{Pr}\left(\left(\text{Lemma 5.4.5 does not hold}|\mathcal{E}_{(5.4.2)}\right)\right. \\
&\qquad\qquad \left.\textbf{or } \left(\text{Lemma 5.4.9 does not hold}|\mathcal{E}_{(5.4.2)}\right)\right) \\
&\geqslant 1 - \mathbf{Pr}\left[\left(\text{Lemma 5.4.5 does not hold}|\mathcal{E}_{(5.4.2)}\right)\right] \\
&\qquad\qquad - \mathbf{Pr}\left[\left(\text{Lemma 5.4.9 does not hold}|\mathcal{E}_{(5.4.2)}\right)\right] \\
&\geqslant 1 - .05 - .1 = .85.
\end{aligned}$$

In the above, $\overline{\mathcal{E}}$ denotes the complement of event $\mathcal{E}$. In the first inequality we used the union bound and in the second inequality we leveraged the bounds for the failure probabilities of Lemmas 5.4.5 and 5.4.9, given that Eqn. (5.4.2) holds. We now let $\mathcal{E}$ denote the event that both Lemmas 5.4.5 and 5.4.9 hold, without any a priori conditioning on event $\mathcal{E}_{(5.4.2)}$; we will bound $\mathbf{Pr}[\mathcal{E}]$ as follows:

$$\begin{aligned}
\mathbf{Pr}[\mathcal{E}] &= \mathbf{Pr}\left[\mathcal{E}|\mathcal{E}_{(5.4.2)}\right] \cdot \mathbf{Pr}\left[\mathcal{E}_{(5.4.2)}\right] + \mathbf{Pr}\left[\mathcal{E}|\overline{\mathcal{E}_{(5.4.2)}}\right] \cdot \mathbf{Pr}\left[\overline{\mathcal{E}_{(5.4.2)}}\right] \\
&\geqslant \mathbf{Pr}\left[\mathcal{E}|\mathcal{E}_{(5.4.2)}\right] \cdot \mathbf{Pr}\left[\mathcal{E}_{(5.4.2)}\right] \\
&= \mathbf{Pr}\left[\mathcal{E}_{5.4.5,5.4.9|(5.4.2)}|\mathcal{E}_{(5.4.2)}\right] \cdot \mathbf{Pr}\left[\mathcal{E}_{(5.4.2)}\right] \\
&\geqslant .85 \cdot .95 \geqslant .8.
\end{aligned}$$

In the first inequality we used the fact that all probabilities are positive. The above derivation immediately bounds the success probability of Theorem 5.2.2. Combining Lemmas 5.4.5 and 5.4.9 with the structural results of Lemma 5.3.5 and setting r as in Eqn. (5.2.3) concludes the proof of the accuracy guarantees of Theorem 5.2.2.

5.5. The running time of the RandLeastSquares algorithm. We now discuss the running time of the RandLeastSquares algorithm. First of all, by the construction of $\mathbf{S}$, the number of non-zero entries in $\mathbf{S}$ is r. In Step 6 we need to compute the products $\mathbf{S}^T\mathbf{HDA}$ and $\mathbf{S}^T\mathbf{HDb}$. Recall that $\mathbf{A}$ has d columns and thus the running time of computing both products is equal to the time needed to apply $\mathbf{S}^T\mathbf{HD}$ on $(d+1)$ vectors. In order to apply $\mathbf{D}$ on $(d+1)$ vectors in $\mathbb{R}^n$, $n(d+1)$ operations suffice. In order to estimate how many operations are needed to apply $\mathbf{S}^T\mathbf{H}$ on $(d+1)$ vectors, we use the following analysis that was first proposed in [2, Section 7].

Let $\mathbf{x}$ be any vector in $\mathbb{R}^n$; multiplying $\mathbf{H}$ by $\mathbf{x}$ can be done as follows:

$$\begin{pmatrix} \mathbf{H}_{n/2} & \mathbf{H}_{n/2} \\ \mathbf{H}_{n/2} & -\mathbf{H}_{n/2} \end{pmatrix} \begin{pmatrix} \mathbf{x}_1 \\ \mathbf{x}_2 \end{pmatrix} = \begin{pmatrix} \mathbf{H}_{n/2}(\mathbf{x}_1 + \mathbf{x}_2) \\ \mathbf{H}_{n/2}(\mathbf{x}_1 - \mathbf{x}_2) \end{pmatrix}.$$

Let $T(n)$ be the number of operations required to perform this operation for n-dimensional vectors. Then,

$$T(n) = 2T(n/2) + n,$$

and thus $T(n) = O(n \log n)$. We can now include the sub-sampling matrix $\mathbf{S}$ to get

$$\begin{pmatrix} \mathbf{S}_1 & \mathbf{S}_2 \end{pmatrix} \begin{pmatrix} \mathbf{H}_{n/2} & \mathbf{H}_{n/2} \\ \mathbf{H}_{n/2} & -\mathbf{H}_{n/2} \end{pmatrix} \begin{pmatrix} \mathbf{x}_1 \\ \mathbf{x}_2 \end{pmatrix} = \mathbf{S}_1 \mathbf{H}_{n/2}(\mathbf{x}_1 + \mathbf{x}_2) + \mathbf{S}_2 \mathbf{H}_{n/2}(\mathbf{x}_1 - \mathbf{x}_2).$$

Let $\mathrm{nnz}(\cdot)$ denote the number of non-zero entries of its argument. Then,

$$T(n, \mathrm{nnz}(\mathbf{S})) = T(n/2, \mathrm{nnz}(\mathbf{S}_1)) + T(n/2, \mathrm{nnz}(\mathbf{S}_2)) + n.$$

From standard methods in the analysis of recursive algorithms, we can now use the fact that $r = \mathrm{nnz}(\mathbf{S}) = \mathrm{nnz}(\mathbf{S}_1) + \mathrm{nnz}(\mathbf{S}_2)$ to prove that

$$T(n, r) \leqslant 2n \log_2(r+1).$$

Towards that end, let $r_1 = \mathrm{nnz}(\mathbf{S}_1)$ and let $r_2 = \mathrm{nnz}(\mathbf{S}_2)$. Then,

$$\begin{aligned} T(n, r) &= T(n/2, r_1) + T(n/2, r_2) + n \\ &\leqslant 2\frac{n}{2}\log_2(r_1+1) + 2\frac{n}{2}\log_2(r_2+1) + n\log_2 2 \\ &= n\log_2(2(r_1+1)(r_2+1)) \\ &\leqslant n\log_2(r+1)^2 \\ &= 2n\log_2(r+1). \end{aligned}$$

The last inequality follows from simple algebra using $r = r_1 + r_2$. Thus, at most $2n(d+1)\log_2(r+1)$ operations are needed to apply $\mathbf{S}^T\mathbf{H}\mathbf{D}$ on $d+1$ vectors. After this preprocessing, the RANDLEASTSQUARES algorithm must compute the pseudoinverse of an $r \times d$ matrix, or, equivalently, solve a least-squares problem on r constraints and d variables. This operation can be performed in $O(rd^2)$ time since $r \geqslant d$. Thus, the entire algorithm runs in time

$$n(d+1) + 2n(d+1)\log_2(r+1) + \mathcal{O}\left(rd^2\right).$$

5.6. References. Our presentation in this chapter follows closely the derivations in [11]; see [11] for a detailed discussion of prior work on this topic. We also refer the interested reader to [3,30] for followup work on randomized solvers for least-squares problems.

6. A RandNLA Algorithm for Low-rank Matrix Approximation

In this section, we will present a simple randomized matrix algorithm for low-rank matrix approximation. Algorithms to compute low-rank approximations to matrices have been of paramount importance historically in scientific computing (see, for example, [23] for traditional numerical methods based on subspace iteration and Krylov subspaces to compute such approximations) as well as more recently in machine learning and data analysis. RandNLA has pioneered an alternative approach, by applying random sampling and random projection algorithms to construct such low-rank approximations with provable accuracy guarantees; see [7] for early work on the topic and [14, 17, 18, 30] for overviews of more recent approaches. In this section, we will present and analyze a simple algorithm to approximate the top k left singular vectors of a matrix $\mathbf{A} \in \mathbb{R}^{m \times n}$. Many RandNLA methods for low-rank approximation boil down to variants of this basic technique; see, e.g., the chapter by Martinsson in this volume [19]. Unlike the previous section on RandNLA algorithms for regression problems, no particular assumptions will be imposed on m and n; indeed, $\mathbf{A}$ could be a square matrix.

6.1. The main algorithm and main theorem. Our main algorithm is quite simple and again leverages the Randomized Hadamard Tranform of Section 5.1. Indeed, let the matrix product $\mathbf{HD}$ denote the $n \times n$ Randomized Hadamard Transform. First, we *postmultiply* the input matrix $\mathbf{A} \in \mathbb{R}^{m \times n}$ by $(\mathbf{HD})^T$, thus forming a new matrix $\mathbf{ADH} \in \mathbb{R}^{m \times n}$.[1] Then, we sample (uniformly at random) c columns from the matrix $\mathbf{ADH}$, thus forming a *smaller* matrix $\mathbf{C} \in \mathbb{R}^{m \times c}$. Finally, we use a Ritz-Rayleigh type procedure to construct approximations $\tilde{\mathbf{U}}_k \in \mathbb{R}^{m \times k}$ to the top k left singular vectors of $\mathbf{A}$ from $\mathbf{C}$; these approximations lie within the column space of $\mathbf{C}$.

See the RandLowRank algorithm (Algorithm 6.1.4) for a detailed description of this procedure, using a sampling-and-rescaling matrix $\mathbf{S} \in \mathbb{R}^{n \times c}$ to form the matrix $\mathbf{C}$. Theorem 6.1.1 is our main quality-of-approximation result for the RandLowRank algorithm.

Theorem 6.1.1. Let $\mathbf{A} \in \mathbb{R}^{m \times n}$, let k be a rank parameter, and let $\epsilon \in (0, 1/2]$. If we set

$$c \geqslant c_0 \frac{k \ln n}{\epsilon^2} \left(\ln \frac{k}{\epsilon^2} + \ln \ln n \right), \tag{6.1.2}$$

(for a fixed constant c_0) then, with probability at least .85, the RandLowRank algorithm returns a matrix $\tilde{\mathbf{U}}_k \in \mathbb{R}^{m \times k}$ such that

$$\|\mathbf{A} - \tilde{\mathbf{U}}_k \tilde{\mathbf{U}}_k^T \mathbf{A}\|_F \leqslant (1+\epsilon)\|\mathbf{A} - \mathbf{U}_k \mathbf{U}_k^T \mathbf{A}\|_F = (1+\epsilon)\|\mathbf{A} - \mathbf{A}_k\|_F. \tag{6.1.3}$$

(Here, $\mathbf{U}_k \in \mathbb{R}^{m \times k}$ contains the top k left singular vectors of $\mathbf{A}$). The running time of the RandLowRank algorithm is $O(mnc)$.

We discuss the dimensions of the matrices in steps 6-9 of the RandLowRank algorithm. One can think of the matrix $\mathbf{C} \in \mathbb{R}^{m \times c}$ as a "sketch" of the input matrix $\mathbf{A}$. Notice that c is $O(k \ln k)$ (up to $\ln \ln$ factors and ignoring constant

[1] Alternatively, we could *premultiply* $\mathbf{A}^T$ by $\mathbf{HD}$. The reader should become comfortable going back and forth with such manipulations.

terms like ϵ and δ) and that the rank of $\mathbf{C}$ (denoted by ρ_C) is at least k, i.e., $\rho_C \geqslant k$. The matrix $\mathbf{U}_C$ has dimensions $m \times \rho_C$ and the matrix $\mathbf{W}$ has dimensions $\rho_C \times n$. Finally, the matrix $\mathbf{U}_{W,k}$ has dimensions $\rho_C \times k$ (by our assumption on the rank of $\mathbf{W}$).

Recall that the *best* rank-k approximation to $\mathbf{A}$ is equal to $\mathbf{A}_k = \mathbf{U}_k\mathbf{U}_k^T\mathbf{A}$. In words, Theorem 6.1.1 argues that the RandLowRank algorithm returns a set of k orthonormal vectors that are excellent approximations to the top k left singular vectors of $\mathbf{A}$, in the sense that projecting $\mathbf{A}$ on the subspace spanned by $\tilde{\mathbf{U}}_k$ returns a matrix that has residual error that is close to that of $\mathbf{A}_k$.

Input: $\mathbf{A} \in \mathbb{R}^{m\times n}$, a rank parameter $k \ll \min\{m, n\}$, and an error parameter $\epsilon \in (0, 1/2)$.

Output: $\tilde{\mathbf{U}}_k \in \mathbb{R}^{m\times k}$.

(1) Let c assume the value of Eqn. (6.1.2).

(2) Let $\mathbf{S}$ be an empty matrix.

(3) **For** $t = 1, \ldots, c$ (i.i.d. trials with replacement) **select uniformly at random** an integer from $\{1, 2, \ldots, n\}$. **If** i is selected, **then** append the column vector $\left(\sqrt{n/c}\right)\mathbf{e}_i$ to $\mathbf{S}$, where $\mathbf{e}_i \in \mathbb{R}^n$ is the i-th canonical vector.

(4) Let $\mathbf{H} \in \mathbb{R}^{n\times n}$ be the normalized Hadamard transform matrix.

(5) Let $\mathbf{D} \in \mathbb{R}^{n\times n}$ be a diagonal matrix with

$$\mathbf{D}_{ii} = \begin{cases} +1 & \text{, with probability } 1/2 \\ -1 & \text{, with probability } 1/2 \end{cases}$$

(6) Compute $\mathbf{C} = \mathbf{ADHS} \in \mathbb{R}^{m\times c}$.

(7) Compute $\mathbf{U}_C$, a basis for the column space of $\mathbf{C}$.

(8) Compute $\mathbf{W} = \mathbf{U}_C^T\mathbf{A}$ and (assuming that its rank is at least k), compute its top k left singular vectors $\mathbf{U}_{W,k}$.

(9) Return $\tilde{\mathbf{U}}_k = \mathbf{U}_C\mathbf{U}_{W,k} \in \mathbb{R}^{m\times k}$.

Algorithm 6.1.4: The RandLowRank algorithm

Remark 6.1.5. We stress that the $O(mnc)$ running time of the RandLowRank algorithm is due to the Ritz-Rayleigh type procedure in steps (7)-(9). These steps guarantee that the proposed algorithm returns a matrix $\tilde{\mathbf{U}}_k$ with *exactly* k columns that approximates the top k left singular vectors of $\mathbf{A}$. The results of [19] focus (in our parlance) on the matrix $\mathbf{C}$, which can be constructed much faster (see Section 6.5), in $O(mn\log_2 c)$ time, but has *more than* k columns. By bounding the error term $\|\mathbf{A} - \mathbf{C}\mathbf{C}^{\dagger}\mathbf{A}\|_F = \|\mathbf{A} - \mathbf{U}_C\mathbf{U}_C^T\mathbf{A}\|_F$, one can prove that the column span of $\mathbf{C}$ contains good approximations to the top k left singular vectors of $\mathbf{A}$.

Remark 6.1.6. Repeating the RANDLOWRANK algorithm $\lceil \ln(1/\delta)/\ln 5 \rceil$ times and keeping the matrix $\tilde{\mathbf{U}}_k$ that minimizes the error $\|\mathbf{A} - \tilde{\mathbf{U}}_k\tilde{\mathbf{U}}_k^T\mathbf{A}\|_F$ reduces the failure probability of the algorithm to at most $1-\delta$, for any $\delta \in (0,1)$.

Remark 6.1.7. As with the sampling process in the RANDLEASTSQUARES algorithm, the operation represented by $\mathbf{DHS}$ in the RANDLOWRANK algorithm can be viewed in one of two equivalent ways: either as a random preconditioning followed by a uniform sampling operation; or as a Johnson-Lindenstrauss style random projection. (In particular, informally, the RANDLOWRANK algorithm "works" for the following reason. If a matrix is well-approximated by a low-rank matrix, then there is redundancy in the columns (and/or rows), and thus random sampling "should" be successful at selecting a good set of columns. That said, just as with the RANDLEASTSQUARES algorithm, there may be some columns that are more important to select, e.g., that have high leverage. Thus, using a random projection, which transforms the input to a new basis where the leverage scores of different columns are uniformized, amounts to preconditioning the input such that uniform sampling is appropriate.)

Remark 6.1.8. The value c is essentially[2] equal to $O((k/\epsilon^2)\ln(k/\epsilon)\ln n)$. For constant ϵ, this grows as a function of $k \ln k$ and $\ln n$.

Remark 6.1.9. Similar bounds can be proven for many other random projection algorithms (using different values for c) and not just the Randomized Hadamard Transform. Well-known alternatives include random Gaussian matrices, the Randomized Discrete Cosine Transform, sparsity-preserving random projections, etc. Which variant is most appropriate in a given situation depends on the sparsity structure of the matrix, the noise properties of the data, the model of data access, etc. See [17,30] for an overview of similar results.

Remark 6.1.10. One can generalize the RANDLOWRANK algorithm to work with the matrix $(\mathbf{A}\mathbf{A}^T)^t\mathbf{ADHS}$ for integer $t \geqslant 0$. This would result in subspace iteration. If all intermediate iterates (for $t = 0, 1, \ldots$) are kept, the Krylov subspace would be formed. See [8,21] and references therein for a detailed treatment and analysis of such methods. (See also the chapter by Martinsson in this volume [19] for related results.)

The remainder of this section will focus on the proof of Theorem 6.1.1. Our proof strategy will consist of three steps. First (Section 6.2), we we will prove that:

$$\|\mathbf{A} - \tilde{\mathbf{U}}_k\tilde{\mathbf{U}}_k^T\mathbf{A}\|_F^2 \leqslant \|\mathbf{A}_k - \mathbf{U}_C\mathbf{U}_C^T\mathbf{A}_k\|_F^2 + \|\mathbf{A}_{k,\perp}\|_F^2.$$

This inequality allows us to focus on the easier-to-bound term $\|\mathbf{A}_k - \mathbf{U}_C\mathbf{U}_C^T\mathbf{A}_k\|_F^2$ instead of the term $\|\mathbf{A} - \tilde{\mathbf{U}}_k\tilde{\mathbf{U}}_k^T\mathbf{A}\|_F^2$. Second (Section 6.3), to bound this term, we will use a structural inequality that is central (in this form or mild variations) in

[2]We omit the $\ln \ln n$ term from this qualitative remark. Recall that $\ln \ln n$ goes to infinity with dignity and therefore, quoting Stan Eisenstat, $\ln \ln n$ is for all practical purposes essentially a constant; see https://rjlipton.wordpress.com/2011/01/19/we-believe-a-lot-but-can-prove-little/.

many RandNLA low-rank approximation algorithms and their analyses. Indeed, we will argue that

$$\|\mathbf{A}_k - \mathbf{U}_C\mathbf{U}_C^T\mathbf{A}_k\|_F^2 \leqslant 2\|(\mathbf{A}-\mathbf{A}_k)\mathbf{DHS}((\mathbf{V}_k^T\mathbf{DHS})^\dagger - (\mathbf{V}_k^T\mathbf{DHS})^T)\|_F^2 \\ + 2\|(\mathbf{A}-\mathbf{A}_k)\mathbf{DHS}(\mathbf{V}_k^T\mathbf{DHS})^T\|_F^2.$$

Third (Section 6.4), we will use results from Section 4 to bound the two terms at the right hand side of the above inequality.

6.2. An alternative expression for the error. The RANDLOWRANK algorithm approximates the top k left singular vectors of $\mathbf{A}$, i.e., the matrix $\mathbf{U}_k \in \mathbb{R}^{m\times k}$, by the orthonormal matrix $\tilde{\mathbf{U}}_k \in \mathbb{R}^{m\times k}$. Bounding $\|\mathbf{A} - \tilde{\mathbf{U}}_k\tilde{\mathbf{U}}_k^T\mathbf{A}\|_F$ directly seems hard, so we present an alternative expression that is easier to analyze and that also reveals an interesting insight for $\tilde{\mathbf{U}}_k$. We will prove that the matrix $\tilde{\mathbf{U}}_k\tilde{\mathbf{U}}_k\mathbf{A}$ is the best rank-k approximation to $\mathbf{A}$ (with respect to the Frobenius norm[3]) that lies within the column space of the matrix $\mathbf{C}$. This optimality property is guaranteed by the Ritz-Rayleigh type procedure implemented in Steps 7-9 of the RANDLOWRANK algorithm.

Lemma 6.2.1. Let $\mathbf{U}_C$ be a basis for the column span of $\mathbf{C}$ and let $\tilde{\mathbf{U}}_k$ be the output of the RANDLOWRANK algorithm. Then

$$\mathbf{A} - \tilde{\mathbf{U}}_k\tilde{\mathbf{U}}_k^T\mathbf{A} = \mathbf{A} - \mathbf{U}_C\left(\mathbf{U}_C^T\mathbf{A}\right)_k. \tag{6.2.2}$$

In addition, $\mathbf{U}_C(\mathbf{U}_C^T\mathbf{A})_k$ is the best rank-k approximation to $\mathbf{A}$, with respect to the Frobenius norm, that lies within the column span of the matrix $\mathbf{C}$, namely

$$\|\mathbf{A} - \mathbf{U}_C(\mathbf{U}_C^T\mathbf{A})_k\|_F^2 = \min_{\mathrm{rank}(\mathbf{Y})\leqslant k} \|\mathbf{A} - \mathbf{U}_C\mathbf{Y}\|_F^2. \tag{6.2.3}$$

Proof. Recall that $\tilde{\mathbf{U}}_k = \mathbf{U}_C\mathbf{U}_{W,k}$, where $\mathbf{U}_{W,k}$ is the matrix of the top k left singular vectors of $\mathbf{W} = \mathbf{U}_C^T\mathbf{A}$. Thus, $\mathbf{U}_{W,k}$ spans the same range as $\mathbf{W}_k$, the best rank-k approximation to $\mathbf{W}$, i.e., $\mathbf{U}_{W,k}\mathbf{U}_{W,k}^T = \mathbf{W}_k\mathbf{W}_k^\dagger$. Therefore

$$\begin{aligned}\mathbf{A} - \tilde{\mathbf{U}}_k\tilde{\mathbf{U}}_k^T\mathbf{A} &= \mathbf{A} - \mathbf{U}_C\mathbf{U}_{W,k}\mathbf{U}_{W,k}^T\mathbf{U}_C^T\mathbf{A}\\ &= \mathbf{A} - \mathbf{U}_C\mathbf{W}_k\mathbf{W}_k^\dagger\mathbf{W} = \mathbf{A} - \mathbf{U}_C\mathbf{W}_k.\end{aligned}$$

The last equality follows from $\mathbf{W}_k\mathbf{W}_k^\dagger$ being the orthogonal projector onto the range of $\mathbf{W}_k$. In order to prove the optimality property of the lemma, we simply observe that

$$\begin{aligned}\|\mathbf{A} - \mathbf{U}_C(\mathbf{U}_C^T\mathbf{A})_k\|_F^2 &= \|\mathbf{A} - \mathbf{U}_C\mathbf{U}_C^T\mathbf{A} + \mathbf{U}_C\mathbf{U}_C^T\mathbf{A} - \mathbf{U}_C(\mathbf{U}_C^T\mathbf{A})_k\|_F^2\\ &= \|(\mathbf{I} - \mathbf{U}_C\mathbf{U}_C^T)\mathbf{A} + \mathbf{U}_C(\mathbf{U}_C^T\mathbf{A} - (\mathbf{U}_C^T\mathbf{A})_k)\|_F^2\\ &= \|(\mathbf{I} - \mathbf{U}_C\mathbf{U}_C^T)\mathbf{A}\|_F^2 + \|\mathbf{U}_C(\mathbf{U}_C^T\mathbf{A} - (\mathbf{U}_C^T\mathbf{A})_k)\|_F^2\\ &= \|(\mathbf{I} - \mathbf{U}_C\mathbf{U}_C^T)\mathbf{A}\|_F^2 + \|\mathbf{U}_C^T\mathbf{A} - (\mathbf{U}_C^T\mathbf{A})_k\|_F^2.\end{aligned}$$

[3]This is not true for other unitarily invariant norms, e.g., the two-norm; see [6] for a detailed discussion.

The second to last equality follows from Matrix Pythagoras (Lemma 2.5.2) and the last equality follows from the orthonormality of the columns of $\mathbf{U}_C$. The second statement of the lemma is now immediate since $(\mathbf{U}_C^T\mathbf{A})_k$ is the best rank-k approximation to $\mathbf{U}_C^T\mathbf{A}$ and thus any other matrix $\mathbf{Y}$ of rank at most k would result in a larger Frobenius norm error. □

Lemma 6.2.1 shows that Eqn. (6.1.3) in Theorem 6.1.1 can be proven by bounding $\|\mathbf{A} - \mathbf{U}_C\left(\mathbf{U}_C^T\mathbf{A}\right)_k\|_F$. Next, we transition from the best rank-k approximation of the projected matrix $(\mathbf{U}_C^T\mathbf{A})_k$ to the best rank-k approximation $\mathbf{A}_k$ of the original matrix. First (recall the notation introduced in Section 2.6), we split

$$\mathbf{A} = \mathbf{A}_k + \mathbf{A}_{k,\perp}, \text{ where } \mathbf{A}_k = \mathbf{U}_k\mathbf{\Sigma}_k\mathbf{V}_k^T \text{ and } \mathbf{A}_{k,\perp} = \mathbf{U}_{k,\perp}\mathbf{\Sigma}_{k,\perp}\mathbf{V}_{k,\perp}^T. \tag{6.2.4}$$

Lemma 6.2.5. Let $\mathbf{U}_C$ be an orthonormal basis for the column span of the matrix $\mathbf{C}$ and let $\tilde{\mathbf{U}}_k$ be the output of the RandLowRank algorithm. Then,

$$\|\mathbf{A} - \tilde{\mathbf{U}}_k\tilde{\mathbf{U}}_k^T\mathbf{A}\|_F^2 \leqslant \|\mathbf{A}_k - \mathbf{U}_C\mathbf{U}_C^T\mathbf{A}_k\|_F^2 + \|\mathbf{A}_{k,\perp}\|_F^2.$$

Proof. The optimality property in Eqn. (6.2.3) in Lemma 6.2.1 and the fact that $\mathbf{U}_C^T\mathbf{A}_k$ has rank at most k imply

$$\begin{aligned}
\|\mathbf{A} - \tilde{\mathbf{U}}_k\tilde{\mathbf{U}}_k^T\mathbf{A}\|_F^2 &= \|\mathbf{A} - \mathbf{U}_C\left(\mathbf{U}_C^T\mathbf{A}\right)_k\|_F^2 \\
&\leqslant \|\mathbf{A} - \mathbf{U}_C\mathbf{U}_C^T\mathbf{A}_k\|_F^2 \\
&= \|\mathbf{A}_k - \mathbf{U}_C\mathbf{U}_C^T\mathbf{A}_k\|_F^2 + \|\mathbf{A}_{k,\perp}\|_F^2.
\end{aligned}$$

The last equality follows from Lemma 2.5.2. □

6.3. A structural inequality. We now state and prove a structural inequality that will help us bound $\|\mathbf{A}_k - \mathbf{U}_C\mathbf{U}_C^T\mathbf{A}_k\|_F^2$ (the first term in the error bound of Lemma 6.2.5) and that, with minor variants, underlies nearly all RandNLA algorithms for low-rank matrix approximation [18]. Recall that, given a matrix $\mathbf{A} \in \mathbb{R}^{m\times n}$, many RandNLA algorithms seek to construct a "sketch" of $\mathbf{A}$ by post-multiplying $\mathbf{A}$ by some "sketching" matrix $\mathbf{Z} \in \mathbb{R}^{n\times c}$, where c is much smaller than n. (In particular, this is precisely what the RandLowRank algorithm does.) Thus, the resulting matrix $\mathbf{AZ} \in \mathbb{R}^{m\times c}$ is much smaller than the original matrix $\mathbf{A}$, and the interesting question is the approximation guarantees that it offers.

A common approach is to explore how well $\mathbf{AZ}$ spans the principal subspace of $\mathbf{A}$, and one metric of accuracy is a suitably chosen norm of the error matrix $\mathbf{A}_k - (\mathbf{AZ})(\mathbf{AZ})^\dagger\mathbf{A}_k$, where $(\mathbf{AZ})(\mathbf{AZ})^\dagger\mathbf{A}_k$ is the projection of $\mathbf{A}_k$ onto the subspace spanned by the columns of $\mathbf{AZ}$. (See Section 2.9 for the definition of the Moore-Penrose pseudoinverse of a matrix.) The following structural result offers a means to bound the Frobenius norm of the error matrix $\mathbf{A}_k - (\mathbf{AZ})(\mathbf{AZ})^\dagger\mathbf{A}_k$.

Lemma 6.3.1. Given $\mathbf{A} \in \mathbb{R}^{m\times n}$, let $\mathbf{Z} \in \mathbb{R}^{n\times c}$ $(c \geqslant k)$ be any matrix such that $\mathbf{V}_k^T\mathbf{Z} \in \mathbb{R}^{k\times c}$ has rank k. Then,

$$\|\mathbf{A}_k - (\mathbf{AZ})(\mathbf{AZ})^\dagger\mathbf{A}_k\|_F^2 \leqslant \|\,(\mathbf{A} - \mathbf{A}_k)\,\mathbf{Z}(\mathbf{V}_k^T\mathbf{Z})^\dagger\|_F^2. \tag{6.3.2}$$

Remark 6.3.3. Lemma 6.3.1 holds for *any* matrix $\mathbf{Z}$, regardless of whether $\mathbf{Z}$ is constructed deterministically or randomly. In the context of RandNLA, typical constructions of $\mathbf{Z}$ would represent a random sampling or random projection operation, like the the matrix $\mathbf{DHS}$ used in the RandLowRank algorithm.

Remark 6.3.4. The lemma actually holds for any unitarily invariant norm, including the two and the nuclear norm of a matrix [18].

Remark 6.3.5. See [18] for a detailed discussion of such structural inequalities and their history. Lemma 6.3.1 immediately suggests a proof strategy for bounding the error of RandNLA algorithms for low-rank matrix approximation: identify a sketching matrix $\mathbf{Z}$ such that $\mathbf{V}_k^T\mathbf{Z}$ has full rank; and, at the same time, bound the relevant norms of $\left(\mathbf{V}_k^T\mathbf{Z}\right)^\dagger$ and $(\mathbf{A}-\mathbf{A}_k)\mathbf{Z}$.

Proof. (of Lemma 6.3.1) First, note that

$$(\mathbf{AZ})^\dagger\mathbf{A}_k = \operatorname{argmin}_{\mathbf{X}\in\mathbb{R}^{c\times n}}\|\mathbf{A}_k-(\mathbf{AZ})\,\mathbf{X}\|_F^2.$$

The above equation follows by viewing the above optimization problem as least-squares regression with multiple right-hand sides. Interestingly, this property holds for any unitarily invariant norm, but the proof is involved; see Lemma 4.2 of [8] for a detailed discussion. The upshot is that, instead of having to bound $\|\mathbf{A}_k-(\mathbf{AZ})(\mathbf{AZ})^\dagger\mathbf{A}_k\|_F^2$, we can replace $(\mathbf{AZ})^+\,\mathbf{A}_k$ with any other $c\times n$ matrix and the equality with an inequality. In particular, we replace $(\mathbf{AZ})^\dagger\,\mathbf{A}_k$ with $(\mathbf{A}_k\mathbf{Z})^\dagger\,\mathbf{A}_k$:

$$\|\mathbf{A}_k-(\mathbf{AZ})(\mathbf{AZ})^\dagger\mathbf{A}_k\|_F^2 \leqslant \|\mathbf{A}_k-\mathbf{AZ}\,(\mathbf{A}_k\mathbf{Z})^\dagger\,\mathbf{A}_k\|_F^2.$$

This suboptimal choice for $\mathbf{X}$ is essentially the "heart" of our proof: it allows us to manipulate and further decompose the error term, thus making the remainder of the analysis feasible. Use $\mathbf{A}=\mathbf{A}-\mathbf{A}_k+\mathbf{A}_k$ to get

$$\begin{aligned}\|\mathbf{A}_k-(\mathbf{AZ})(\mathbf{AZ})^\dagger\mathbf{A}_k\|_F^2 &\leqslant \|\mathbf{A}_k-(\mathbf{A}-\mathbf{A}_k+\mathbf{A}_k)\mathbf{Z}\,(\mathbf{A}_k\mathbf{Z})^\dagger\,\mathbf{A}_k\|_F^2\\ &= \|\mathbf{A}_k-\mathbf{A}_k\mathbf{Z}(\mathbf{A}_k\mathbf{Z})^\dagger\mathbf{A}_k-(\mathbf{A}-\mathbf{A}_k)\mathbf{Z}(\mathbf{A}_k\mathbf{Z})^\dagger\mathbf{A}_k\|_F^2\\ &= \|(\mathbf{A}-\mathbf{A}_k)\mathbf{Z}(\mathbf{A}_k\mathbf{Z})^\dagger\mathbf{A}_k\|_F^2.\end{aligned}$$

To derive the last inequality, we used

$$\begin{aligned}\mathbf{A}_k-\mathbf{A}_k\mathbf{Z}(\mathbf{A}_k\mathbf{Z})^\dagger\mathbf{A}_k &= \mathbf{A}_k-\mathbf{U}_k\boldsymbol{\Sigma}_k\mathbf{V}_k^T\mathbf{Z}(\mathbf{U}_k\boldsymbol{\Sigma}_k\mathbf{V}_k^T\mathbf{Z})^\dagger\mathbf{A}_k\\ &= \mathbf{A}_k-\mathbf{U}_k\boldsymbol{\Sigma}_k(\mathbf{V}_k^T\mathbf{Z})(\mathbf{V}_k^T\mathbf{Z})^\dagger\boldsymbol{\Sigma}_k^{-1}\mathbf{U}_k^T\mathbf{A}_k \qquad (6.3.6)\\ &= \mathbf{A}_k-\mathbf{U}_k\mathbf{U}_k^T\mathbf{A}_k=\mathbf{0}. \qquad (6.3.7)\end{aligned}$$

In Eqn. (6.3.6), we used Eqn. (2.9.1) and the fact that both matrices $\mathbf{V}_k^T\mathbf{Z}$ and $\mathbf{U}_k\boldsymbol{\Sigma}_k$ have rank k. The latter fact also implies that $(\mathbf{V}_k^T\mathbf{Z})(\mathbf{V}_k^T\mathbf{Z})^\dagger=\mathbf{I}_k$, which derives Eqn. (6.3.7). Finally, the fact that $\mathbf{U}_k\mathbf{U}_k^T\mathbf{A}_k=\mathbf{A}_k$ concludes the derivation and the proof of the lemma. □

6.4. Completing the proof of Theorem 6.1.1. In order to complete the proof of the relative error guarantee of Theorem 6.1.1, we will complete the strategy outlined at the end of Section 6.1. First, recall that from Lemma 6.2.5 it suffices to bound

$$\|\mathbf{A} - \tilde{\mathbf{U}}_k \tilde{\mathbf{U}}_k^T \mathbf{A}\|_F^2 \leqslant \|\mathbf{A}_k - \mathbf{U}_C \mathbf{U}_C^T \mathbf{A}_k\|_F^2 + \|\mathbf{A} - \mathbf{A}_k\|_F^2. \tag{6.4.1}$$

Then, to bound the first term in the right-hand side of the above inequality, we will apply the structural result of Lemma 6.3.1 on the matrix

$$\mathbf{\Phi} = \mathbf{ADH},$$

with $\mathbf{Z} = \mathbf{S}$, where the matrices $\mathbf{D}$, $\mathbf{H}$, and $\mathbf{S}$ are constructed as described in the RandLowRank algorithm. If $\mathbf{V}_{\Phi,k}^T \mathbf{S}$ has rank k, then Lemma 6.3.1 gives the estimate that

$$\|\mathbf{\Phi}_k - (\mathbf{\Phi S})(\mathbf{\Phi S})^\dagger \mathbf{\Phi}_k\|_F^2 \leqslant \|(\mathbf{\Phi} - \mathbf{\Phi}_k)\mathbf{S}(\mathbf{V}_{\Phi,k}^T \mathbf{S})^\dagger\|_F^2. \tag{6.4.2}$$

Here, we used $\mathbf{V}_{\Phi,k} \in \mathbb{R}^{n \times k}$ to denote the matrix of the top k right singular vectors of $\mathbf{\Phi}$.

Recall from Section 5.1 that $\mathbf{DH}$ is an orthogonal matrix and thus the left singular vectors and the singular values of the matrices $\mathbf{A}$ and $\mathbf{\Phi} = \mathbf{ADH}$ are identical. The right singular vectors of the matrix $\mathbf{\Phi}$ are simply the right singular vectors of $\mathbf{A}$, rotated by $\mathbf{DH}$, namely

$$\mathbf{V}_{\Phi}^T = \mathbf{V}^T \mathbf{DH},$$

where $\mathbf{V}$ (respectively, $\mathbf{V}_{\Phi}$) denotes the matrix of the right singular vectors of $\mathbf{A}$ (respectively, $\mathbf{\Phi}$). Thus, we have $\mathbf{\Phi}_k = \mathbf{A}_k \mathbf{DH}$, $\mathbf{\Phi} - \mathbf{\Phi}_k = (\mathbf{A} - \mathbf{A}_k)\mathbf{DH}$, and $\mathbf{V}_{\Phi,k} = \mathbf{V}_k \mathbf{DH}$. Using all the above, we can rewrite Eqn. (6.4.2) as follows:

$$\|\mathbf{A}_k - (\mathbf{ADHS})(\mathbf{ADHS})^\dagger \mathbf{A}_k\|_F^2 \leqslant \|(\mathbf{A} - \mathbf{A}_k)\mathbf{DHS}(\mathbf{V}_k^T \mathbf{DHS})^\dagger\|_F^2. \tag{6.4.3}$$

In the above derivation, we used unitary invariance to drop a $\mathbf{DH}$ term from the Frobenius norm. Recall that $\mathbf{A}_{k,\perp} = \mathbf{A} - \mathbf{A}_k$; we now proceed to manipulate the right-hand side of the above inequality as follows[4]:

$$\begin{aligned}
&\|\mathbf{A}_{k,\perp}\mathbf{DHS}(\mathbf{V}_k^T \mathbf{DHS})^\dagger\|_F^2 \\
&\qquad = \|\mathbf{A}_{k,\perp}\mathbf{DHS}((\mathbf{V}_k^T \mathbf{DHS})^\dagger - (\mathbf{V}_k^T \mathbf{DHS})^T + (\mathbf{V}_k^T \mathbf{DHS})^T)\|_F^2 \\
&\qquad \leqslant 2\|\mathbf{A}_{k,\perp}\mathbf{DHS}((\mathbf{V}_k^T \mathbf{DHS})^\dagger - (\mathbf{V}_k^T \mathbf{DHS})^T)\|_F^2 && (6.4.4) \\
&\qquad\qquad + 2\|\mathbf{A}_{k,\perp}\mathbf{DHS}(\mathbf{V}_k^T \mathbf{DHS})^T\|_F^2. && (6.4.5)
\end{aligned}$$

We now proceed to bound the terms in (6.4.4) and (6.4.5) separately. Our first order of business, however, will be to quantify the manner in which the Randomized Hadamard Transform approximately uniformizes information in the top k right singular vectors of $\mathbf{A}$.

[4]We use the following easy-to-prove version of the triangle inequality for the Frobenius norm: for any two matrices $\mathbf{X}$ and $\mathbf{Y}$ that have the same dimensions, $\|\mathbf{X} + \mathbf{Y}\|_F^2 \leqslant 2\|\mathbf{X}\|_F^2 + 2\|\mathbf{Y}\|_F^2$.

The effect of the Randomized Hadamard Transform. Here, we state a lemma that quantifies the manner in which $\mathbf{HD}$ (premultiplying $\mathbf{V}_k$, or $\mathbf{DH}$ postmultiplying $\mathbf{V}_k^T$) approximately "uniformizes" information in the right singular subspace of the matrix $\mathbf{A}$, thus allowing us to apply our matrix multiplication results from Section 4 in order to bound (6.4.4) and (6.4.5). This is completely analogous to our discussion in Section 5.4 regarding the RandLeastSquares algorithm.

Lemma 6.4.6. Let $\mathbf{V}_k$ be an $n \times k$ matrix with orthonormal columns and let the product $\mathbf{HD}$ be the $n \times n$ Randomized Hadamard Transform of Section 5.1. Then, with probability at least .95,

$$\|(\mathbf{HDV}_k)_{i*}\|_2^2 \leqslant \frac{2k\ln(40nk)}{n}, \qquad \text{for all } i = 1, \ldots, n. \tag{6.4.7}$$

The proof of the above lemma is identical to the proof of Lemma 5.4.1, with $\mathbf{V}_k$ instead of $\mathbf{U}$ and k instead of d.

6.4.1. Bounding Expression (6.4.4). To bound the term in Expression (6.4.4), we first use the strong submultiplicativity of the Frobenius norm (see Section 2.5) to get

$$\begin{aligned} &\|\mathbf{A}_{k,\perp}\mathbf{DHS}((\mathbf{V}_k^T\mathbf{DHS})^\dagger - (\mathbf{V}_k^T\mathbf{DHS})^T)\|_F^2 \\ &\leqslant \|\mathbf{A}_{k,\perp}\mathbf{DHS}\|_F^2\|(\mathbf{V}_k^T\mathbf{DHS})^\dagger - (\mathbf{V}_k^T\mathbf{DHS})^T\|_2^2. \end{aligned} \tag{6.4.8}$$

Our first lemma bounds the term $\|(\mathbf{A}-\mathbf{A}_k)\mathbf{DHS}\|_F^2 = \|\mathbf{A}_{k,\perp}\mathbf{DHS}\|_F^2$. We actually prove the result for any matrix $\mathbf{X}$ and for our choice for the matrix $\mathbf{S}$ in the RandLowRank algorithm.

Lemma 6.4.9. Let the sampling matrix $\mathbf{S} \in \mathbb{R}^{n\times c}$ be constructed as in the RandLowRank algorithm. Then, for any matrix $\mathbf{X} \in \mathbb{R}^{m\times n}$,

$$\mathbf{E}\left[\|\mathbf{XS}\|_F^2\right] = \|\mathbf{X}\|_F^2,$$

and, from Markov's inequality (see Section 3.11), with probability at least 0.95,

$$\|\mathbf{XS}\|_F^2 \leqslant 20\|\mathbf{X}\|_F^2.$$

Remark 6.4.10. The above lemma holds even if the sampling of the canonical vectors $\mathbf{e}_i$ to be included in $\mathbf{S}$ is not done uniformly at random, but with respect to any set of probabilities $\{p_1, \ldots, p_n\}$ summing up to one, as long as the selected canonical vector at the t-th trial (say the i_t-th canonical vector $\mathbf{e}_{i_t}$) is rescaled by $\sqrt{1/cp_{i_t}}$. Thus, even for nonuniform sampling, $\mathbf{XS}$ is an unbiased estimator for the Frobenius norm of the matrix $\mathbf{X}$.

Proof. (of Lemma 6.4.9) We compute the expectation of $\|\mathbf{XS}\|_F^2$ from first principles as follows:

$$\mathbf{E}\left[\|\mathbf{XS}\|_F^2\right] = \sum_{t=1}^{c}\sum_{j=1}^{n}\frac{1}{n}\cdot\left\|\sqrt{\frac{n}{c}}\mathbf{X}_{*j}\right\|_F^2 = \frac{1}{c}\sum_{t=1}^{c}\sum_{j=1}^{n}\|\mathbf{X}_{*j}\|_F^2 = \|\mathbf{X}\|_F^2.$$

The lemma now follows by applying Markov's inequality. □

We can now prove the following lemma, assuming that Eqn. (6.4.7) holds.

Lemma 6.4.11. Assume that Eqn. (6.4.7) holds. If c satisfies

$$c \geqslant \frac{192k\ln(40nk)}{\epsilon^2}\ln\left(\frac{192\sqrt{20}k\ln(40nk)}{\epsilon^2}\right), \tag{6.4.12}$$

then with probability at least .95,

$$\|\left(\mathbf{V}_k^T\mathbf{DHS}\right)^\dagger - \left(\mathbf{V}_k^T\mathbf{DHS}\right)^T\|_2^2 \leqslant 2\epsilon^2.$$

Proof. Let σ_i denote the i-th singular value of the matrix $\mathbf{V}_k^T\mathbf{DHS}$. Conditioned on Eqn. (6.4.7) holding, we can replicate the proof of Lemma 5.4.5 to argue that if c satisfies Eqn. (6.4.12), then, with probability at least .95,

$$\left|1-\sigma_i^2\right| \leqslant \epsilon \tag{6.4.13}$$

holds for all i. (Indeed, we can replicate the proof of Lemma 5.4.5 using $\mathbf{V}_k$ instead of $\mathbf{U}_A$ and k instead of d; we also evaluate the bound for arbitrary ϵ instead of fixing it.) We now observe that the matrices

$$\left(\mathbf{V}_k^T\mathbf{DHS}\right)^\dagger \quad \text{and} \quad \left(\mathbf{V}_k^T\mathbf{DHS}\right)^T$$

have the same left and right singular vectors[5]. Recall that in this lemma we used σ_i to denote the singular values of the matrix $\mathbf{V}_k^T\mathbf{DHS}$. Then, the singular values of the matrix $\left(\mathbf{V}_k^T\mathbf{DHS}\right)^T$ are equal to the σ_i's, while the singular values of the matrix $\left(\mathbf{V}_k^T\mathbf{DHS}\right)^\dagger$ are equal to σ_i^{-1}. Thus,

$$\|\left(\mathbf{V}_k^T\mathbf{DHS}\right)^\dagger - \left(\mathbf{V}_k^T\mathbf{DHS}\right)^T\|_2^2 = \max_i\left|\sigma_i^{-1}-\sigma_i\right|^2 = \max_i\left|(1-\sigma_i^2)^2\sigma_i^{-2}\right|.$$

Combining with Eqn. (6.4.13) and using the fact that $\epsilon \leqslant 1/2$,

$$\|\left(\mathbf{V}_k^T\mathbf{DHS}\right)^\dagger - \left(\mathbf{V}_k^T\mathbf{DHS}\right)^T\|_2^2 = \max_i\left|(1-\sigma_i^2)\sigma_i^{-2}\right| \leqslant (1-\epsilon)^{-1}\epsilon^2 \leqslant 2\epsilon^2. \quad \square$$

Lemma 6.4.14. Assume that Eqn. (6.4.7) holds. If c satisfies Eqn. (6.4.12), then, with probability at least .9,

$$2\|\mathbf{A}_{k,\perp}\mathbf{DHS}((\mathbf{V}_k^T\mathbf{DHS})^\dagger - (\mathbf{V}_k^T\mathbf{DHS})^T)\|_F^2 \leqslant 80\epsilon^2\|\mathbf{A}_{k,\perp}\|_F^2.$$

Proof. We combine Eqn. (6.4.8) with Lemmas 6.4.9 (applied to $\mathbf{X} = \mathbf{A}_{k,\perp}$) and Lemma 6.4.11 to get that, conditioned on Eqn. (6.4.7) holding, with probability at least 1-0.05-0.05=0.9,

$$\|\mathbf{A}_{k,\perp}\mathbf{DHS}((\mathbf{V}_k^T\mathbf{DHS})^\dagger - (\mathbf{V}_k^T\mathbf{DHS})^T)\|_F^2 \leqslant 40\epsilon^2\|\mathbf{A}_{k,\perp}\|_F^2.$$

The aforementioned failure probability follows from a simple union bound on the failure probabilities of Lemmas 6.4.9 and 6.4.11. $\square$

Bounding Expression (6.4.5). Our bound for Expression (6.4.5) will be conditioned on Eqn. (6.4.7) holding; then, we will use our matrix multiplication results

[5]Given any matrix $\mathbf{X}$ with thin SVD $\mathbf{X} = \mathbf{U}_X\mathbf{\Sigma}_X\mathbf{V}_X^T$ its transpose is $\mathbf{X}^T = \mathbf{V}_X\mathbf{\Sigma}_X\mathbf{U}_X^T$ and its pseudoinverse is $\mathbf{X} = \mathbf{V}_X\mathbf{\Sigma}_X^{-1}\mathbf{U}_X^T$.

from Section 4 to derive our bounds. Our discussion is completely analogous to the proof of Lemma 5.4.9. We will prove the following lemma.

Lemma 6.4.15. Assume that Eqn. (6.4.7) holds. If $c \geqslant 40k\ln(40nk)/\epsilon$, then, with probability at least .95,

$$\|\mathbf{A}_{k,\perp}\mathbf{DHS}(\mathbf{V}_k^T\mathbf{DHS})^T\|_F^2 \leqslant \epsilon\|\mathbf{A}_{k,\perp}\|_F^2.$$

Proof. To prove the lemma, we first observe that

$$\|\mathbf{A}_{k,\perp}\mathbf{DHS}(\mathbf{V}_k^T\mathbf{DHS})^T\|_F^2 = \|\mathbf{A}_{k,\perp}\mathbf{DHSS}^T\mathbf{H}^T\mathbf{DV}_k - \mathbf{A}_{k,\perp}\mathbf{DHH}^T\mathbf{DV}_k\|_F^2,$$

since $\mathbf{DHH}^T\mathbf{D} = \mathbf{I}_n$ and $\mathbf{A}_{k,\perp}\mathbf{V}_k = \mathbf{0}$. Thus, we can view $\mathbf{A}_{k,\perp}\mathbf{DHSS}^T\mathbf{H}^T\mathbf{DV}_k$ as approximating the product of two matrices, $\mathbf{A}_{k,\perp}\mathbf{DH}$ and $\mathbf{H}^T\mathbf{DV}_k$, by randomly sampling columns from the first matrix and the corresponding rows from the second matrix, with uniform sampling probabilities that do not depend on the two matrices involved in the product. We will apply the bounds of Eqn. (4.2.6), after arguing that the assumptions of Eqn. (4.2.5) are satisfied. Indeed, since we condition on Eqn. (6.4.7) holding, the rows of $\mathbf{H}^T\mathbf{DV}_k = \mathbf{HDV}_k$ satisfy

$$\frac{1}{n} \geqslant \beta\frac{\|(\mathbf{HDV}_k)_{i*}\|_2^2}{k}, \qquad \text{for all } i = 1,\ldots,n, \tag{6.4.16}$$

for $\beta = (2\ln(40nk))^{-1}$. Thus, Eqn. (4.2.6) implies

$$\begin{aligned}\mathbf{E}\left[\|\mathbf{A}_{k,\perp}\mathbf{DHSS}^T\mathbf{H}^T\mathbf{DV}_k - \mathbf{A}_{k,\perp}\mathbf{DHH}^T\mathbf{DV}_k\|_F^2\right] &\leqslant \frac{1}{\beta c}\|\mathbf{A}_{k,\perp}\mathbf{DH}\|_F^2\|\mathbf{HDV}_k\|_F^2 \\ &= \frac{k}{\beta c}\|\mathbf{A}_{k,\perp}\|_F^2.\end{aligned}$$

In the above we used $\|\mathbf{HDV}_k\|_F^2 = k$. Markov's inequality now implies that with probability at least .95,

$$\|\mathbf{A}_{k,\perp}\mathbf{DHSS}^T\mathbf{H}^T\mathbf{DV}_k - \mathbf{A}_{k,\perp}\mathbf{DHH}^T\mathbf{DV}_k\|_F^2 \leqslant \frac{20k}{\beta c}\|\mathbf{A}_{k,\perp}\|_F^2.$$

Setting $r \geqslant 20k/(\beta\epsilon)$ and using the value of β specified above concludes the proof of the lemma. □

Concluding the proof of Theorem 6.1.1. We are now ready to conclude the proof, and therefore we revert back to using $\mathbf{A}_{k,\perp} = \mathbf{A} - \mathbf{A}_k$. We first state the following lemma.

Lemma 6.4.17. Assume that Eqn. (6.4.7) holds. There exists a constant c_0 such that, if

$$c \geqslant c_0\frac{k\ln n}{\epsilon^2}\ln\left(\frac{k\ln n}{\epsilon^2}\right), \tag{6.4.18}$$

then with probability at least .85,

$$\|\mathbf{A} - \tilde{\mathbf{U}}_k\tilde{\mathbf{U}}_k^T\mathbf{A}\|_F \leqslant (1+\epsilon)\|\mathbf{A} - \mathbf{A}_k\|_F.$$

Proof. Combining Lemma 6.2.5 with Expressions (6.4.4) and (6.4.5), and Lemmas 6.4.14 and 6.4.15, we get

$$\|\mathbf{A} - \tilde{\mathbf{U}}_k\tilde{\mathbf{U}}_k^T\mathbf{A}\|_F^2 \leqslant (1+\epsilon+80\epsilon^2)\|\mathbf{A} - \mathbf{A}_k\|_F^2 \leqslant (1+41\epsilon)\|\mathbf{A} - \mathbf{A}_k\|_F^2.$$

The last inequality follows by using $\epsilon \leqslant 1/2$. Taking square roots of both sides and using $\sqrt{1+41\epsilon} \leqslant 1+7\epsilon$, we get

$$\|\mathbf{A} - \tilde{\mathbf{U}}_k \tilde{\mathbf{U}}_k^T \mathbf{A}\|_F \leqslant (1+7\epsilon)\|\mathbf{A} - \mathbf{A}_k\|_F.$$

Observe that c has to be set to the maximum of the values used in Lemmas 6.4.14 and 6.4.15, which is the value of Eqn. (6.4.12). Adjusting ϵ to $\epsilon/21$ and appropriately adjusting the constants in the expression of c gives the lemma. (We made no special effort to compute or optimize the constant c_0 in the expression of c.)

The failure probability follows by a union bound on the failure probabilities of Lemmas 6.4.14 and 6.4.15 conditioned on Eqn. (6.4.7). □

To conclude the proof of Theorem 6.1.1, we simply need to remove the conditional probability from Lemma 6.4.17. Towards that end, we follow the same strategy as in Section 5.4, to conclude that the success probability of the overall approach is at least $0.85 \cdot 0.95 \geqslant 0.8$.

6.5. Running time. The RandLowRank Algorithm 6.1.4 computes the product $\mathbf{C} = \mathbf{AHDS}$ using the ideas of Section 5.5, thus taking $2n(m+1)\log_2(c+1)$ time. Step 7 takes $O(mc^2)$; step 8 takes $O(mnc + nc^2)$ time; step 9 takes $O(mck)$ time. Overall, the running time is, asymptotically, dominated by the $O(mnc)$ term is step 8, with c as in Eqn. (6.4.18).

6.6. References. Our presentation in this chapter follows the derivations in [8]. We also refer the interested reader to [21, 29] for related work.

Acknowledgements. The authors would like to thank Ilse Ipsen for allowing them to use her slides for the introductory linear algebra lecture delivered at the PCMI Summer School on which the first section of this chapter is heavily based. The authors would also like to thank Aritra Bose, Eugenia-Maria Kontopoulou, and Fred Roosta for their help in proofreading early drafts of this manuscript.

References

[1] N. Ailon and B. Chazelle, *The fast Johnson-Lindenstrauss transform and approximate nearest neighbors*, SIAM J. Comput. **39** (2009), no. 1, 302–322, DOI 10.1137/060673096. MR2506527 ←25

[2] N. Ailon and E. Liberty, *Fast dimension reduction using Rademacher series on dual BCH codes*, Discrete Comput. Geom. **42** (2009), no. 4, 615–630, DOI 10.1007/s00454-008-9110-x. MR2556458 ←35

[3] H. Avron, P. Maymounkov, and S. Toledo, *Blendenpik: supercharging Lapack's least-squares solver*, SIAM J. Sci. Comput. **32** (2010), no. 3, 1217–1236, DOI 10.1137/090767911. MR2639236 ←27, 28, 36

[4] Rajendra Bhatia, *Matrix analysis*, Graduate Texts in Mathematics, vol. 169, Springer-Verlag, New York, 1997. MR1477662 ←11

[5] Å. Björck, *Numerical Methods in Matrix Computations*, Springer, Heidelberg, 2015. ←10, 11

[6] C. Boutsidis, P. Drineas, and M. Magdon-Ismail, *Near-optimal column-based matrix reconstruction*, SIAM J. Comput. **43** (2014), no. 2, 687–717, DOI 10.1137/12086755X. MR3504679 ←40

[7] Petros Drineas, Ravi Kannan, and Michael W. Mahoney, *Fast Monte Carlo algorithms for matrices. II. Computing a low-rank approximation to a matrix*, SIAM J. Comput. **36** (2006), no. 1, 158–183, DOI 10.1137/S0097539704442696. MR2231644 ←37

[8] Petros Drineas, Ilse C. F. Ipsen, Eugenia-Maria Kontopoulou, and Malik Magdon-Ismail, *Structural Convergence Results for Approximation of Dominant Subspaces from Block Krylov Spaces*, SIAM J. Matrix Anal. Appl. **39** (2018), no. 2, 567–586, DOI 10.1137/16M1091745. MR3782400 ←39, 42, 47

[9] Petros Drineas, Ravi Kannan, and Michael W. Mahoney, *Fast Monte Carlo algorithms for matrices. I. Approximating matrix multiplication*, SIAM J. Comput. **36** (2006), no. 1, 132–157, DOI 10.1137/S0097539704442684. MR2231643 ←18, 20, 21, 24

[10] P. Drineas and M. W. Mahoney, *RandNLA: Randomized Numerical Linear Algebra*, Communications of the ACM **59** (2016), no. 6, 80–90. ←3

[11] P. Drineas, M. W. Mahoney, S. Muthukrishnan, and T. Sarlós, *Faster least squares approximation*, Numer. Math. **117** (2011), no. 2, 219–249, DOI 10.1007/s00211-010-0331-6. MR2754850 ←29, 36

[12] John C. Duchi, *Introductory lectures on stochastic optimization*, The Mathematics of Data, IAS/Park City Math. Ser., vol. 25, Amer. Math. Soc., Providence, RI, 2018. ←3

[13] G. H. Golub and C. F. Van Loan, *Matrix computations*, 3rd ed., Johns Hopkins Studies in the Mathematical Sciences, Johns Hopkins University Press, Baltimore, MD, 1996. MR1417720 ←4, 11

[14] N. Halko, P. G. Martinsson, and J. A. Tropp, *Finding structure with randomness: probabilistic algorithms for constructing approximate matrix decompositions*, SIAM Rev. **53** (2011), no. 2, 217–288, DOI 10.1137/090771806. MR2806637 ←37

[15] Wassily Hoeffding, *Probability inequalities for sums of bounded random variables*, J. Amer. Statist. Assoc. **58** (1963), 13–30. MR0144363 ←32

[16] John T. Holodnak and Ilse C. F. Ipsen, *Randomized approximation of the Gram matrix: exact computation and probabilistic bounds*, SIAM J. Matrix Anal. Appl. **36** (2015), no. 1, 110–137, DOI 10.1137/130940116. MR3306014 ←24

[17] M. W. Mahoney, *Randomized algorithms for matrices and data*, Foundations and Trends in Machine Learning, NOW Publishers, Boston, 2011. ←3, 22, 26, 37, 39

[18] Michael W. Mahoney and Petros Drineas, *Structural properties underlying high-quality randomized numerical linear algebra algorithms*, Handbook of big data, Chapman & Hall/CRC Handb. Mod. Stat. Methods, CRC Press, Boca Raton, FL, 2016, pp. 137–154. MR3674816 ←37, 41, 42

[19] Per-Gunnar Martinsson, *Randomized methods for matrix computations*, The Mathematics of Data, IAS/Park City Math. Ser., vol. 25, Amer. Math. Soc., Providence, RI, 2018. ←3, 37, 38, 39

[20] Rajeev Motwani and Prabhakar Raghavan, *Randomized algorithms*, Cambridge University Press, Cambridge, 1995. MR1344451 ←16

[21] C. Musco and C. Musco, *Stronger and Faster Approximate Singular Value Decomposition via the Block Lanczos Method*, Neural Information Processing Systems (NIPS), 2015, available at arXiv:1504.05477. ←39, 47

[22] Roberto Imbuzeiro Oliveira, *Sums of random Hermitian matrices and an inequality by Rudelson*, Electron. Commun. Probab. **15** (2010), 203–212, DOI 10.1214/ECP.v15-1544. MR2653725 ←22

[23] Yousef Saad, *Numerical methods for large eigenvalue problems*, Classics in Applied Mathematics, vol. 66, Society for Industrial and Applied Mathematics (SIAM), Philadelphia, PA, 2011. Revised edition of the 1992 original [1177405]. MR3396212 ←37

[24] Nikhil Srivastava, *Spectral sparsification and restricted invertibility*, ProQuest LLC, Ann Arbor, MI, 2010. Thesis (Ph.D.)–Yale University. MR2941475 ←24

[25] G. W. Stewart and J. G. Sun, *Matrix Perturbation Theory*, Academic Press, New York, 1990. ←11

[26] Gilbert Strang, *Linear algebra and its applications*, 2nd ed., Academic Press [Harcourt Brace Jovanovich, Publishers], New York-London, 1980. MR575349 ←11

[27] L.N. Trefethen and D. Bau III, *Numerical Linear Algebra*, SIAM, Philadelphia, 1997. ←11

[28] Roman Vershynin, *Four lectures on probabilistic methods for data science*, The Mathematics of Data, IAS/Park City Math. Ser., vol. 25, Amer. Math. Soc., Providence, RI, 2018. ←3, 22, 32

[29] S. Wang and Z. Zhang and T. Zhang, *Improved Analyses of the Randomized Power Method and Block Lanczos Method* (2015), 1–22 pp., available at arXiv:1508.06429. ←47

[30] David P. Woodruff, *Sketching as a tool for numerical linear algebra*, Found. Trends Theor. Comput. Sci. **10** (2014), no. 1-2, iv+157. MR3285427 ←36, 37, 39

Purdue University, Computer Science Department, 305 N University Street, West Lafayette, IN 47906.

Email address: pdrineas@purdue.edu

University of California at Berkeley, ICSI and Department of Statistics, 367 Evans Hall, Berkeley, CA 94720.

Email address: mmahoney@stat.berkeley.edu

IAS/Park City Mathematics Series
Volume 25, Pages 49–97
https://doi.org/10.1090/pcms/025/00830

Optimization Algorithms for Data Analysis

Stephen J. Wright

Contents

2010 *Mathematics Subject Classification.* Primary 14Dxx; Secondary 14Dxx.
Key words and phrases. Park City Mathematics Institute.

1. Introduction

In this article, we consider algorithms for solving smooth optimization problems, possibly with simple constraints or structured nonsmooth regularizers. One such canonical formulation is

$$\min_{x \in \mathbb{R}^n} f(x), \tag{1.0.1}$$

where $f : \mathbb{R}^n \to \mathbb{R}$ has at least Lipschitz continuous gradients. Additional assumptions about f, such as convexity and Lipschitz continuity of the Hessian, are introduced as needed. Another formulation we consider is

$$\min_{x \in \mathbb{R}^n} f(x) + \lambda \psi(x), \tag{1.0.2}$$

where f is as in (1.0.1), $\psi : \mathbb{R}^n \to \mathbb{R}$ is a function that is usually convex and usually nonsmooth, and $\lambda \geqslant 0$ is a regularization parameter.[1] We refer to (1.0.2) as a *regularized* minimization problem because the presence of the term involving ψ induces certain structural properties on the solution, that make it more desirable or plausible in the context of the application. We describe iterative algorithms that generate a sequence $\{x^k\}_{k=0,1,2,\ldots}$ of points that, in the case of convex objective functions, converges to the set of solutions. (Some algorithms also generate other "auxiliary" sequences of iterates.)

We are motivated to study problems of the forms (1.0.1) and (1.0.2) by their ubiquity in data analysis applications. Accordingly, Section 2 describes some canonical problems in data analysis and their formulation as optimization problems. After some preliminaries in Section 3, we describe in Section 4 algorithms that take step based on the gradients $\nabla f(x^k)$. Extensions of these methods to

[1] A set S is said to be *convex* if for any pair of points $z', z'' \in S$, we have that $\alpha z' + (1-\alpha) z'' \in S$ for all $\alpha \in [0,1]$. A function $\phi : \mathbb{R}^n \to \mathbb{R}$ is convex if $\phi(\alpha z' + (1-\alpha) z'') \leqslant \alpha \phi(z') + (1-\alpha)\phi(z'')$ for all z', z'' in the (convex) domain of ϕ and all $\alpha \in [0,1]$.

the case (1.0.2) of regularized objectives are described in Section 5. Section 6 describes accelerated gradient methods, which achieve better worst-case complexity than basic gradient methods, while still only using first-derivative information. We discuss Newton's method in Section 7, outlining variants that can guarantee convergence to points that approximately satisfy second-order conditions for a local minimizer of a smooth nonconvex function.

1.1. Omissions Our approach throughout is to give a concise description of some of the most important algorithmic tools for smooth nonlinear optimization and regularized optimization, along with the basic convergence theory for each. (In any given context, we mean by "smooth" that the function is differentiable as many times as is necessary for the discussion to make sense.) In most cases, the theory is elementary enough to include here in its entirety. In the few remaining cases, we provide citations to works in which complete proofs can be found.

Although we allow nonsmoothness in the regularization term in (1.0.2), we do not cover subgradient methods or mirror descent explicitly in this chapter. We also do not discuss stochastic gradient methods, a class of methods that is central to modern machine learning. All these topics are discussed in the contribution of John Duchi to the current volume [22]. Other omissions include the following.

- Coordinate descent methods; see [47] for a recent review.
- Augmented Lagrangian methods, including alternating direction methods of multipliers (ADMM) [23]. The review [5] remains a good reference for the latter topic, especially as it applies to problems from data analysis.
- Semidefinite programming (see [43,45]) and conic optimization (see [6]).
- Methods tailored specifically to linear or quadratic programming, such as the simplex method or interior-point methods (see [46] for a discussion of the latter).
- Quasi-Newton methods, which modify Newton's method by approximating the Hessian or its inverse, thus attaining attractive theoretical and practical performance without using any second-derivative information. For a discussion of these methods, see [36, Chapter 6]. One important method of this class, which is useful in data analysis and many other large-scale problems, is the limited-memory method L-BFGS [30]; see also [36, Section 7.2].

1.2. Notation Our notational conventions in this chapter are as follows. We use upper-case Roman characters (A, L, R, and so on) for matrices and lower-case Roman (x, v, u, and so on) for vectors. (Vectors are assumed to be *column* vectors.) Transposes are indicated by a superscript "T." Elements of matrices and vectors are indicated by subscripts, for example, A_{ij} and x_j. Iteration numbers are indicated by superscripts, for example, x^k. We denote the set of real numbers by $\mathbb{R}$, so that $\mathbb{R}^n$ denotes the Euclidean space of dimension n. The set of symmetric real $n \times n$ matrices is denoted by $S\mathbb{R}^{n\times n}$. Real scalars are usually denoted by

Greek characters, for example, α, β, and so on, though in deference to convention, we sometimes use Roman capitals (for example, L for the Lipschitz constant of a gradient). Where vector norms appear, the type of norm in use is indicated by a subscript (for example $\|x\|_1$), except that when no subscript appears, the Euclidean norm $\|\cdot\|_2$ is assumed. Matrix norms are defined where first used.

2. Optimization Formulations of Data Analysis Problems

In this section, we describe briefly some representative problems in data analysis and machine learning, emphasizing their formulation as optimization problems. Our list is by no means exhaustive. In many cases, there are a number of different ways to formulate a given application as an optimization problem. We do not try to describe all of them. But our list here gives a flavor of the interface between data analysis and optimization.

2.1. Setup Practical data sets are often extremely messy. Data may be mislabeled, noisy, incomplete, or otherwise corrupted. Much of the hard work in data analysis is done by professionals, familiar with the underlying applications, who "clean" the data and prepare it for analysis, while being careful not to change the essential properties that they wish to discern from the analysis. Dasu and Johnson [19] claim out that "80% of data analysis is spent on the process of cleaning and preparing the data." We do not discuss this aspect of the process, focusing instead on the part of the data analysis pipeline in which the problem is formulated and solved.

The data set in a typical analysis problem consists of m objects:

$$D := \{(a_j, y_j),\ j = 1, 2, \ldots, m\}, \tag{2.1.1}$$

where a_j is a vector (or matrix) of *features* and y_j is an *observation* or *label*. (Each pair (a_j, y_j) has the same size and shape for all $j = 1, 2, \ldots, m$.) The analysis task then consists of discovering a function ϕ such that $\phi(a_j) \approx y_j$ holds for most $j = 1, 2, \ldots, m$. The process of discovering the mapping ϕ is often called "learning" or "training."

The function ϕ is often defined in terms of a vector or matrix of parameters, which we denote by x or X. (Other notation also appears below.) With these parametrizations, the problem of identifying ϕ becomes a data-fitting problem: "Find the parameters x defining ϕ such that $\phi(a_j) \approx y_j$, $j = 1, 2, \ldots, m$ in some optimal sense." Once we come up with a definition of the term "optimal," we have an optimization problem. Many such optimization formulations have objective functions of the "summation" type

$$L_D(x) := \sum_{j=1}^{m} \ell(a_j, y_j; x), \tag{2.1.2}$$

where the jth term $\ell(a_j, y_j; x)$ is a measure of the mismatch between $\phi(a_j)$ and y_j, and x is the vector of parameters that determines ϕ.

One use of ϕ is to make predictions about future data items. Given another previously unseen item of data $\hat{a}$ of the same type as a_j, $j = 1, 2, \ldots, m$, we predict that the label $\hat{y}$ associated with $\hat{a}$ would be $\phi(\hat{a})$. The mapping may also expose other structure and properties in the data set. For example, it may reveal that only a small fraction of the features in a_j are needed to reliably predict the label y_j. (This is known as *feature selection*.) The function ϕ or its parameter x may also reveal important structure in the data. For example, X could reveal a low-dimensional subspace that contains most of the a_j, or X could reveal a matrix with particular structure (low-rank, sparse) such that observations of X prompted by the feature vectors a_j yield results close to y_j.

Examples of labels y_j include the following.

- A real number, leading to a *regression* problem.
- A label, say $y_j \in \{1, 2, \ldots, M\}$ indicating that a_j belongs to one of M classes. This is a *classification* problem. We have $M = 2$ for binary classification and $M > 2$ for multiclass classification.
- Null. Some problems only have feature vectors a_j and no labels. In this case, the data analysis task may consist of grouping the a_j into clusters (where the vectors within each cluster are deemed to be functionally similar), or identifying a low-dimensional subspace (or a collection of low-dimensional subspaces) that approximately contains the a_j. Such problems require the labels y_j to be learned, alongside the function ϕ. For example, in a clustering problem, y_j could represent the cluster to which a_j is assigned.

Even after cleaning and preparation, the setup above may contain many complications that need to be dealt with in formulating the problem in rigorous mathematical terms. The quantities (a_j, y_j) may contain noise, or may be otherwise corrupted. We would like the mapping ϕ to be robust to such errors. There may be *missing data*: parts of the vectors a_j may be missing, or we may not know all the labels y_j. The data may be arriving in *streaming* fashion rather than being available all at once. In this case, we would learn ϕ in an *online* fashion.

One particular consideration is that we wish to avoid *overfitting* the model to the data set D in (2.1.1). The particular data set D available to us can often be thought of as a finite sample drawn from some underlying larger (often infinite) collection of data, and we wish the function ϕ to perform well on the unobserved data points as well as the observed subset D. In other words, we want ϕ to be not too sensitive to the particular sample D that is used to define empirical objective functions such as (2.1.2). The optimization formulation can be modified in various ways to achieve this goal, by the inclusion of constraints or penalty terms that limit some measure of "complexity" of the function (such techniques are called *generalization* or *regularization*). Another approach is to terminate the optimization algorithm early, the rationale being that overfitting occurs mainly in the later stages of the optimization process.

2.2. Least Squares Probably the oldest and best-known data analysis problem is linear least squares. Here, the data points (a_j, y_j) lie in $\mathbb{R}^n \times \mathbb{R}$, and we solve

$$\min_x \frac{1}{2m} \sum_{j=1}^{m} (a_j^T x - y_j)^2 = \frac{1}{2m} \|Ax - y\|_2^2, \tag{2.2.1}$$

where A is the matrix whose rows are a_j^T, $j = 1, 2, \ldots, m$ and $y = (y_1, y_2, \ldots, y_m)^T$. In the terminology above, the function ϕ is defined by $\phi(a) := a^T x$. (We could also introduce a nonzero intercept by adding an extra parameter $\beta \in \mathbb{R}$ and defining $\phi(a) := a^T x + \beta$.) This formulation can be motivated statistically, as a maximum-likelihood estimate of x when the observations y_j are exact but for i.i.d. Gaussian noise. Randomized linear algebra methods for large-scale instances of this problem are discussed in Section 5 of the lectures of Drineas and Mahoney [20] in this volume.

Various modifications of (2.2.1) impose desirable structure on x and hence on ϕ. For example, Tikhonov regularization with a squared ℓ_2-norm, which is

$$\min_x \frac{1}{2m} \|Ax - y\|_2^2 + \lambda \|x\|_2^2, \quad \text{for some parameter } \lambda > 0,$$

yields a solution x with less sensitivity to perturbations in the data (a_j, y_j). The LASSO formulation

$$\min_x \frac{1}{2m} \|Ax - y\|_2^2 + \lambda \|x\|_1 \tag{2.2.2}$$

tends to yield solutions x that are sparse, that is, containing relatively few nonzero components [42]. This formulation performs feature selection: The locations of the nonzero components in x reveal those components of a_j that are instrumental in determining the observation y_j. Besides its statistical appeal — predictors that depend on few features are potentially simpler and more comprehensible than those depending on many features — feature selection has practical appeal in making predictions about future data. Rather than gathering all components of a new data vector $\hat{a}$, we need to find only the "selected" features, since only these are needed to make a prediction. The LASSO formulation (2.2.2) is an important prototype for many problems in data analysis, in that it involves a regularization term $\lambda \|x\|_1$ that is nonsmooth and convex, but with relatively simple structure that can potentially be exploited by algorithms.

2.3. Matrix Completion Matrix completion is in one sense a natural extension of least-squares to problems in which the data a_j are naturally represented as matrices rather than vectors. Changing notation slightly, we suppose that each A_j is an $n \times p$ matrix, and we seek another $n \times p$ matrix X that solves

$$\min_X \frac{1}{2m} \sum_{j=1}^{m} (\langle A_j, X \rangle - y_j)^2, \tag{2.3.1}$$

where $\langle A, B \rangle := \text{trace}(A^T B)$. Here we can think of the A_j as "probing" the unknown matrix X. Commonly considered types of observations are random linear combinations (where the elements of A_j are selected i.i.d. from some distribution) or single-element observations (in which each A_j has 1 in a single location and zeros elsewhere). A regularized version of (2.3.1), leading to solutions X that are low-rank, is

$$\min_X \frac{1}{2m} \sum_{j=1}^{m} (\langle A_j, X\rangle - y_j)^2 + \lambda \|X\|_*, \tag{2.3.2}$$

where $\|X\|_*$ is the nuclear norm, which is the sum of singular values of X [39]. The nuclear norm plays a role analogous to the ℓ_1 norm in (2.2.2). Although the nuclear norm is a somewhat complex nonsmooth function, it is at least convex, so that the formulation (2.3.2) is also convex. This formulation can be shown to yield a statistically valid solution when the true X is low-rank and the observation matrices A_j satisfy a "restricted isometry" property, commonly satisfied by random matrices, but not by matrices with just one nonzero element. The formulation is also valid in a different context, in which the true X is incoherent (roughly speaking, it does not have a few elements that are much larger than the others), and the observations A_j are of single elements [10].

In another form of regularization, the matrix X is represented explicitly as a product of two "thin" matrices L and R, where $L \in \mathbb{R}^{n \times r}$ and $R \in \mathbb{R}^{p \times r}$, with $r \ll \min(n, p)$. We set $X = LR^T$ in (2.3.1) and solve

$$\min_{L,R} \frac{1}{2m} \sum_{j=1}^{m} (\langle A_j, LR^T\rangle - y_j)^2. \tag{2.3.3}$$

In this formulation, the rank r is "hard-wired" into the definition of X, so there is no need to include a regularizing term. This formulation is also typically much more compact than (2.3.2); the total number of elements in (L, R) is $(n + p)r$, which is much less than np. A disadvantage is that it is nonconvex. An active line of current research, pioneered in [9] and also drawing on statistical sources, shows that the nonconvexity is benign in many situations, and that under certain assumptions on the data (A_j, y_j), $j = 1, 2, \ldots, m$ and careful choice of algorithmic strategy, good solutions can be obtained from the formulation (2.3.3). A clue to this good behavior is that although this formulation is nonconvex, it is in some sense an approximation to a tractable problem: If we have a complete observation of X, then a rank-r approximation can be found by performing a singular value decomposition of X, and defining L and R in terms of the r leading left and right singular vectors.

2.4. Nonnegative Matrix Factorization Some applications in computer vision, chemometrics, and document clustering require us to find factors L and R like those in (2.3.3) in which all elements are nonnegative. If the full matrix $Y \in \mathbb{R}^{n \times p}$ is observed, this problem has the form

$$\min_{L,R} \|LR^T - Y\|_F^2, \quad \text{subject to } L \geqslant 0,\ R \geqslant 0.$$

2.5. Sparse Inverse Covariance Estimation In this problem, the labels y_j are null, and the vectors $a_j \in \mathbb{R}^n$ are viewed as independent observations of a random vector $A \in \mathbb{R}^n$, which has zero mean. The sample covariance matrix constructed from these observations is

$$S = \frac{1}{m-1} \sum_{j=1}^{m} a_j a_j^T.$$

The element S_{il} is an estimate of the covariance between the ith and lth elements of the random variable vector A. Our interest is in calculating an estimate X of the *inverse* covariance matrix that is *sparse*. The structure of X yields important information about A. In particular, if $X_{il} = 0$, we can conclude that the i and l components of A are *conditionally independent*. (That is, they are independent given knowledge of the values of the other $n-2$ components of A.) Stated another way, the nonzero locations in X indicate the arcs in the dependency graph whose nodes correspond to the n components of A.

One optimization formulation that has been proposed for estimating the inverse sparse covariance matrix X is the following:

$$\min_{X \in \mathbb{SR}^{n \times n},\, X \succeq 0} \langle S, X \rangle - \log\det(X) + \lambda \|X\|_1, \tag{2.5.1}$$

where $\mathbb{SR}^{n \times n}$ is the set of $n \times n$ symmetric matrices, $X \succeq 0$ indicates that X is positive definite, and $\|X\|_1 := \sum_{i,l=1}^{n} |X_{il}|$ (see [17,25]).

2.6. Sparse Principal Components The setup for this problem is similar to the previous section, in that we have a sample covariance matrix S that is estimated from a number of observations of some underlying random vector. The *principal components* of this matrix are the eigenvectors corresponding to the largest eigenvalues. It is often of interest to find *sparse* principal components, approximations to the leading eigenvectors that also contain few nonzeros. An explicit optimization formulation of this problem is

$$\max_{v \in \mathbb{R}^n} v^T S v \quad \text{s.t. } \|v\|_2 = 1, \;\; \|v\|_0 \leqslant k, \tag{2.6.1}$$

where $\|\cdot\|_0$ indicates the cardinality of v (that is, the number of nonzeros in v) and k is a user-defined parameter indicating a bound on the cardinality of v. The problem (2.6.1) is NP-hard, so exact formulations (for example, as a quadratic program with binary variables) are intractable. We consider instead a relaxation, due to [18], which replaces vv^T by a positive semidefinite proxy $M \in \mathbb{SR}^{n \times n}$:

$$\max_{M \in \mathbb{SR}^{n \times n}} \langle S, M \rangle \quad \text{s.t. } M \succeq 0, \; \langle I, M \rangle = 1, \; \|M\|_1 \leqslant \rho, \tag{2.6.2}$$

for some parameter $\rho > 0$ that can be adjusted to attain the desired sparsity. This formulation is a convex optimization problem, in fact, a semidefinite programming problem.

This formulation can be generalized to find the leading $r > 1$ sparse principal components. Ideally, we would obtain these from a matrix $V \in \mathbb{R}^{n \times r}$ whose columns are mutually orthogonal and have at most k nonzeros each. We can write a convex relaxation of this problem, once again a semidefinite program, as

$$\max_{M \in S\mathbb{R}^{n \times n}} \langle S, M \rangle \quad \text{s.t. } 0 \preceq M \preceq I,\ \langle I, M \rangle = 1,\ \|M\|_1 \leqslant \rho . \tag{2.6.3}$$

A more compact (but nonconvex) formulation is

$$\max_{F \in \mathbb{R}^{n \times r}} \langle S, FF^T \rangle \quad \text{s.t. } \|F\|_2 \leqslant 1,\ \|F\|_{2,1} \leqslant \bar{R},$$

where $\|F\|_{2,1} := \sum_{i=1}^{n} \|F_{i\cdot}\|_2$ [15]. The latter regularization term is often called a "group-sparse" or "group-LASSO" regularizer. (An early use of this type of regularizer was described in [44].)

2.7. Sparse Plus Low-Rank Matrix Decomposition Another useful paradigm is to decompose a partly or fully observed $n \times p$ matrix Y into the sum of a sparse matrix and a low-rank matrix. A convex formulation of the fully-observed problem is

$$\min_{M,S} \|M\|_* + \lambda \|S\|_1 \quad \text{s.t. } Y = M + S,$$

where $\|S\|_1 := \sum_{i=1}^{n} \sum_{j=1}^{p} |S_{ij}|$ [11, 14]. Compact, nonconvex formulations that allow noise in the observations include the following:

$$\min_{L,R,S} \frac{1}{2} \|LR^T + S - Y\|_F^2 \quad \text{(fully observed)}$$

$$\min_{L,R,S} \frac{1}{2} \|P_\Phi(LR^T + S - Y)\|_F^2 \quad \text{(partially observed)},$$

where Φ represents the locations of the observed entries of Y and P_Φ is projection onto this set [15, 48].

One application of these formulations is to robust PCA, where the low-rank part represents principal components and the sparse part represents "outlier" observations. Another application is to foreground-background separation in video processing. Here, each column of Y represents the pixels in one frame of video, whereas each row of Y shows the evolution of one pixel over time.

2.8. Subspace Identification In this application, the $a_j \in \mathbb{R}^n$, $j = 1, 2, \dots, m$ are vectors that lie (approximately) in a low-dimensional subspace. The aim is to identify this subspace, expressed as the column subspace of a matrix $X \in \mathbb{R}^{n \times r}$. If the a_j are fully observed, an obvious way to solve this problem is to perform a singular value decomposition of the $n \times m$ matrix $A = [a_j]_{j=1}^m$, and take X to be the leading r right singular vectors. In interesting variants of this problem, however, the vectors a_j may be arriving in streaming fashion and may be only partly observed, for example in indices $\Phi_j \subset \{1, 2, \dots, n\}$. We would thus need to identify a matrix X and vectors $s_j \in \mathbb{R}^r$ such that

$$P_{\Phi_j}(a_j - Xs_j) \approx 0, \quad j = 1, 2, \dots, m.$$

The algorithm for identifying X, described in [1], is a manifold-projection scheme that takes steps in incremental fashion for each a_j in turn. Its validity relies on incoherence of the matrix X with respect to the principal axes, that is, the matrix X should not have a few elements that are much larger than the others. A local convergence analysis of this method is given in [2].

2.9. Support Vector Machines Classification via support vector machines (SVM) is a classical paradigm in machine learning. This problem takes as input data (a_j, y_j) with $a_j \in \mathbb{R}^n$ and $y_j \in \{-1, 1\}$, and seeks a vector $x \in \mathbb{R}^n$ and a scalar $\beta \in \mathbb{R}$ such that

$$a_j^T x - \beta \geqslant 1 \quad \text{when } y_j = +1; \tag{2.9.1a}$$

$$a_j^T x - \beta \leqslant -1 \quad \text{when } y_j = -1. \tag{2.9.1b}$$

Any pair (x, β) that satisfies these conditions defines a *separating hyperplane* in $\mathbb{R}^n$, that separates the "positive" cases $\{a_j \,|\, y_j = +1\}$ from the "negative" cases $\{a_j \,|\, y_j = -1\}$. (In the language of Section 2.1, we could define the function ϕ as $\phi(a_j) = \text{sign}(a_j^T x - \beta)$.) Among all separating hyperplanes, the one that minimizes $\|x\|^2$ is the one that maximizes the *margin* between the two classes, that is, the hyperplane whose distance to the nearest point a_j of either class is greatest.

We can formulate the problem of finding a separating hyperplane as an optimization problem by defining an objective with the summation form (2.1.2):

$$H(x, \beta) = \frac{1}{m} \sum_{j=1}^{m} \max(1 - y_j(a_j^T x - \beta), 0). \tag{2.9.2}$$

Note that the jth term in this summation is zero if the conditions (2.9.1) are satisfied, and positive otherwise. Even if no pair (x, β) exists with $H(x, \beta) = 0$, the pair (x, β) that minimizes (2.1.2) will be the one that comes as close as possible to satisfying (2.9.1), in a suitable sense. A term $\lambda\|x\|_2^2$, where λ is a small positive parameter, is often added to (2.9.2), yielding the following regularized version:

$$H(x, \beta) = \frac{1}{m} \sum_{j=1}^{m} \max(1 - y_j(a_j^T x - \beta), 0) + \frac{1}{2}\lambda\|x\|_2^2. \tag{2.9.3}$$

If λ is sufficiently small (but positive), and if separating hyperplanes exist, the pair (x, β) that minimizes (2.9.3) is the maximum-margin separating hyperplane. The maximum-margin property is consistent with the goals of generalizability and robustness. For example, if the observed data (a_j, y_j) is drawn from an underlying "cloud" of positive and negative cases, the maximum-margin solution usually does a reasonable job of separating other empirical data samples drawn from the same clouds, whereas a hyperplane that passes close by several of the observed data points may not do as well (see Figure 2.9.4).

The problem of minimizing (2.9.3) can be written as a convex quadratic program — having a convex quadratic objective and linear constraints — by introducing variables s_j, $j = 1, 2, \ldots, m$ to represent the residual terms. Then,

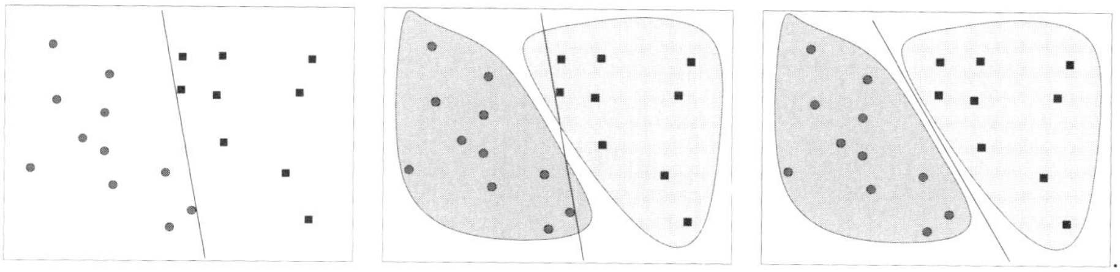

FIGURE 2.9.4. Linear support vector machine classification, with one class represented by circles and the other by squares. One possible choice of separating hyperplane is shown at left. If the observed data is an empirical sample drawn from a cloud of underlying data points, this plane does not do well in separating the two clouds (middle). The maximum-margin separating hyperplane does better (right).

$$\min_{x,\beta,s} \frac{1}{m}\mathbf{1}^T s + \frac{1}{2}\lambda\|x\|_2^2, \tag{2.9.5a}$$

$$\text{subject to} \quad s_j \geqslant 1 - y_j(a_j^T x - \beta), \quad s_j \geqslant 0, \quad j = 1,2,\dots,m, \tag{2.9.5b}$$

where $\mathbf{1} = (1,1,\dots,1)^T \in \mathbb{R}^m$.

Often it is not possible to find a hyperplane that separates the positive and negative cases well enough to be useful as a classifier. One solution is to transform all of the raw data vectors a_j by a mapping ζ into a higher-dimensional Euclidean space, then perform the support-vector-machine classification on the vectors $\zeta(a_j)$, $j = 1,2,\dots,m$.

The conditions (2.9.1) would thus be replaced by

$$\zeta(a_j)^T x - \beta \geqslant 1 \quad \text{when } y_j = +1; \tag{2.9.6a}$$

$$\zeta(a_j)^T x - \beta \leqslant -1 \quad \text{when } y_j = -1, \tag{2.9.6b}$$

leading to the following analog of (2.9.3):

$$H(x,\beta) = \frac{1}{m}\sum_{j=1}^{m} \max(1 - y_j(\zeta(a_j)^T x - \beta), 0) + \frac{1}{2}\lambda\|x\|_2^2. \tag{2.9.7}$$

When transformed back to $\mathbb{R}^m$, the surface $\{a \mid \zeta(a)^T x - \beta = 0\}$ is nonlinear and possibly disconnected, and is often a much more powerful classifier than the hyperplanes resulting from (2.9.3).

We can formulate (2.9.7) as a convex quadratic program in exactly the same manner as we derived (2.9.5) from (2.9.3). By taking the dual of this quadratic program, we obtain another convex quadratic program, in m variables:

$$\min_{\alpha\in\mathbb{R}^m} \frac{1}{2}\alpha^T Q\alpha - \mathbf{1}^T\alpha \quad \text{subject to } 0 \leqslant \alpha \leqslant \frac{1}{\lambda}\mathbf{1}, \ y^T\alpha = 0, \tag{2.9.8}$$

where

$$Q_{kl} = y_k y_l \zeta(a_k)^T \zeta(a_l), \quad y = (y_1, y_2, \dots, y_m)^T, \quad \mathbf{1} = (1,1,\dots,1)^T \in \mathbb{R}^m.$$

Interestingly, problem (2.9.8) can be formulated and solved without any explicit knowledge or definition of the mapping ζ. We need only a technique to define the elements of Q. This can be done with the use of a *kernel function* $K : \mathbb{R}^n \times \mathbb{R}^n \to \mathbb{R}$, where $K(a_k, a_l)$ replaces $\zeta(a_k)^T\zeta(a_l)$ [4, 16]. This is the so-called "kernel trick." (The kernel function K can also be used to construct a classification function ϕ from the solution of (2.9.8).) A particularly popular choice of kernel is the Gaussian kernel:

$$K(a_k, a_l) := \exp(-\|a_k - a_l\|^2/(2\sigma)),$$

where σ is a positive parameter.

2.10. Logistic Regression Logistic regression can be viewed as a variant of binary support-vector machine classification, in which rather than the classification function ϕ giving a unqualified prediction of the class in which a new data vector a lies, it returns an estimate of the *odds* of a belonging to one class or the other. We seek an "odds function" p parametrized by a vector $x \in \mathbb{R}^n$ as follows:

$$p(a; x) := (1 + \exp(a^T x))^{-1}, \tag{2.10.1}$$

and aim to choose the parameter x so that

$$p(a_j; x) \approx 1 \quad \text{when } y_j = +1; \tag{2.10.2a}$$

$$p(a_j; x) \approx 0 \quad \text{when } y_j = -1. \tag{2.10.2b}$$

(Note the similarity to (2.9.1).) The optimal value of x can be found by maximizing a log-likelihood function:

$$L(x) := \frac{1}{m}\left[\sum_{j:y_j=-1} \log(1 - p(a_j; x)) + \sum_{j:y_j=1} \log p(a_j; x)\right]. \tag{2.10.3}$$

We can perform feature selection using this model by introducing a regularizer $\lambda\|x\|_1$, as follows:

$$\max_x \frac{1}{m}\left[\sum_{j:y_j=-1} \log(1 - p(a_j; x)) + \sum_{j:y_j=1} \log p(a_j; x)\right] - \lambda\|x\|_1, \tag{2.10.4}$$

where $\lambda > 0$ is a regularization parameter. (Note that we *subtract* rather than add the regularization term $\lambda\|x\|_1$ to the objective, because this problem is formulated as a maximization rather than a minimization.) As we see later, this term has the effect of producing a solution in which few components of x are nonzero, making it possible to evaluate $p(a; x)$ by knowing only those components of a that correspond to the nonzeros in x.

An important extension of this technique is to *multiclass* (or *multinomial*) logistic regression, in which the data vectors a_j belong to more than two classes. Such applications are common in modern data analysis. For example, in a speech recognition system, the M classes could each represent a *phoneme* of speech, one of the potentially thousands of distinct elementary sounds that can be uttered by humans in a few tens of milliseconds. A multinomial logistic regression problem requires a distinct odds function p_k for each class $k \in \{1, 2, \dots, M\}$. These functions are parametrized by vectors $x_{[k]} \in \mathbb{R}^n$, $k = 1, 2, \dots, M$, defined as follows:

$$p_k(a; X) := \frac{\exp(a^T x_{[k]})}{\sum_{l=1}^{M} \exp(a^T x_{[l]})}, \quad k = 1, 2, \dots, M, \tag{2.10.5}$$

where we define $X := \{x_{[k]} \mid k = 1, 2, \dots, M\}$. Note that for all a and for all $k = 1, 2, \dots, M$, we have $p_k(a) \in (0, 1)$ and also $\sum_{k=1}^{M} p_k(a) = 1$. The operation in (2.10.5) is referred to as a "softmax" on the quantities $\{a^T x_{[l]} \mid l = 1, 2, \dots, M\}$. If one of these inner products dominates the others, that is, $a^T x_{[k]} \gg a^T x_{[l]}$ for all $l \neq k$, the formula (2.10.5) will yield $p_k(a; X) \approx 1$ and $p_l(a; X) \approx 0$ for all $l \neq k$.

In the setting of multiclass logistic regression, the labels y_j are vectors in R^M, whose elements are defined as follows:

$$y_{jk} = \begin{cases} 1 & \text{when } a_j \text{ belongs to class } k, \\ 0 & \text{otherwise.} \end{cases} \tag{2.10.6}$$

Similarly to (2.10.2), we seek to define the vectors $x_{[k]}$ so that

$$p_k(a_j; X) \approx 1 \quad \text{when } y_{jk} = 1 \tag{2.10.7a}$$

$$p_k(a_j; X) \approx 0 \quad \text{when } y_{jk} = 0. \tag{2.10.7b}$$

The problem of finding values of $x_{[k]}$ that satisfy these conditions can again be formulated as one of maximizing a log-likelihood:

$$L(X) := \frac{1}{m} \sum_{j=1}^{m} \left[\sum_{\ell=1}^{M} y_{j\ell}(x_{[\ell]}^T a_j) - \log \left(\sum_{\ell=1}^{M} \exp(x_{[\ell]}^T a_j) \right) \right]. \tag{2.10.8}$$

"Group-sparse" regularization terms can be included in this formulation to select a set of features in the vectors a_j, common to each class, that distinguish effectively between the classes.

2.11. Deep Learning Deep neural networks are often designed to perform the same function as multiclass logistic regression, that is, to classify a data vector a into one of M possible classes, where $M \geqslant 2$ is large in some key applications. The difference is that the data vector a undergoes a series of structured transformations before being passed through a multiclass logistic regression classifier of the type described in the previous subsection.

The simple neural network shown in Figure 2.11.1 illustrates the basic ideas. In this figure, the data vector a_j enters at the bottom of the network, each node in the bottom layer corresponding to one component of a_j. The vector then moves upward through the network, undergoing a structured nonlinear transformation as it moves from one layer to the next. A typical form of this transformation, which converts the vector a_j^{l-1} at layer $l-1$ to input vector a_j^l at layer l, is

$$a_j^l = \sigma(W^l a_j^{l-1} + g^l), \quad l = 1, 2, \dots, D,$$

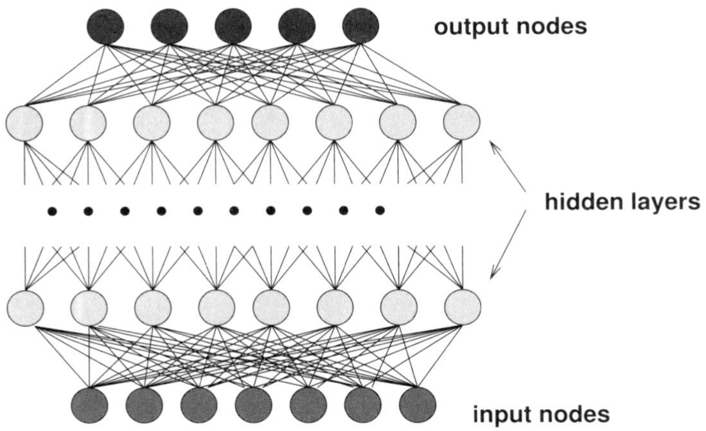

FIGURE 2.11.1. A deep neural network, showing connections between adjacent layers.

where W^l is a matrix of dimension $|a_j^l| \times |a_j^{l-1}|$ and g^l is a vector of length $|a_j^l|$, σ is a *componentwise* nonlinear transformation, and D is the number of *hidden layers*, defined as the layers situated strictly between the bottom and top layers. Each arc in Figure 2.11.1 represents one of the elements of a transformation matrix W^l. We define a_j^0 to be the "raw" input vector a_j, and let a_j^D be the vector formed by the nodes at the topmost hidden layer in Figure 2.11.1. Typical forms of the function σ include the following, acting identically on each component $t \in \mathbb{R}$ of its input vector:

- Logistic function: $t \to 1/(1+e^{-t})$;
- Hinge loss: $t \to \max(t, 0)$;
- Bernoulli: a random function that outputs 1 with probability $1/(1+e^{-t})$ and 0 otherwise.

Each node in the top layer corresponds to a particular class, and the output of each node corresponds to the odds of the input vector belonging to each class. As mentioned, the "softmax" operator is typically used to convert the transformed input vector in the second-top layer (layer D) to a set of odds at the top layer. Associated with each input vector a_j are labels y_{jk}, defined as in (2.10.6) to indicate which of the M classes that a_j belongs to.

The parameters in this neural network are the matrix-vector pairs (W^l, g^l), $l = 1, 2, \ldots, D$ that transform the input vector a_j into its form a_j^D at the topmost hidden layer, together with the parameters X of the multiclass logistic regression operation that takes place at the very top stage, where X is defined exactly as in the discussion of Section 2.10. We aim to choose all these parameters so that the network does a good job on classifying the training data correctly. Using the notation w for the hidden layer transformations, that is,

$$w := (W^1, g^1, W^2, g^2, \ldots, W^D, g^D), \tag{2.11.2}$$

and defining $X := \{x_{[k]} \mid k = 1, 2, \dots, M\}$ as in Section 2.10, we can write the loss function for deep learning as follows:

$$L(w, X) := \frac{1}{m} \sum_{j=1}^{m} \left[\sum_{\ell=1}^{M} y_{j\ell}(x_{[\ell]}^T a_j^D(w)) - \log \left(\sum_{\ell=1}^{M} \exp(x_{[\ell]}^T a_j^D(w)) \right) \right]. \tag{2.11.3}$$

Note that this is exactly the function (2.10.8) applied to the output of the top hidden layer $a_j^D(w)$. We write $a_j^D(w)$ to make explicit the dependence of a_j^D on the parameters w of (2.11.2), as well as on the input vector a_j. (We can view multiclass logistic regression (2.10.8) as a special case of deep learning in which there are no hidden layers, so that $D = 0$, w is null, and $a_j^D = a_j$, $j = 1, 2, \dots, m$.)

Neural networks in use for particular applications (in image recognition and speech recognition, for example, where they have been very successful) include many variants on the basic design above. These include restricted connectivity between layers (that is, enforcing structure on the matrices W^l, $l = 1, 2, \dots, D$), layer arrangements that are more complex than the linear layout illustrated in Figure 2.11.1, with outputs coming from different levels, connections across non-adjacent layers, different componentwise transformations σ at different layers, and so on. Deep neural networks for practical applications are highly engineered objects.

The loss function (2.11.3) shares with many other applications the "summation" form (2.1.2), but it has several features that set it apart from the other applications discussed above. First, and possibly most important, it is *nonconvex* in the parameters w. There is reason to believe that the "landscape" of L is complex, with the global minimizer being exceedingly difficult to find. Second, the total number of parameters in (w, X) is usually very large. The most popular algorithms for minimizing (2.11.3) are of stochastic gradient type, which like most optimization methods come with no guarantee for finding the minimizer of a nonconvex function. Effective training of deep learning classifiers typically requires a great deal of data and computation power. Huge clusters of powerful computers, often using multicore processors, GPUs, and even specially architected processing units, are devoted to this task. Efficiency also requires many heuristics in the formulation and the algorithm (for example, in the choice of regularization functions and in the steplengths for stochastic gradient).

3. Preliminaries

We discuss here some foundations for the analysis of subsequent sections. These include useful facts about smooth and nonsmooth convex functions, Taylor's theorem and some of its consequences, optimality conditions, and proximal operators.

In the discussion of this section, our basic assumption is that f is a mapping from $\mathbb{R}^n$ to $\mathbb{R} \cup \{+\infty\}$, continuous on its effective domain $\mathcal{D} := \{x \mid f(x) < \infty\}$. Further assumptions of f are introduced as needed.

3.1. Solutions Consider the problem of minimizing f (1.0.1). We have the following terminology:

- x^* is a *local minimizer* of f if there is a neighborhood N of x^* such that $f(x) \geqslant f(x^*)$ for all $x \in N$.
- x^* is a *global minimizer* of f if $f(x) \geqslant f(x^*)$ for all $x \in \mathbb{R}^n$.
- x^* is a *strict local minimizer* if it is a local minimizer on some neighborhood N and in addition $f(x) > f(x^*)$ for all $x \in N$ with $x \neq x^*$.
- x^* is an *isolated local minimizer* if there is a neighborhood N of x^* such that $f(x) \geqslant f(x^*)$ for all $x \in N$ and in addition, N contains no local minimizers other than x^*.

3.2. Convexity and Subgradients A convex set $\Omega \subset \mathbb{R}^n$ has the property that

$$x, y \in \Omega \;\Rightarrow\; (1-\alpha)x + \alpha y \in \Omega \;\text{ for all } \alpha \in [0,1]. \tag{3.2.1}$$

We usually deal with *closed* convex sets in this article. For a convex set $\Omega \subset \mathbb{R}^n$ we define the *indicator function* $I_\Omega(x)$ as follows:

$$I_\Omega(x) = \begin{cases} 0 & \text{if } x \in \Omega \\ +\infty & \text{otherwise.} \end{cases}$$

Indicator functions are useful devices for deriving optimality conditions for constrained problems, and even for developing algorithms. The constrained optimization problem

$$\min_{x \in \Omega} f(x) \tag{3.2.2}$$

can be restated equivalently as follows:

$$\min f(x) + I_\Omega(x). \tag{3.2.3}$$

We noted already that a convex function $\phi : \mathbb{R}^n \to \mathbb{R} \cup \{+\infty\}$ has the following defining property:

(3.2.4)

$$\phi((1-\alpha)x + \alpha y) \leqslant (1-\alpha)\phi(x) + \alpha\phi(y), \quad \text{for all } x, y \in \mathbb{R}^n \text{ and all } \alpha \in [0,1].$$

The concepts of "minimizer" are simpler in the case of convex objective functions than in the general case. In particular, the distinction between "local" and "global" minimizers disappears. For f convex in (1.0.1), we have the following.

(a) Any local minimizer of (1.0.1) is also a global minimizer.
(b) The set of global minimizers of (1.0.1) is a convex set.

If there exists a value $\gamma > 0$ such that

$$\phi((1-\alpha)x + \alpha y) \leqslant (1-\alpha)\phi(x) + \alpha\phi(y) - \frac{1}{2}\gamma\alpha(1-\alpha)\|x-y\|_2^2 \tag{3.2.5}$$

for all x and y in the domain of ϕ and $\alpha \in [0,1]$, we say that ϕ is *strongly convex with modulus of convexity* γ.

We summarize some definitions and results about subgradients of convex functions here. For a more extensive discussion, see [22].

Definition 3.2.6. A vector $v \in \mathbb{R}^n$ is a *subgradient* of f at a point x if

$$f(x+d) \geqslant f(x) + v^T d. \quad \text{for all } d \in \mathbb{R}^n.$$

The *subdifferential*, denoted $\partial f(x)$, is the set of all subgradients of f at x.

Subdifferentials satisfy a *monotonicity* property, as we show now.

Lemma 3.2.7. *If* $a \in \partial f(x)$ *and* $b \in \partial f(y)$, *we have* $(a-b)^T(x-y) \geqslant 0$.

Proof. From the convexity of f and the definitions of a and b, we deduce that $f(y) \geqslant f(x) + a^T(y-x)$ and $f(x) \geqslant f(y) + b^T(x-y)$. The result follows by adding these two inequalities. □

We can easily characterize a minimum in terms of the subdifferential.

Theorem 3.2.8. *The point* x^* *is the minimizer of a convex function* f *if and only if* $0 \in \partial f(x^*)$.

Proof. Suppose that $0 \in \partial f(x^*)$, we have by substituting $x = x^*$ and $v = 0$ into Definition 3.2.6 that $f(x^*+d) \geqslant f(x^*)$ for all $d \in \mathbb{R}^n$, which implies that x^* is a minimizer of f.

The converse follows trivially by showing that $v = 0$ satisfies Definition 3.2.6 when x^* is a minimizer. □

The subdifferential is the generalization to nonsmooth convex functions of the concept of derivative of a smooth function.

Theorem 3.2.9. *If* f *is convex and differentiable at* x, *then* $\partial f(x) = \{\nabla f(x)\}$.

A converse of this result is also true. Specifically, if the subdifferential of a convex function f at x contains a single subgradient, then f is differentiable with gradient equal to this subgradient (see [40, Theorem 25.1]).

3.3. Taylor's Theorem Taylor's theorem is a foundational result for optimization of smooth nonlinear functions. It shows how smooth functions can be approximated locally by low-order (linear or quadratic) functions.

Theorem 3.3.1. *Given a continuously differentiable function* $f : \mathbb{R}^n \to \mathbb{R}$, *and given* $x, p \in \mathbb{R}^n$, *we have that*

$$f(x+p) = f(x) + \int_0^1 \nabla f(x+\xi p)^T p \, d\xi, \tag{3.3.2}$$

$$f(x+p) = f(x) + \nabla f(x+\xi p)^T p, \quad \textit{some } \xi \in (0,1). \tag{3.3.3}$$

If f *is twice continuously differentiable, we have*

$$\nabla f(x+p) = \nabla f(x) + \int_0^1 \nabla^2 f(x+\xi p) p \, d\xi, \tag{3.3.4}$$

$$f(x+p) = f(x) + \nabla f(x)^T p + \frac{1}{2} p^T \nabla^2 f(x+\xi p) p, \quad \textit{for some } \xi \in (0,1). \tag{3.3.5}$$

We can derive an important consequence of this theorem when f is *Lipschitz* continuously differentiable with constant L, that is,

$$\|\nabla f(x) - \nabla f(y)\| \leqslant L\|x - y\|, \quad \text{for all } x, y \in \mathbb{R}^n. \tag{3.3.6}$$

We have by setting $y = x + p$ in (3.3.2) and subtracting the term $\nabla f(x)^T(y - x)$ from both sides that

$$f(y) - f(x) - \nabla f(x)^T(y - x) = \int_0^1 [\nabla f(x + \xi(y - x)) - \nabla f(x)]^T (y - x)\, d\xi.$$

By using (3.3.6), we have

$$\begin{aligned}[\nabla f(x + \xi(y - x)) - \nabla f(x)]^T (y - x) &\leqslant \|\nabla f(x + \xi(y - x)) - \nabla f(x)\| \|y - x\| \\ &\leqslant L\xi \|y - x\|^2.\end{aligned}$$

By substituting this bound into the previous integral, we obtain

$$f(y) - f(x) - \nabla f(x)^T (y - x) \leqslant \frac{L}{2}\|y - x\|^2. \tag{3.3.7}$$

For the remainder of Section 3.3, we assume that f is continuously differentiable and also *convex*. The definition of convexity (3.2.4) and the fact that $\partial f(x) = \{\nabla f(x)\}$ implies that

$$f(y) \geqslant f(x) + \nabla f(x)^T (y - x), \quad \text{for all } x, y \in \mathbb{R}^n. \tag{3.3.8}$$

We defined "strong convexity with modulus γ" in (3.2.5). When f is differentiable, we have the following equivalent definition, obtained by rearranging (3.2.5) and letting $\alpha \downarrow 0$.

$$f(y) \geqslant f(x) + \nabla f(x)^T (y - x) + \frac{\gamma}{2}\|y - x\|^2. \tag{3.3.9}$$

By combining this expression with (3.3.7), we have the following result.

Lemma 3.3.10. *Given convex* f *satisfying* (3.2.5), *with* ∇f *uniformly Lipschitz continuous with constant* L, *we have for any* x, y *that*

$$\frac{\gamma}{2}\|y - x\|^2 \leqslant f(y) - f(x) - \nabla f(x)^T (y - x) \leqslant \frac{L}{2}\|y - x\|^2. \tag{3.3.11}$$

For later convenience, we define a *condition number* κ as follows:

$$\kappa := \frac{L}{\gamma}. \tag{3.3.12}$$

When f is *twice* continuously differentiable, we can characterize the constants γ and L in terms of the eigenvalues of the Hessian $\nabla f(x)$. Specifically, we can show that (3.3.11) is equivalent to

$$\gamma I \preceq \nabla^2 f(x) \preceq LI, \quad \text{for all } x. \tag{3.3.13}$$

When f is strictly convex and quadratic, κ defined in (3.3.12) is the condition number of the (constant) Hessian, in the usual sense of linear algebra.

Strongly convex functions have unique minimizers, as we now show.

Theorem 3.3.14. *Let* f *be differentiable and strongly convex with modulus* $\gamma > 0$. *Then the minimizer* x^* *of* f *exists and is unique.*

Proof. We show first that for any point x^0, the level set $\{x \mid f(x) \leqslant f(x^0)\}$ is closed and bounded, and hence compact. Suppose for contradiction that there is a sequence $\{x^\ell\}$ such that $\|x^\ell\| \to \infty$ and

$$f(x^\ell) \leqslant f(x^0). \tag{3.3.15}$$

By strong convexity of f, we have for some $\gamma > 0$ that

$$f(x^\ell) \geqslant f(x^0) + \nabla f(x^0)^T (x^\ell - x^0) + \frac{\gamma}{2}\|x^\ell - x^0\|^2.$$

By rearranging slightly, and using (3.3.15), we obtain

$$\frac{\gamma}{2}\|x^\ell - x^0\|^2 \leqslant -\nabla f(x^0)^T (x^\ell - x^0) \leqslant \|\nabla f(x^0)\| \|x^\ell - x^0\|.$$

By dividing both sides by $(\gamma/2)\|x^\ell - x^0\|$, we obtain $\|x^\ell - x^0\| \leqslant (2/\gamma)\|\nabla f(x^0)\|$ for all ℓ, which contradicts unboundedness of $\{x^\ell\}$. Thus, the level set is bounded. Since it is also closed (by continuity of f), it is compact.

Since f is continuous, it attains its minimum on the compact level set, which is also the solution of $\min_x f(x)$, and we denote it by x^*. Suppose for contradiction that the minimizer is not unique, so that we have two points x_1^* and x_2^* that minimize f. Obviously, these points must attain equal objective values, so that $f(x_1^*) = f(x_2^*) = f^*$ for some f^*. By taking (3.2.5) and setting $\phi = f^*$, $x = x_1^*$, $y = x_2^*$, and $\alpha = 1/2$, we obtain

$$f((x_1^* + x_2^*)/2) \leqslant \frac{1}{2}(f(x_1^*) + f(x_2^*)) - \frac{1}{8}\gamma\|x_1^* - x_2^*\|^2 < f^*,$$

so the point $(x_1^* + x_2^*)/2$ has a smaller function value than both x_1^* and x_2^*, contradicting our assumption that x_1^* and x_2^* are both minimizers. Hence, the minimizer x^* is unique. □

3.4. Optimality Conditions for Smooth Functions We consider the case of a smooth (twice continuously differentiable) function f that is not necessarily convex. Before designing algorithms to find a minimizer of f, we need to identify properties of f and its derivatives at a point $\bar{x}$ that tell us whether or not $\bar{x}$ is a minimizer, of one of the types described in Subsection 3.1. We call such properties *optimality conditions*.

A *first-order necessary* condition for optimality is that $\nabla f(\bar{x}) = 0$. More precisely, if $\bar{x}$ is a local minimizer, then $\nabla f(\bar{x}) = 0$. We can prove this by using Taylor's theorem. Supposing for contradiction that $\nabla f(\bar{x}) \neq 0$, we can show by setting $x = \bar{x}$ and $p = -\alpha\nabla f(\bar{x})$ for $\alpha > 0$ in (3.3.3) that $f(\bar{x} - \alpha\nabla f(\bar{x})) < f(\bar{x})$ for all $\alpha > 0$ sufficiently small. Thus *any* neighborhood of $\bar{x}$ will contain points x with a $f(x) < f(\bar{x})$, so $\bar{x}$ cannot be a local minimizer.

If f is convex, as well as smooth, the condition $\nabla f(\bar{x}) = 0$ is *sufficient* for $\bar{x}$ to be a *global* solution. This claim follows immediately from Theorems 3.2.8 and 3.2.9.

A *second-order necessary condition* for $\bar{x}$ to be a local solution is that $\nabla f(\bar{x}) = 0$ and $\nabla^2 f(\bar{x})$ is positive semidefinite. The proof is by an argument similar to that of the first-order necessary condition, but using the second-order Taylor series expansion (3.3.5) instead of (3.3.3). A *second-order sufficient condition* is that $\nabla f(\bar{x}) = 0$

and $\nabla^2 f(\bar{x})$ is positive definite. This condition guarantees that $\bar{x}$ is a *strict* local minimizer, that is, there is a neighborhood of $\bar{x}$ such that $\bar{x}$ has a strictly smaller function value than all other points in this neighborhood. Again, the proof makes use of (3.3.5).

We call $\bar{x}$ a *stationary point* for smooth f if it satisfies the first-order necessary condition $\nabla f(\bar{x}) = 0$. Stationary points are not necessarily local minimizers. In fact, local *maximizers* satisfy the same condition. More interestingly, stationary points can be *saddle points*. These are points for which there exist directions u and v such that $f(\bar{x} + \alpha u) < f(\bar{x})$ and $f(\bar{x} + \alpha v) > f(\bar{x})$ for all positive α sufficiently small. When the Hessian $\nabla^2 f(\bar{x})$ has both strictly positive and strictly negative eigenvalues, it follows from (3.3.5) that $\bar{x}$ is a saddle point. When $\nabla^2 f(\bar{x})$ is positive semidefinite or negative semidefinite, second derivatives alone are insufficient to classify $\bar{x}$; higher-order derivative information is needed.

3.5. Proximal Operators and the Moreau Envelope Here we present some analysis for analyzing the convergence of algorithms for the regularized problem (1.0.2), where the objective is the sum of a smooth function and a convex (usually nonsmooth) function.

We start with a formal definition.

Definition 3.5.1. For a closed proper convex function h and a positive scalar λ, the *Moreau envelope* is

$$M_{\lambda,h}(x) := \inf_u \left\{ h(u) + \frac{1}{2\lambda}\|u - x\|^2 \right\} = \frac{1}{\lambda} \inf_u \left\{ \lambda h(u) + \frac{1}{2}\|u - x\|^2 \right\}. \tag{3.5.2}$$

The proximal operator of the function λh is the value of u that achieves the infimum in (3.5.2), that is,

$$\operatorname{prox}_{\lambda h}(x) := \arg\min_u \left\{ \lambda h(u) + \frac{1}{2}\|u - x\|^2 \right\}. \tag{3.5.3}$$

From optimality properties for (3.5.3) (see Theorem 3.2.8), we have

$$0 \in \lambda \partial h(\operatorname{prox}_{\lambda h}(x)) + (\operatorname{prox}_{\lambda h}(x) - x). \tag{3.5.4}$$

The Moreau envelope can be viewed as a kind of smoothing or regularization of the function h. It has a finite value for all x, even when h takes on infinite values for some $x \in \mathbb{R}^n$. In fact, it is differentiable everywhere, with gradient

$$\nabla M_{\lambda,h}(x) = \frac{1}{\lambda}(x - \operatorname{prox}_{\lambda h}(x)).$$

Moreover, x^* is a minimizer of h if and only if it is a minimizer of $M_{\lambda,h}$.

The proximal operator satisfies a nonexpansiveness property. From the optimality conditions (3.5.4) at two points x and y, we have

$$x - \operatorname{prox}_{\lambda h}(x) \in \lambda \partial(\operatorname{prox}_{\lambda h}(x)), \quad y - \operatorname{prox}_{\lambda h}(y) \in \lambda \partial(\operatorname{prox}_{\lambda h}(y)).$$

By applying monotonicity (Lemma 3.2.7), we have

$$(1/\lambda)\big((x - \operatorname{prox}_{\lambda h}(x)) - (y - \operatorname{prox}_{\lambda h}(y))\big)^T (\operatorname{prox}_{\lambda h}(x) - \operatorname{prox}_{\lambda h}(y)) \geqslant 0,$$

Rearranging this and applying the Cauchy-Schwartz inequality yields

$$\begin{aligned}\|\text{prox}_{\lambda h}(x) - \text{prox}_{\lambda h}(y)\|^2 &\leqslant (x-y)^T(\text{prox}_{\lambda h}(x) - \text{prox}_{\lambda h}(y)) \\ &\leqslant \|x-y\| \, \|\text{prox}_{\lambda h}(x) - \text{prox}_{\lambda h}(y)\|,\end{aligned}$$

from which we obtain $\|\text{prox}_{\lambda h}(x) - \text{prox}_{\lambda h}(y)\| \leqslant \|x-y\|$, as claimed.

We list the prox operator for several instances of h that are common in data analysis applications. These definitions are useful in implementing the prox-gradient algorithms of Section 5.

- $h(x) = 0$ for all x, for which we have $\text{prox}_{\lambda h}(x) = 0$. (This observation is useful in proving that the prox-gradient method reduces to the familiar steepest descent method when the objective contains no regularization term.)
- $h(x) = I_\Omega(x)$, the indicator function for a closed convex set Ω. In this case, we have for any $\lambda > 0$ that

$$\text{prox}_{\lambda I_\Omega}(x) = \arg\min_u \left\{ \lambda I_\Omega(u) + \frac{1}{2}\|u-x\|^2 \right\} = \arg\min_{u \in \Omega} \frac{1}{2}\|u-x\|^2,$$

  which is simply the projection of x onto the set Ω.
- $h(x) = \|x\|_1$. By substituting into definition (3.5.3) we see that the minimization separates into its n separate components, and that the ith component of $\text{prox}_{\lambda\|\cdot\|_1}(x)$ is

$$\left[\text{prox}_{\lambda\|\cdot\|_1}(x)\right]_i = \arg\min_{u_i} \left\{ \lambda|u_i| + \frac{1}{2}(u_i - x_i)^2 \right\}.$$

  We can thus verify that

$$[\text{prox}_{\lambda\|\cdot\|_1}(x)]_i = \begin{cases} x_i - \lambda & \text{if } x_i > \lambda; \\ 0 & \text{if } x_i \in [-\lambda, \lambda]; \\ x_i + \lambda & \text{if } x_i < -\lambda, \end{cases} \tag{3.5.5}$$

  an operation that is known as *soft-thresholding*.
- $h(x) = \|x\|_0$, where $\|x\|_0$ denotes the *cardinality* of the vector x, its number of nonzero components. Although this h is not a convex function (as we can see by considering convex combinations of the vectors $(0,1)^T$ and $(1,0)^T$ in $\mathbb{R}^2$), its proximal operator is well defined, and is known as *hard thresholding*:

$$[\text{prox}_{\lambda\|\cdot\|_0}(x)]_i = \begin{cases} x_i & \text{if } |x_i| \geqslant \sqrt{2\lambda}; \\ 0 & \text{if } |x_i| < \sqrt{2\lambda}. \end{cases}$$

  As in (3.5.5), the definition (3.5.3) separates into n individual components.

3.6. Convergence Rates An important measure for evaluating algorithms is the rate of convergence to zero of some measure of error. For smooth f, we may be interested in how rapidly the sequence of gradient norms $\{\|\nabla f(x^k)\|\}$ converges to zero. For nonsmooth convex f, a measure of interest may be convergence to zero of $\{\text{dist}(0, \partial f(x^k))\}$ (the sequence of distances from 0 to the subdifferential $\partial f(x^k)$). Other error measures for which we may be able to prove convergence rates include $\|x^k - x^*\|$ (where x^* is a solution) and $f(x^k) - f^*$ (where f^* is the optimal value of the objective function f). For generality, we denote by $\{\phi_k\}$ the sequence of nonnegative scalars whose rate of convergence to 0 we wish to find.

We say that *linear* convergence holds if there is some $\sigma \in (0, 1)$ such that

$$\phi_{k+1}/\phi_k \leqslant 1 - \sigma, \quad \text{for all k sufficiently large.} \tag{3.6.1}$$

(This property is sometimes also called *geometric* or *exponential* convergence, but the term *linear* is standard in the optimization literature, so we use it here.) It follows from (3.6.1) that there is some positive constant C such that

$$\phi_k \leqslant C(1 - \sigma)^k, \quad k = 1, 2, \ldots . \tag{3.6.2}$$

While (3.6.1) implies (3.6.2), the converse does not hold. The sequence

$$\phi_k = \begin{cases} 2^{-k} & \text{k even} \\ 0 & \text{k odd,} \end{cases}$$

satisfies (3.6.2) with $C = 1$ and $\sigma = .5$, but does not satisfy (3.6.1). To distinguish between these two slightly different definitions, (3.6.1) is sometimes called *Q-linear* while (3.6.2) is called *R-linear*.

Sublinear convergence is, as its name suggests, slower than linear. Several varieties of sublinear convergence are encountered in optimization algorithms for data analysis, including the following

$$\phi_k \leqslant C/\sqrt{k}, \quad k = 1, 2, \ldots, \tag{3.6.3a}$$

$$\phi_k \leqslant C/k, \quad k = 1, 2, \ldots, \tag{3.6.3b}$$

$$\phi_k \leqslant C/k^2, \quad k = 1, 2, \ldots, \tag{3.6.3c}$$

where in each case, C is some positive constant.

Superlinear convergence occurs when the constant $\sigma \in (0, 1)$ in (3.6.1) can be chosen arbitrarily close to 1. Specifically, we say that the sequence $\{\phi_k\}$ converges *Q-superlinearly* to 0 if

$$\lim_{k \to \infty} \phi_{k+1}/\phi_k = 0. \tag{3.6.4}$$

Q-Quadratic convergence occurs when

$$\phi_{k+1}/\phi_k^2 \leqslant C, \quad k = 1, 2, \ldots, \tag{3.6.5}$$

for some sufficiently large C. We say that the convergence is *R-superlinear* if there is a Q-superlinearly convergent sequence $\{\nu_k\}$ that dominates $\{\phi_k\}$ (that is, $0 \leqslant \phi_k \leqslant \nu_k$ for all k). R-quadratic convergence is defined similarly. Quadratic and superlinear rates are associated with higher-order methods, such as Newton and quasi-Newton methods.

When a convergence rate applies *globally*, from any reasonable starting point, it can be used to derive a complexity bound for the algorithm, which takes the form of a bound on the number of iterations K required to reduce ϕ_k below some specified tolerance ϵ. For a sequence satisfying the R-linear convergence condition (3.6.2) a sufficient condition for $\phi_K \leqslant \epsilon$ is $C(1-\sigma)^K \leqslant \epsilon$. By using the estimate $\log(1-\sigma) \leqslant -\sigma$ for all $\sigma \in (0,1)$, we have that

$$C(1-\sigma)^K \leqslant \epsilon \quad \Leftrightarrow \quad K\log(1-\sigma) \leqslant \log(\epsilon/C) \quad \Leftarrow \quad K \geqslant \log(C/\epsilon)/\sigma.$$

It follows that for linearly convergent algorithms, the number of iterations required to converge to a tolerance ϵ depends logarithmically on $1/\epsilon$ and inversely on the rate constant σ. For an algorithm that satisfies the sublinear rate (3.6.3a), a sufficient condition for $\phi_K \leqslant \epsilon$ is $C/\sqrt{K} \leqslant \epsilon$, which is equivalent to $K \geqslant (C/\epsilon)^2$, so the complexity is $O(1/\epsilon^2)$. Similar analyses for (3.6.3b) reveal complexity of $O(1/\epsilon)$, while for (3.6.3c), we have complexity $O(1/\sqrt{\epsilon})$.

For quadratically convergent methods, the complexity is doubly logarithmic in ϵ (that is, $O(\log\log(1/\epsilon))$). Once the algorithm enters a neighborhood of quadratic convergence, just a few additional iterations are required for convergence to a solution of high accuracy.

4. Gradient Methods

We consider here iterative methods for solving the unconstrained smooth problem (1.0.1) that make use of the gradient ∇f (see also, [22] which describes subgradient methods for nonsmooth convex functions.) We consider mostly methods that generate an iteration sequence $\{x^k\}$ via the formula

$$x^{k+1} = x^k + \alpha_k d^k, \tag{4.0.1}$$

where d^k is the search direction and α_k is a steplength.

We consider the steepest descent method, which searches along the negative gradient direction $d^k = -\nabla f(x^k)$, proving convergence results for nonconvex functions, convex functions, and strongly convex functions. In Subsection 4.5, we consider methods that use more general descent directions d^k, proving convergence of methods that make careful choices of the line search parameter α_k at each iteration. In Subsection 4.6, we consider the conditional gradient method for minimization of a smooth function f over a compact set.

4.1. Steepest Descent The simplest stepsize protocol is the short-step variant of steepest descent. We assume here that f is differentiable, with gradient ∇f satisfying the Lipschitz continuity condition (3.3.6) with constant L. We choose the search direction $d^k = -\nabla f(x^k)$ in (4.0.1), and set the steplength α_k to be the constant $1/L$, to obtain the iteration

$$x^{k+1} = x^k - \frac{1}{L}\nabla f(x^k), \quad k = 0,1,2,\ldots. \tag{4.1.1}$$

To estimate the amount of decrease in f obtained at each iterate of this method, we use Taylor's theorem. From (3.3.7), we have

$$f(x+\alpha d) \leqslant f(x) + \alpha\nabla f(x)^T d + \alpha^2\frac{L}{2}\|d\|^2, \tag{4.1.2}$$

For $x = x^k$ and $d = -\nabla f(x^k)$, the value of α that minimizes the expression on the right-hand side is $\alpha = 1/L$. By substituting these values, we obtain

$$f(x^{k+1}) = f(x^k - (1/L)\nabla f(x^k)) \leqslant f(x^k) - \frac{1}{2L}\|\nabla f(x^k)\|^2. \tag{4.1.3}$$

This expression is one of the foundational inequalities in the analysis of optimization methods. Depending on the assumptions about f, we can derive a variety of different convergence rates from this basic inequality.

4.2. General Case We consider first a function f that is Lipschitz continuously differentiable and bounded below, but that need not necessarily be convex. Using (4.1.3) alone, we can prove a sublinear convergence result for the steepest descent method.

Theorem 4.2.1. *Suppose that* f *is Lipschitz continuously differentiable, satisfying* (3.3.6), *and that* f *is bounded below by a constant* $\bar{f}$. *Then for the steepest descent method with constant steplength* $\alpha_k \equiv 1/L$, *applied from a starting point* x^0, *we have for any integer* $T \geqslant 1$ *that*

$$\min_{0 \leqslant k \leqslant T-1} \|\nabla f(x^k)\| \leqslant \sqrt{\frac{2L[f(x^0) - f(x^T)]}{T}} \leqslant \sqrt{\frac{2L[f(x^0) - \bar{f}]}{T}}.$$

Proof. Rearranging (4.1.3) and summing over the first $T-1$ iterates, we have

$$\sum_{k=0}^{T-1} \|\nabla f(x^k)\|^2 \leqslant 2L \sum_{k=0}^{T-1} [f(x^k) - f(x^{k+1})] = 2L[f(x^0) - f(x^T)]. \tag{4.2.2}$$

(Note the telescoping sum.) Since f is bounded below by $\bar{f}$, the right-hand side is bounded above by the constant $2L[f(x^0) - \bar{f}]$. We also have that

$$\min_{0 \leqslant k \leqslant T-1} \|\nabla f(x^k)\| = \sqrt{\min_{0 \leqslant k \leqslant T-1} \|\nabla f(x^k)\|^2} \leqslant \sqrt{\frac{1}{T}\sum_{k=0}^{T-1} \|\nabla f(x^k)\|^2}.$$

The result is obtained by combining this bound with (4.2.2). □

This result shows that within the first $T-1$ steps of steepest descent, at least one of the iterates has gradient norm less than $\sqrt{2L[f(x^0) - \bar{f}]/T}$, which represents sublinear convergence of type (3.6.3a). It follows too from (4.2.2) that for f bounded below, any accumulation point of the sequence $\{x^k\}$ is stationary.

4.3. Convex Case When f is also convex, we have the following stronger result for the steepest descent method.

Theorem 4.3.1. *Suppose that* f *is convex and Lipschitz continuously differentiable, satisfying* (3.3.6), *and that* (1.0.1) *has a solution* x^*. *Then the steepest descent method with stepsize* $\alpha_k \equiv 1/L$ *generates a sequence* $\{x^k\}_{k=0}^{\infty}$ *that satisfies*

$$f(x^T) - f^* \leqslant \frac{L}{2T}\|x^0 - x^*\|^2. \tag{4.3.2}$$

Proof. By convexity of f, we have $f(x^*) \geqslant f(x^k) + \nabla f(x^k)^T(x^* - x^k)$, so by substituting into (4.1.3), we obtain for $k = 0, 1, 2, \ldots$ that

$$\begin{aligned} f(x^{k+1}) &\leqslant f(x^*) + \nabla f(x^k)^T(x^k - x^*) - \frac{1}{2L}\|\nabla f(x^k)\|^2 \\ &= f(x^*) + \frac{L}{2}\left(\|x^k - x^*\|^2 - \left\|x^k - x^* - \frac{1}{L}\nabla f(x^k)\right\|^2\right) \\ &= f(x^*) + \frac{L}{2}\left(\|x^k - x^*\|^2 - \|x^{k+1} - x^*\|^2\right). \end{aligned}$$

By summing over $k = 0, 1, 2, \ldots, T-1$, and noting the telescoping sum, we have

$$\begin{aligned} \sum_{k=0}^{T-1}(f(x^{k+1}) - f^*) &\leqslant \frac{L}{2}\sum_{k=0}^{T-1}\left(\|x^k - x^*\|^2 - \|x^{k+1} - x^*\|^2\right) \\ &= \frac{L}{2}\left(\|x^0 - x^*\|^2 - \|x^T - x^*\|^2\right) \\ &\leqslant \frac{L}{2}\|x^0 - x^*\|^2. \end{aligned}$$

Since $\{f(x^k)\}$ is a nonincreasing sequence, we have, as required,

$$f(x^T) - f(x^*) \leqslant \frac{1}{T}\sum_{k=0}^{T-1}(f(x^{k+1}) - f^*) \leqslant \frac{L}{2T}\|x^0 - x^*\|^2. \qquad \square$$

4.4. Strongly Convex Case Recall that the definition (3.3.9) of strong convexity shows that f can be bounded below by a quadratic with Hessian γI. A strongly convex f with L-Lipschitz gradients is also bounded *above* by a similar quadratic (see (3.3.7)) differing only in the quadratic term, which becomes LI. From this "sandwich" effect, we derive a linear convergence rate for the gradient method, stated formally in the following theorem.

Theorem 4.4.1. *Suppose that f is Lipschitz continuously differentiable, satisfying* (3.3.6), *and strongly convex, satisfying* (3.2.5) *with modulus of convexity* γ. *Then f has a unique minimizer* x^*, *and the steepest descent method with stepsize* $\alpha_k \equiv 1/L$ *generates a sequence* $\{x^k\}_{k=0}^{\infty}$ *that satisfies*

$$f(x^{k+1}) - f(x^*) \leqslant \left(1 - \frac{\gamma}{L}\right)(f(x^k) - f(x^*)), \quad k = 0, 1, 2, \ldots.$$

Proof. Existence of the unique minimizer x^* follows from Theorem 3.3.14. Minimizing both sides of the inequality (3.3.9) with respect to y, we find that the minimizer on the left side is attained at $y = x^*$, while on the right side it is attained at $x - \nabla f(x)/\gamma$. Plugging these optimal values into (3.3.9), we obtain

$$\begin{aligned} &\min_y f(y) \geqslant \min_y f(x) + \nabla f(x)^T(y - x) + \frac{\gamma}{2}\|y - x\|^2 \\ &\Rightarrow \quad f(x^*) \geqslant f(x) - \nabla f(x)^T\left(\frac{1}{\gamma}\nabla f(x)\right) + \frac{\gamma}{2}\left\|\frac{1}{\gamma}\nabla f(x)\right\|^2 \\ &\Rightarrow \quad f(x^*) \geqslant f(x) - \frac{1}{2\gamma}\|\nabla f(x)\|^2. \end{aligned}$$

By rearrangement, we obtain

$$\|\nabla f(x)\|^2 \geqslant 2\gamma[f(x) - f(x^*)]. \tag{4.4.2}$$

By substituting (4.4.2) into our basic inequality (4.1.3), we obtain

$$f(x^{k+1}) = f\left(x^k - \frac{1}{L}\nabla f(x^k)\right) \leqslant f(x^k) - \frac{1}{2L}\|\nabla f(x^k)\|^2 \leqslant f(x^k) - \frac{\gamma}{L}(f(x^k) - f^*).$$

Subtracting f^* from both sides of this inequality yields the result. □

Note that After T steps, we have

$$f(x^T) - f^* \leqslant \left(1 - \frac{\gamma}{L}\right)^T (f(x^0) - f^*), \tag{4.4.3}$$

which is convergence of type (3.6.2) with constant $\sigma = \gamma/L$.

4.5. General Case: Line-Search Methods Returning to the case in which f has Lipschitz continuous gradients but is possibly nonconvex, we consider algorithms that take steps of the form (4.0.1), where d^k is a *descent direction*, that is, it makes a positive inner product with the negative gradient $-\nabla f(x^k)$, so that $\nabla f(x^k)^T d^k < 0$. This condition ensures that $f(x^k + \alpha d^k) < f(x^k)$ for sufficiently small positive values of step length α — we obtain improvement in f by taking small steps along d^k. (This claim follows from (3.3.3).) *Line-search methods* are built around this fundamental observation. By introducing additional conditions on d^k and α_k, that can be verified in practice with reasonable effort, we can establish a bound on decrease similar to (4.1.3) on each iteration, and thus a conclusion similar to that of Theorem 4.2.1.

We assume that d^k satisfies the following for some $\eta > 0$:

$$\nabla f(x^k)^T d^k \leqslant -\eta \|\nabla f(x^k)\| \|d^k\|. \tag{4.5.1}$$

For the steplength α_k, we assume the following *weak Wolfe* conditions hold, for some constants c_1 and c_2 with $0 < c_1 < c_2 < 1$:

$$f(x^k + \alpha_k d^k) \leqslant f(x^k) + c_1 \alpha_k \nabla f(x^k)^T d^k \tag{4.5.2a}$$

$$\nabla f(x^k + \alpha_k d^k)^T d^k \geqslant c_2 \nabla f(x^k)^T d^k. \tag{4.5.2b}$$

Condition (4.5.2a) is called "sufficient decrease;" it ensures descent at each step of at least a small fraction c_1 of the amount promised by the first-order Taylor-series expansion (3.3.3). Condition (4.5.2b) ensures that the directional derivative of f along the search direction d^k is significantly less negative at the chosen steplength α_k than at $\alpha = 0$. This condition ensures that the step is "not too short." It can be shown that it is always possible to find α_k that satisfies both conditions (4.5.2) simultaneously.

Line-search procedures, which are specialized optimization procedures for minimizing functions of one variable, have been devised to find such values efficiently; see [36, Chapter 3] for details.

For line-search methods of this type, we have the following generalization of Theorem 4.2.1.

Theorem 4.5.3. *Suppose that* f *is Lipschitz continuously differentiable, satisfying* (3.3.6), *and that* f *is bounded below by a constant* $\bar{f}$. *Consider the method that takes steps of the form* (4.0.1), *where* d^k *satisfies* (4.5.1) *for some* $\eta > 0$ *and the conditions* (4.5.2) *hold at all* k, *for some constants* c_1 *and* c_2 *with* $0 < c_1 < c_2 < 1$. *Then for any integer* $T \geqslant 1$, *we have*

$$\min_{0 \leqslant k \leqslant T-1} \|\nabla f(x^k)\| \leqslant \sqrt{\frac{L}{\eta^2 c_1 (1-c_2)}} \sqrt{\frac{f(x^0) - \bar{f}}{T}}.$$

Proof. By combining the Lipschitz property (3.3.6) with (4.5.2b), we have

$$-(1-c_2)\nabla f(x^k)^T d^k \leqslant [\nabla f(x^k + \alpha_k d^k) - \nabla f(x^k)]^T d^k \leqslant L\alpha_k \|d^k\|^2.$$

By comparing the first and last terms in these inequalities, we obtain the following lower bound on α_k:

$$\alpha_k \geqslant -\frac{(1-c_2)}{L} \frac{\nabla f(x^k)^T d^k}{\|d^k\|^2}.$$

By substituting this bound into (4.5.2a), and using (4.5.1) and the step definition (4.0.1), we obtain

$$\begin{aligned} f(x^{k+1}) = f(x^k + \alpha_k d^k) &\leqslant f(x^k) + c_1 \alpha_k \nabla f(x^k)^T d^k \\ &\leqslant f(x^k) - \frac{c_1(1-c_2)}{L} \frac{(\nabla f(x^k)^T d^k)^2}{\|d^k\|^2} \\ &\leqslant f(x^k) - \frac{c_1(1-c_2)}{L} \eta^2 \|\nabla f(x^k)\|^2, \end{aligned} \tag{4.5.4}$$

which by rearrangement yields

$$\|\nabla f(x^k)\|^2 \leqslant \frac{L}{c_1(1-c_2)\eta^2} \left(f(x^k) - f(x^{k+1}) \right). \tag{4.5.5}$$

The result now follows as in the proof of Theorem 4.2.1. □

It follows by taking limits on both sides of (4.5.5) that

$$\lim_{k \to \infty} \|\nabla f(x^k)\| = 0, \tag{4.5.6}$$

and therefore all accumulation points $\bar{x}$ of the sequence $\{x^k\}$ generated by the algorithm (4.0.1) have $\nabla f(\bar{x}) = 0$. In the case of f convex, this condition guarantees that $\bar{x}$ is a solution of (1.0.1). When f is nonconvex, $\bar{x}$ may be a local minimum, but it may also be a saddle point or a local maximum.

The paper [29] uses the stable manifold theorem to show that line-search gradient methods are highly unlikely to converge to stationary points $\bar{x}$ at which some eigenvalues of the Hessian $\nabla^2 f(\bar{x})$ are negative. Although it is easy to construct examples for which such bad behavior occurs, it requires special choices of starting point x^0. Possibly the most obvious example is where $f(x_1, x_2) = x_1^2 - x_2^2$ starting from $x^0 = (1, 0)^T$, where $d^k = -\nabla f(x^k)$ at each k. For this example, all iterates have $x_2^k = 0$ and, under appropriate conditions, converge to the saddle point $\bar{x} = 0$. Any starting point with $x_2^0 \neq 0$ cannot converge to 0, in fact, it is easy to see that x_2^k diverges away from 0.

4.6. Conditional Gradient Method The conditional gradient approach, often known as "Frank-Wolfe" after the authors who devised it [24], is a method for convex nonlinear optimization over compact convex sets. This is the problem

$$\min_{x \in \Omega} f(x), \tag{4.6.1}$$

(see earlier discussion around (3.2.2)), where Ω is a compact convex set and f is a convex function whose gradient is Lipschitz continuously differentiable in a neighborhood of Ω, with Lipschitz constant L. We assume that Ω has diameter D, that is, $\|x-y\| \leqslant D$ for all $x, y \in \Omega$.

The conditional gradient method replaces the objective in (4.6.1) at each iteration by a linear Taylor-series approximation around the current iterate x^k, and minimizes this linear objective over the original constraint set Ω. It then takes a step from x^k towards the minimizer of this linearized subproblem. The full method is as follows:

$$v^k := \arg\min_{v \in \Omega} v^T \nabla f(x^k); \tag{4.6.2a}$$

$$x^{k+1} := x^k + \alpha_k (v^k - x^k), \quad \alpha_k := \frac{2}{k+2}. \tag{4.6.2b}$$

The method has a sublinear convergence rate, as we show below, and indeed requires many iterations in practice to obtain an accurate solution. Despite this feature, it makes sense in many interesting applications, because the subproblems (4.6.2a) can be solved very cheaply in some settings, and because highly accurate solutions are not required in some applications.

We have the following result for sublinear convergence of the conditional gradient method.

Theorem 4.6.3. *Under the conditions above, where* L *is the Lipschitz constant for* ∇f *on an open neighborhood of* Ω *and* D *is the diameter of* Ω*, the conditional gradient method* (4.6.2) *applied to* (4.6.1) *satisfies*

$$f(x^k) - f(x^*) \leqslant \frac{2LD^2}{k+2}, \quad k = 1, 2, \ldots, \tag{4.6.4}$$

where x^* *is any solution of* (4.6.1).

Proof. Setting $x = x^k$ and $y = x^{k+1} = x^k + \alpha_k (v^k - x^k)$ in (3.3.7), we have

$$\begin{aligned} f(x^{k+1}) &\leqslant f(x^k) + \alpha_k \nabla f(x^k)^T (v^k - x^k) + \frac{1}{2}\alpha_k^2 L \|v^k - x^k\|^2 \\ &\leqslant f(x^k) + \alpha_k \nabla f(x^k)^T (v^k - x^k) + \frac{1}{2}\alpha_k^2 L D^2, \end{aligned} \tag{4.6.5}$$

where the second inequality comes from the definition of D. For the first-order term, we have since v^k solves (4.6.2a) and x^* is feasible for (4.6.2a) that

$$\nabla f(x^k)^T (v^k - x^k) \leqslant \nabla f(x^k)^T (x^* - x^k) \leqslant f(x^*) - f(x^k).$$

By substituting in (4.6.5) and subtracting $f(x^*)$ from both sides, we obtain

$$f(x^{k+1}) - f(x^*) \leqslant (1 - \alpha_k)[f(x^k) - f(x^*)] + \frac{1}{2}\alpha_k^2 L D^2. \tag{4.6.6}$$

We now apply an inductive argument. For $k = 0$, we have $\alpha_0 = 1$ and

$$f(x^1) - f(x^*) \leqslant \frac{1}{2}LD^2 < \frac{2}{3}LD^2,$$

so that (4.6.4) holds in this case. Supposing that (4.6.4) holds for some value of k, we aim to show that it holds for $k+1$ too. We have

$$\begin{aligned}
&f(x^{k+1}) - f(x^*) \\
&\leqslant \left(1 - \frac{2}{k+2}\right)[f(x^k) - f(x^*)] + \frac{1}{2}\frac{4}{(k+2)^2}LD^2 && \text{from (4.6.6), (4.6.2b)} \\
&\leqslant LD^2\left[\frac{2k}{(k+2)^2} + \frac{2}{(k+2)^2}\right] && \text{from (4.6.4)} \\
&= 2LD^2\frac{(k+1)}{(k+2)^2} \\
&= 2LD^2\frac{k+1}{k+2}\frac{1}{k+2} \\
&\leqslant 2LD^2\frac{k+2}{k+3}\frac{1}{k+2} = \frac{2LD^2}{k+3},
\end{aligned}$$

as required. □

5. Prox-Gradient Methods

We now describe an elementary but powerful approach for solving the regularized optimization problem

$$\min_{x \in \mathbb{R}^n} \phi(x) := f(x) + \lambda\psi(x), \tag{5.0.1}$$

where f is a smooth convex function, ψ is a convex regularization function (known simply as the "regularizer"), and $\lambda \geqslant 0$ is a regularization parameter. The technique we describe here is a natural extension of the steepest-descent approach, in that it reduces to the steepest-descent method analyzed in Theorems 4.3.1 and 4.4.1 applied to f when the regularization term is not present ($\lambda = 0$). It is useful when the regularizer ψ has a simple structure that is easy to account for explicitly, as is true for many regularizers that arise in data analysis, such as the ℓ_1 function ($\psi(x) = \|x\|_1$) of the indicator function for a simple set Ω ($\psi(x) = I_\Omega(x)$), such as a box $\Omega = [l_1, u_1] \otimes [l_2, u_2] \otimes \ldots \otimes [l_n, u_n]$. For such regularizers, the proximal operators can be computed explicitly and efficiently.[2]

Each step of the algorithm is defined as follows:

$$x^{k+1} := \text{prox}_{\alpha_k\lambda\psi}(x^k - \alpha_k \nabla f(x^k)), \tag{5.0.2}$$

for some steplength $\alpha_k > 0$, and the prox operator defined in (3.5.3). By substituting into this definition, we can verify that x^{k+1} is the solution of an approximation to the objective ϕ of (5.0.1), namely:

$$x^{k+1} := \arg\min_z \nabla f(x^k)^T(z - x^k) + \frac{1}{2\alpha_k}\|z - x^k\|^2 + \lambda\psi(z). \tag{5.0.3}$$

[2]For the analysis of this section I am indebted to class notes of L. Vandenberghe, from 2013-14.

One way to verify this equivalence is to note that the objective in (5.0.3) can be written as

$$\frac{1}{\alpha_k}\left\{\frac{1}{2}\left\|z-(x^k-\alpha_k\nabla f(x^k))\right\|^2+\alpha_k\lambda\psi(x)\right\},$$

(modulo a term $\alpha_k\|\nabla f(x^k)\|^2$ that does not involve z). The subproblem objective in (5.0.3) consists of a linear term $\nabla f(x^k)^T(z-x^k)$ (the first-order term in a Taylor-series expansion), a proximality term $\frac{1}{2\alpha_k}\|z-x^k\|^2$ that becomes more strict as $\alpha_k \downarrow 0$, and the regularization term $\lambda\psi(x)$ in unaltered form. When $\lambda = 0$, we have $x^{k+1} = x^k - \alpha_k\nabla f(x^k)$, so the iteration (5.0.2) (or (5.0.3)) reduces to the usual steepest-descent approach discussed in Section 4 in this case. It is useful to continue thinking of α_k as playing the role of a line-search parameter, though here the line search is expressed implicitly through a proximal term.

We will demonstrate convergence of the method (5.0.2) at a sublinear rate, for functions f whose gradients satisfy a Lipschitz continuity property with Lipschitz constant L (see (3.3.6)), and for the constant steplength choice $\alpha_k = 1/L$. The proof makes use of a "gradient map" defined by

$$G_\alpha(x) := \frac{1}{\alpha}\left(x-\text{prox}_{\alpha\lambda\psi}(x-\alpha\nabla f(x))\right). \tag{5.0.4}$$

By comparing with (5.0.2), we see that this map defines the step taken at iteration k as:

$$x^{k+1}=x^k-\alpha_k G_{\alpha_k}(x^k) \quad\Leftrightarrow\quad G_{\alpha_k}(x^k)=\frac{1}{\alpha_k}(x^k-x^{k+1}). \tag{5.0.5}$$

The following technical lemma reveals some useful properties of $G_\alpha(x)$.

Lemma 5.0.6. *Suppose that in problem* (5.0.1), ψ *is a closed convex function and that* f *is convex with Lipschitz continuous gradient on* $\mathbb{R}^n$, *with Lipschitz constant* L. *Then for the definition* (5.0.4) *with* $\alpha > 0$, *the following claims are true.*

(a) $G_\alpha(x) \in \nabla f(x) + \lambda\partial\psi(x-\alpha G_\alpha(x))$.

(b) *For any* z, *and any* $\alpha \in (0, 1/L]$, *we have that*

$$\phi(x-\alpha G_\alpha(x)) \leqslant \phi(z) + G_\alpha(x)^T(x-z) - \frac{\alpha}{2}\|G_\alpha(x)\|^2.$$

Proof. For part (a), we use the optimality property (3.5.4) of the prox operator, and make the following substitutions: $x-\alpha\nabla f(x)$ for "x", $\alpha\lambda$ for "λ", and ψ for "h" to obtain

$$0 \in \alpha\lambda\partial\psi(\text{prox}_{\alpha\lambda\psi}(x-\alpha\nabla f(x))) + (\text{prox}_{\alpha\lambda\psi}(x-\alpha\nabla f(x)) - (x-\alpha\nabla f(x)).$$

We make the substitution $\text{prox}_{\alpha\lambda\psi}(x-\alpha\nabla f(x)) = x-\alpha G_\alpha(x)$, using definition (5.0.4), to obtain

$$0 \in \alpha\lambda\partial\psi(x-\alpha G_\alpha(x)) - \alpha(G_\alpha(x)-\nabla f(x)),$$

and the result follows when we divide by α.

For (b), we start with the following consequence of Lipschitz continuity of ∇f, from Lemma 3.3.10:

$$f(y) \leqslant f(x) + \nabla f(x)^T(y-x) + \frac{L}{2}\|y-x\|^2.$$

By setting $y = x - \alpha G_\alpha(x)$, for any $\alpha \in (0, 1/L]$, we have

$$\text{(5.0.7)} \quad \begin{aligned} f(x - \alpha G_\alpha(x)) &\leqslant f(x) - \alpha G_\alpha(x)^T \nabla f(x) + \frac{L\alpha^2}{2} \|G_\alpha(x)\|^2 \\ &\leqslant f(x) - \alpha G_\alpha(x)^T \nabla f(x) + \frac{\alpha}{2} \|G_\alpha(x)\|^2. \end{aligned}$$

(The second inequality uses $\alpha \in (0, 1/L]$.) We also have by convexity of f and ψ that for any z and any $v \in \partial\psi(x - \alpha G_\alpha(x)$ the following are true:

$$\text{(5.0.8)} \quad \begin{aligned} f(z) &\geqslant f(x) + \nabla f(x)^T (z - x), \\ \psi(z) &\geqslant \psi(x - \alpha G_\alpha(x)) + v^T (z - (x - \alpha G_\alpha(x))). \end{aligned}$$

From part (a) that $v = (G_\alpha(x) - \nabla f(x))/\lambda \in \partial\psi(x - \alpha G_\alpha(x))$. Making this choice of v in (5.0.8) and using (5.0.7) we have for any $\alpha \in (0, 1/L]$ that

$$\begin{aligned} &\phi(x - \alpha G_\alpha(x)) \\ &= f(x - \alpha G_\alpha(x)) + \lambda\psi(x - \alpha G_\alpha(x)) \\ &\leqslant f(x) - \alpha G_\alpha(x)^T \nabla f(x) + \frac{\alpha}{2} \|G_\alpha(x)\|^2 + \lambda\psi(x - \alpha G_\alpha(x)) \quad \text{(from (5.0.7))} \\ &\leqslant f(z) + \nabla f(x)^T (x - z) - \alpha G_\alpha(x)^T \nabla f(x) + \frac{\alpha}{2} \|G_\alpha(x)\|^2 \\ &\qquad + \lambda\psi(z) + (G_\alpha(x) - \nabla f(x))^T (x - \alpha G_\alpha(x) - z) \quad \text{(from (5.0.8))} \\ &= f(z) + \lambda\psi(z) + G_\alpha(x)^T (x - z) - \frac{\alpha}{2} \|G_\alpha(x)\|^2, \end{aligned}$$

where the last equality follows from cancellation of several terms in the previous line. Thus (b) is proved. □

Theorem 5.0.9. *Suppose that in problem* (5.0.1), ψ *is a closed convex function and that* f *is convex with Lipschitz continuous gradient on* $\mathbb{R}^n$, *with Lipschitz constant* L. *Suppose that* (5.0.1) *attains a minimizer* x^* *(not necessarily unique) with optimal objective value* ϕ^*. *Then if* $\alpha_k = 1/L$ *for all* k *in* (5.0.2), *we have*

$$\phi(x^k) - \phi^* \leqslant \frac{L\|x^0 - x^*\|^2}{2k}, \quad k = 1, 2, \ldots.$$

Proof. Since $\alpha_k = 1/L$ satisfies the conditions of Lemma 5.0.6, we can use part (b) of this result to show that the sequence $\{\phi(x^k)\}$ is decreasing and that the distance to the optimum x^* also decreases at each iteration. Setting $x = z = x^k$ and $\alpha = \alpha_k$ in Lemma 5.0.6, and recalling (5.0.5), we have

$$\phi(x^{k+1}) = \phi(x^k - \alpha_k G_{\alpha_k}(x^k)) \leqslant \phi(x^k) - \frac{\alpha_k}{2} \|G_{\alpha_k}(x^k)\|^2,$$

justifying the first claim. For the second claim, we have by setting $x = x^k$, $\alpha = \alpha_k$, and $z = x^*$ in Lemma 5.0.6 that

$$\text{(5.0.10)} \quad \begin{aligned} 0 \leqslant \phi(x^{k+1}) - \phi^* &= \phi(x^k - \alpha_k G_{\alpha_k}(x^k)) - \phi^* \\ &\leqslant G_{\alpha_k}(x^k)^T (x^k - x^*) - \frac{\alpha_k}{2} \|G_{\alpha_k}(x^k)\|^2 \\ &= \frac{1}{2\alpha_k} \left(\|x^k - x^*\|^2 - \|x^k - x^* - \alpha_k G_{\alpha_k}(x^k)\|^2 \right) \\ &= \frac{1}{2\alpha_k} \left(\|x^k - x^*\|^2 - \|x^{k+1} - x^*\|^2 \right), \end{aligned}$$

from which $\|x^{k+1} - x^*\| \leqslant \|x^k - x^*\|$ follows.

By setting $\alpha_k = 1/L$ in (5.0.10), and summing over $k = 0, 1, 2, \dots, K-1$, we obtain from a telescoping sum on the right-hand side that

$$\sum_{k=0}^{K-1} (\phi(x^{k+1}) - \phi^*) \leqslant \frac{L}{2}\left(\|x^0 - x^*\|^2 - \|x^K - x^*\|^2\right) \leqslant \frac{L}{2}\|x^0 - x^*\|^2.$$

By monotonicity of $\{\phi(x^k)\}$, we have

$$K(\phi(x^K) - \phi^*) \leqslant \sum_{k=0}^{K-1} (\phi(x^{k+1}) - \phi^*).$$

The result follows immediately by combining these last two expressions. □

6. Accelerating Gradient Methods

We showed in Section 4 that the basic steepest descent method for solving (1.0.1) for smooth f converges sublinearly at a $1/k$ rate when f is convex, and linearly at a rate of $(1 - \gamma/L)$ when f is strongly convex, satisfying (3.3.13) for positive γ and L. We show in this section that by using the gradient information in a more clever way, faster convergence rates can be attained.

The key idea is *momentum*. In iteration k of a momentum method, we tend to continue moving along the *previous* search direction at each iteration, making a small adjustment toward the negative gradient $-\nabla f$ evaluated at x^k or a nearby point. (Steepest descent simply uses $-\nabla f(x^k)$ as the search direction.) Although not obvious at first, there is some intuition behind the momentum idea. The step taken at the previous iterate x^{k-1} was based on negative gradient information at that iteration, along with the search direction from the iteration prior to that one, namely, x^{k-2}. By continuing this line of reasoning backwards, we see that the previous step is a linear combination of all the gradient information that we have encountered at all iterates so far, going back to the initial iterate x^0. If this information is aggregated properly, it can produce a richer overall picture of the function than the latest negative gradient alone, and thus has the potential to yield better convergence.

Sure enough, several intricate methods that use the momentum idea have been proposed, and have been widely successful. These methods are often called *accelerated gradient methods*. A major contributor in this area is Yuri Nesterov, dating to his seminal contribution in 1983 [33] and explicated further in his book [34] and other publications. Another key contribution is [3], which derived an accelerated method for the regularized case (1.0.2).

6.1. Heavy-Ball Method Possibly the most elementary method of momentum type is the *heavy-ball* method of Polyak [37]; see also [38]. Each iteration of this method has the form

$$x^{k+1} = x^k - \alpha_k \nabla f(x^k) + \beta_k (x^k - x^{k-1}), \tag{6.1.1}$$

where α_k and β_k are positive scalars. That is, a momentum term $\beta_k(x^k - x^{k-1})$ is added to the usual steepest descent update. Although this method can be applied to any smooth convex f (and even to nonconvex functions), the convergence analysis is most straightforward for the special case of strongly convex *quadratic* functions (see [38]). (This analysis also suggests appropriate values for the step lengths α_k and β_k.) Consider the function

$$\min_{x \in \mathbb{R}^n} f(x) := \frac{1}{2}x^T A x - b^T x, \tag{6.1.2}$$

where the (constant) Hessian A has eigenvalues in the range $[\gamma, L]$, with $0 < \gamma \leqslant L$. For the following constant choices of steplength parameters:

$$\alpha_k = \alpha := \frac{4}{(\sqrt{L} + \sqrt{\gamma})^2}, \quad \beta_k = \beta := \frac{\sqrt{L} - \sqrt{\gamma}}{\sqrt{L} + \sqrt{\gamma}},$$

it can be shown that $\|x^k - x^*\| \leqslant C\beta^k$, for some (possibly large) constant C. We can use (3.3.7) to translate this into a bound on the function error, as follows:

$$f(x^k) - f(x^*) \leqslant \frac{L}{2}\|x^k - x^*\|^2 \leqslant \frac{LC^2}{2}\beta^{2k},$$

allowing a direct comparison with the rate (4.4.3) for the steepest descent method. If we suppose that $L \gg \gamma$, we have

$$\beta \approx 1 - 2\sqrt{\frac{\gamma}{L}},$$

so that we achieve approximate convergence $f(x^k) - f(x^*) \leqslant \epsilon$ (for small positive ϵ) in $O(\sqrt{L/\gamma}\log(1/\epsilon))$ iterations, compared with $O((L/\gamma)\log(1/\epsilon))$ for steepest descent — a significant improvement.

The heavy-ball method is fundamental, but several points should be noted. First, the analysis for convex quadratic f is based on linear algebra arguments, and does not generalize to general strongly convex nonlinear functions. Second, the method requires knowledge of γ and L, for the purposes of defining parameters α and β. Third, it is not a descent method; we usually have $f(x^{k+1}) > f(x^k)$ for many k. These properties are not specific to the heavy-ball method — some of them are shared by other methods that use momentum.

6.2. Conjugate Gradient The conjugate gradient method for solving linear systems $Ax = b$ (or, equivalently, minimizing the convex quadratic (6.1.2)) where A is symmetric positive definite, is one of the most important algorithms in computational science. Though invented earlier than the other algorithms discussed in this section (see [27]) and motivated in a different way, conjugate gradient clearly makes use of momentum. Its steps have the form

$$x^{k+1} = x^k + \alpha_k p^k, \quad \text{where } p^k = -\nabla f(x^k) + \xi_k p^{k-1}, \tag{6.2.1}$$

for some choices of α_k and ξ_k, which is identical to (6.1.1) when we define β_k appropriately. For convex, strongly quadratic problems (6.1.2), conjugate gradient has excellent properties. It does not require prior knowledge of the range $[\gamma, L]$ of the eigenvalue spectrum of A, choosing the steplengths α_k and ξ_k in an adaptive fashion. (In fact, α_k is chosen to be the exact minimizer along the search direction p_k.) The main arithmetic operation per iteration is one matrix-vector multiplication involving A, the same cost as a gradient evaluation for f in (6.1.2). Most importantly, there is a rich convergence theory, that characterizes convergence in terms of the properties of the full spectrum of A (not just its extreme elements), showing in particular that good approximate solutions can be obtained quickly if the eigenvalues are clustered. Convergence to an exact solution of (6.1.2) in at most n iterations is guaranteed (provided, naturally, that the arithmetic is carried out exactly).

There has been much work over the years on extending the conjugate gradient method to general smooth functions f. Few of the theoretical properties for the quadratic case carry over to the nonlinear setting, though several results are known; see [36, Chapter 5], for example. Such "nonlinear" conjugate gradient methods vary in the accuracy with which they perform the line search for α_k in (6.2.1) and — more fundamentally — in the choice of ξ_k. The latter is done in a way that ensures that each search direction p^k is a descent direction. In some methods, ξ_k is set to zero on some iterations, which causes the method to take a steepest descent step, effectively "restarting" the conjugate gradient method at the latest iterate.

Despite these qualifications, nonlinear conjugate gradient is quite commonly used in practice, because of its minimal storage requirements and the fact that it requires only one gradient evaluation per iteration. Its popularity has been eclipsed in recent years by the limited-memory quasi-Newton method L-BFGS [30], [36, Section 7.2], which requires more storage (though still $O(n)$) and is similarly economical and easy to implement.

6.3. Nesterov's Accelerated Gradient: Weakly Convex Case We now describe Nesterov's method for (1.0.1) and prove its convergence — sublinear at a $1/k^2$ rate — for the case of f convex with Lipschitz continuous gradients satisfying (3.3.6). Each iteration of this method has the form

$$x^{k+1} = x^k - \alpha_k \nabla f\left(x^k + \beta_k(x^k - x^{k-1})\right) + \beta_k(x^k - x^{k-1}), \tag{6.3.1}$$

for choices of the parameters α_k and β_k to be defined. Note immediately the similarity to the heavy-ball formula (6.1.1). The only difference is that the extrapolation step $x^k \to x^k + \beta_k(x^k - x^{k-1})$ is taken before evaluation of the gradient ∇f in (6.3.1), whereas in (6.1.1) the gradient is simply evaluated at x^k. It is convenient for purposes of analysis (and implementation) to introduce an auxiliary sequence $\{y^k\}$, fix $\alpha_k \equiv 1/L$, and rewrite the update (6.3.1) as follows:

$$x^{k+1} = y^k - \frac{1}{L}\nabla f(y^k), \tag{6.3.2a}$$

$$y^{k+1} = x^{k+1} + \beta_{k+1}(x^{k+1} - x^k), \quad k = 0, 1, 2, \ldots, \tag{6.3.2b}$$

where we initialize at an arbitrary y^0 and set $x^0 = y^0$. We define β_k with reference to another scalar sequence λ_k in the following manner:

$$\lambda_0 = 0, \quad \lambda_{k+1} = \frac{1}{2}\left(1 + \sqrt{1 + 4\lambda_k^2}\right), \quad \beta_k = \frac{\lambda_k - 1}{\lambda_{k+1}}. \tag{6.3.3}$$

Since $\lambda_k \geqslant 1$ for $k = 1, 2, \ldots$, we have $\beta_{k+1} \geqslant 0$ for $k = 0, 1, 2, \ldots$. It also follows from the definition of λ_{k+1} that

$$\lambda_{k+1}^2 - \lambda_{k+1} = \lambda_k^2. \tag{6.3.4}$$

We have the following result for convergence of Nesterov's scheme on general convex functions. We prove it using an argument from [3], as reformulated in [7, Section 3.7]. The analysis is famously technical, and intuition is hard to come by. Some recent progress has been made in deriving algorithms similar to (6.3.2) that have a plausible geometric or algebraic motivation; see [8, 21].

Theorem 6.3.5. *Suppose that* f *in* (1.0.1) *is convex, with* ∇f *Lipschitz continuously differentiable with constant* L *(as in* (3.3.6)*) and that the minimum of* f *is attained at* x^**, with* $f^* := f(x^*)$*. Then the method defined by* (6.3.2), (6.3.3) *with* $x^0 = y^0$ *yields an iteration sequence* $\{x^k\}$ *with the following property:*

$$f(x^T) - f^* \leqslant \frac{2L\|x^0 - x^*\|^2}{(T+1)^2}, \quad T = 1, 2, \ldots.$$

Proof. From convexity of f and (3.3.7), we have for any x and y that

$$\begin{aligned}
& f(y - \nabla f(y)/L) - f(x) \\
&\leqslant f(y - \nabla f(y)/L) - f(y) + \nabla f(y)^T(y - x) \\
&\leqslant \nabla f(y)^T(y - \nabla f(y)/L - y) + \frac{L}{2}\|y - \nabla f(y)/L - y\|^2 + \nabla f(y)^T(y - x) \\
&= -\frac{1}{2L}\|\nabla f(y)\|^2 + \nabla f(y)^T(y - x).
\end{aligned} \tag{6.3.6}$$

Setting $y = y^k$ and $x = x^k$ in this bound, we obtain

$$\begin{aligned}
f(x^{k+1}) - f(x^k) &= f(y^k - \nabla f(y^k)/L) - f(x^k) \\
&\leqslant -\frac{1}{2L}\|\nabla f(y^k)\|^2 + \nabla f(y^k)^T(y^k - x^k) \\
&= -\frac{L}{2}\|x^{k+1} - y^k\|^2 - L(x^{k+1} - y^k)^T(y^k - x^k).
\end{aligned} \tag{6.3.7}$$

We now set $y = y^k$ and $x = x^*$ in (6.3.6), and use (6.3.2a) to obtain

$$f(x^{k+1}) - f(x^*) \leqslant -\frac{L}{2}\|x^{k+1} - y^k\|^2 - L(x^{k+1} - y^k)^T(y^k - x^*). \tag{6.3.8}$$

Introducing notation $\delta_k := f(x^k) - f(x^*)$, we multiply (6.3.7) by $\lambda_{k+1} - 1$ and add it to (6.3.8) to obtain

$$\begin{aligned}
& (\lambda_{k+1} - 1)(\delta_{k+1} - \delta_k) + \delta_{k+1} \\
&\leqslant -\frac{L}{2}\lambda_{k+1}\|x^{k+1} - y^k\|^2 - L(x^{k+1} - y^k)^T(\lambda_{k+1}y^k - (\lambda_{k+1} - 1)x^k - x^*).
\end{aligned}$$

We multiply this bound by λ_{k+1}, and use (6.3.4) to obtain

$$\begin{aligned}
(6.3.9)\quad & \lambda_{k+1}^2\delta_{k+1} - \lambda_k^2\delta_k \\
& \leqslant -\frac{L}{2}\left[\|\lambda_{k+1}(x^{k+1}-y^k)\|^2 + 2\lambda_{k+1}(x^{k+1}-y^k)^T(\lambda_{k+1}y^k - (\lambda_{k+1}-1)x^k - x^*)\right] \\
& = -\frac{L}{2}\left[\|\lambda_{k+1}x^{k+1} - (\lambda_{k+1}-1)x^k - x^*\|^2 - \|\lambda_{k+1}y^k - (\lambda_{k+1}-1)x^k - x^*\|^2\right],
\end{aligned}$$

where in the final equality we used the identity $\|a\|^2 + 2a^Tb = \|a+b\|^2 - \|b\|^2$. By multiplying (6.3.2b) by λ_{k+2}, and using $\lambda_{k+2}\beta_{k+1} = \lambda_{k+1} - 1$ from (6.3.3), we have

$$\begin{aligned}
\lambda_{k+2}y^{k+1} &= \lambda_{k+2}x^{k+1} + \lambda_{k+2}\beta_{k+1}(x^{k+1} - x^k) \\
&= \lambda_{k+2}x^{k+1} + (\lambda_{k+1} - 1)(x^{k+1} - x^k).
\end{aligned}$$

By rearranging this equality, we have

$$\lambda_{k+1}x^{k+1} - (\lambda_{k+1} - 1)x^k = \lambda_{k+2}y^{k+1} - (\lambda_{k+2} - 1)x^{k+1}.$$

By substituting into the first term on the right-hand side of (6.3.9), and using the definition

$$u^k := \lambda_{k+1}y^k - (\lambda_{k+1} - 1)x^k - x^*, \tag{6.3.10}$$

we obtain

$$\lambda_{k+1}^2\delta_{k+1} - \lambda_k^2\delta_k \leqslant -\frac{L}{2}(\|u^{k+1}\|^2 - \|u^k\|^2).$$

By summing both sides of this inequality over $k = 0, 1, \ldots, T-1$, and using $\lambda_0 = 0$, we obtain

$$\lambda_T^2\delta_T \leqslant \frac{L}{2}(\|u^0\|^2 - \|u^T\|^2) \leqslant \frac{L}{2}\|x^0 - x^*\|^2,$$

so that

$$\delta_T = f(x^T) - f(x^*) \leqslant \frac{L\|x^0 - x^*\|^2}{2\lambda_T^2}. \tag{6.3.11}$$

A simple induction confirms that $\lambda_k \geqslant (k+1)/2$ for $k = 1, 2, \ldots$, and the claim of the theorem follows by substituting this bound into (6.3.11). □

6.4. Nesterov's Accelerated Gradient: Strongly Convex Case We turn now to Nesterov's approach for smooth strongly convex functions, which satisfy (3.2.5) with $\gamma > 0$. Again, we follow the proof in [7, Section 3.7], which is based on the analysis in [34]. The method uses the same update formula (6.3.2) as in the weakly convex case, and the same initialization, but with a different choice of β_{k+1}, namely:

$$\beta_{k+1} \equiv \frac{\sqrt{L} - \sqrt{\gamma}}{\sqrt{L} + \sqrt{\gamma}} = \frac{\sqrt{\kappa} - 1}{\sqrt{\kappa} + 1}. \tag{6.4.1}$$

The condition measure κ is defined in (3.3.12). We prove the following convergence result.

Theorem 6.4.2. *Suppose that f is such that ∇f is Lipschitz continuously differentiable with constant L, and that it is strongly convex with modulus of convexity γ and unique* *minimizer* x^*. *Then the method* (6.3.2), (6.4.1) *with starting point* $x^0 = y^0$ *satisfies*

$$f(x^T) - f(x^*) \leqslant \frac{L+\gamma}{2}\|x^0 - x^*\|^2 \left(1 - \frac{1}{\sqrt{\kappa}}\right)^T, \quad T = 1, 2, \dots.$$

Proof. The proof makes use of a family of strongly convex functions $\Phi_k(z)$ defined inductively as follows:

$$\Phi_0(z) = f(y^0) + \frac{\gamma}{2}\|z - y^0\|^2, \tag{6.4.3a}$$

$$\Phi_{k+1}(z) = (1 - 1/\sqrt{\kappa})\Phi_k(z) + \frac{1}{\sqrt{\kappa}}\left(f(y^k) + \nabla f(y^k)^T(z - y^k) + \frac{\gamma}{2}\|z - y^k\|^2\right). \tag{6.4.3b}$$

Each $\Phi_k(\cdot)$ is a quadratic, and an inductive argument shows that $\nabla^2 \Phi_k(z) = \gamma I$ for all k and all z. Thus, each Φ_k has the form

$$\Phi_k(z) = \Phi_k^* + \frac{\gamma}{2}\|z - v^k\|^2, \quad k = 0, 1, 2, \dots, \tag{6.4.4}$$

where v^k is the minimizer of $\Phi_k(\cdot)$ and Φ_k^* is its optimal value. (From (6.4.3a), we have $v^0 = y^0$.) We note too that Φ_k becomes a tighter *overapproximation* to f as $k \to \infty$. To show this, we use (3.3.9) to replace the final term in parentheses in (6.4.3b) by $f(z)$, then subtract $f(z)$ from both sides of (6.4.3b) to obtain

$$\Phi_{k+1}(z) - f(z) \leqslant (1 - 1/\sqrt{\kappa})(\Phi_k(z) - f(z)). \tag{6.4.5}$$

In the remainder of the proof, we show that the following bound holds:

$$f(x^k) \leqslant \min_z \Phi_k(z) = \Phi_k^*, \quad k = 0, 1, 2, \dots. \tag{6.4.6}$$

The upper bound in Lemma 3.3.10 for $x = x^*$ gives $f(z) - f(x^*) \leqslant (L/2)\|z - x^*\|^2$. By combining this bound with (6.4.5) and (6.4.6), we have

$$\begin{aligned} f(x^k) - f(x^*) &\leqslant \Phi_k^* - f(x^*) \\ &\leqslant \Phi_k(x^*) - f(x^*) \\ &\leqslant (1 - 1/\sqrt{\kappa})^k(\Phi_0(x^*) - f(x^*)) \\ &\leqslant (1 - 1/\sqrt{\kappa})^k[(\Phi_0(x^*) - f(x^0)) + (f(x^0) - f(x^*))] \\ &\leqslant (1 - 1/\sqrt{\kappa})^k \frac{\gamma + L}{2}\|x^0 - x^*\|^2. \end{aligned} \tag{6.4.7}$$

The proof is completed by establishing (6.4.6), by induction on k. Since $x^0 = y^0$, it holds by definition at $k = 0$. By using step formula (6.3.2a), the convexity property (3.3.8) (with $x = y^k$), and the inductive hypothesis, we have

$$\begin{aligned} &f(x^{k+1}) \\ &\leqslant f(y^k) - \frac{1}{2L}\|\nabla f(y^k)\|^2 \\ &= (1 - 1/\sqrt{\kappa})f(x^k) + (1 - 1/\sqrt{\kappa})(f(y^k) - f(x^k)) + f(y^k)/\sqrt{\kappa} - \frac{1}{2L}\|\nabla f(y^k)\|^2 \\ &\leqslant (1 - 1/\sqrt{\kappa})\Phi_k^* + (1 - 1/\sqrt{\kappa})\nabla f(y^k)^T(y^k - x^k) + f(y^k)/\sqrt{\kappa} - \frac{1}{2L}\|\nabla f(y^k)\|^2. \end{aligned} \tag{6.4.8}$$

Thus the claim is established (and the theorem is proved) if we can show that the right-hand side in (6.4.8) is bounded above by Φ^*_{k+1}.

Recalling the observation (6.4.4), we have by taking derivatives of both sides of (6.4.3b) with respect to z that

$$\nabla\Phi_{k+1}(z) = \gamma(1-1/\sqrt{\kappa})(z-v^k) + \nabla f(y^k)/\sqrt{\kappa} + \gamma(z-y^k)/\sqrt{\kappa}. \tag{6.4.9}$$

Since v^{k+1} is the minimizer of Φ_{k+1} we can set $\nabla\Phi_{k+1}(v^{k+1}) = 0$ in (6.4.9) to obtain

$$v^{k+1} = (1-1/\sqrt{\kappa})v^k + y^k/\sqrt{\kappa} - \nabla f(y^k)/(\gamma\sqrt{\kappa}). \tag{6.4.10}$$

By subtracting y^k from both sides of this expression, and taking $\|\cdot\|^2$ of both sides, we obtain

$$\begin{aligned}\|v^{k+1}-y^k\|^2 = (1-1/\sqrt{\kappa})^2\|y^k-v^k\|^2 + \|\nabla f(y^k)\|^2/(\gamma^2\kappa)\\ -2(1-1/\sqrt{\kappa})/(\gamma\sqrt{\kappa})\nabla f(y^k)^T(v^k-y^k).\end{aligned} \tag{6.4.11}$$

By evaluating Φ_{k+1} at $z = y^k$, using both (6.4.4) and (6.4.3b), we obtain

$$\begin{aligned}&\Phi^*_{k+1} + \frac{\gamma}{2}\|y^k - v^{k+1}\|^2\\ &= (1-1/\sqrt{\kappa})\Phi_k(y^k) + f(y^k)/\sqrt{\kappa}\\ &= (1-1/\sqrt{\kappa})\Phi^*_k + \frac{\gamma}{2}(1-1/\sqrt{\kappa})\|y^k-v^k\|^2 + f(y^k)/\sqrt{\kappa}.\end{aligned} \tag{6.4.12}$$

By substituting (6.4.11) into (6.4.12), we obtain

$$\begin{aligned}\Phi^*_{k+1} &= (1-1/\sqrt{\kappa})\Phi^*_k + f(y^k)/\sqrt{\kappa} + \gamma(1-1/\sqrt{\kappa})/(2\sqrt{\kappa})\|y^k-v^k\|^2\\ &\quad -\frac{1}{2L}\|\nabla f(y^k)\|^2 + (1-1/\sqrt{\kappa})\nabla f(y^k)^T(v^k-y^k)/\sqrt{\kappa}\\ &\geqslant (1-1/\sqrt{\kappa})\Phi^*_k + f(y^k)/\sqrt{\kappa}\\ &\quad -\frac{1}{2L}\|\nabla f(y^k)\|^2 + (1-1/\sqrt{\kappa})\nabla f(y^k)^T(v^k-y^k)/\sqrt{\kappa},\end{aligned} \tag{6.4.13}$$

where we simply dropped a nonnegative term from the right-hand side to obtain the inequality. The final step is to show that

$$v^k - y^k = \sqrt{\kappa}(y^k - x^k), \tag{6.4.14}$$

which we do by induction. Note that $v^0 = x^0 = y^0$, so the claim holds for $k = 0$. We have

$$\begin{aligned}v^{k+1} - y^{k+1} &= (1-1/\sqrt{\kappa})v^k + y^k/\sqrt{\kappa} - \nabla f(y^k)/(\gamma\sqrt{\kappa}) - y^{k+1}\\ &= \sqrt{\kappa}y^k - (\sqrt{\kappa}-1)x^k - \sqrt{\kappa}\nabla f(y^k)/L - y^{k+1}\\ &= \sqrt{\kappa}x^{k+1} - (\sqrt{\kappa}-1)x^k - y^{k+1}\\ &= \sqrt{\kappa}(y^{k+1} - x^{k+1}),\end{aligned} \tag{6.4.15}$$

where the first equality is from (6.4.10), the second equality is from the inductive hypothesis, the third equality is from the iteration formula (6.3.2a), and the final equality is from the iteration formula (6.3.2b) with the definition of β_{k+1} from (6.4.1). We have thus proved (6.4.14), and by substituting this equality into (6.4.13), we obtain that Φ_{k+1}^* is an upper bound on the right-hand side of (6.4.8). This establishes (6.4.6) and thus completes the proof of the theorem. □

6.5. Lower Bounds on Rates The term "optimal" in Nesterov's optimal method is used because the convergence rate achieved by the method is the best possible (possibly up to a constant), among algorithms that make use of gradient information at the iterates x^k. This claim can be proved by means of a carefully designed function, for which *no* method that makes use of all gradients observed up to and including iteration k (namely, $\nabla f(x^i)$, $i = 0, 1, 2, \ldots, k$) can produce a sequence $\{x^k\}$ that achieves a rate better than that of Theorem 6.3.5. The function proposed in [32] is a convex quadratic $f(x) = (1/2)x^T A x - e_1^T x$, where

$$A = \begin{bmatrix} 2 & -1 & 0 & 0 & \ldots & \ldots & 0 \\ -1 & 2 & -1 & 0 & \ldots & \ldots & 0 \\ 0 & -1 & 2 & -1 & 0 & \ldots & 0 \\ & & \ddots & \ddots & \ddots & & \\ 0 & \ldots & & 0 & -1 & 2 & -1 \\ 0 & \ldots & & & 0 & -1 & 2 \end{bmatrix}, \quad e_1 = \begin{bmatrix} 1 \\ 0 \\ 0 \\ \vdots \\ 0 \end{bmatrix}.$$

The solution x^* satisfies $Ax^* = e_1$; its components are $x_i^* = 1 - i/(n+1)$, for $i = 1, 2, \ldots, n$. If we use $x^0 = 0$ as the starting point, and construct the iterate x^{k+1} as

$$x^{k+1} = x^k + \sum_{j=0}^{k} \xi_j \nabla f(x^j),$$

for some coefficients ξ_j, $j = 0, 1, \ldots, k$, an elementary inductive argument shows that each iterate x^k can have nonzero entries only in its first k components. It follows that for any such algorithm, we have

$$\|x^k - x^*\|^2 \geqslant \sum_{j=k+1}^{n} (x_j^*)^2 = \sum_{j=k+1}^{n} \left(1 - \frac{j}{n+1}\right)^2. \tag{6.5.1}$$

A little arithmetic shows that

$$\|x^k - x^*\|^2 \geqslant \frac{1}{8}\|x^0 - x^*\|^2, \quad k = 1, 2, \ldots, \frac{n}{2} - 1, \tag{6.5.2}$$

It can be shown further that

$$f(x^k) - f^* \geqslant \frac{3L}{32(k+1)^2}\|x^0 - x^*\|^2, \quad k = 1, 2, \ldots, \frac{n}{2} - 1, \tag{6.5.3}$$

where $L = \|A\|_2$. This lower bound on $f(x^k) - x^*$ is within a constant factor of the upper bound of Theorem 6.3.5.

The restriction $k \leqslant n/2$ in the argument above is not fully satisfying. A more compelling example would show that the lower bound (6.5.3) holds for *all* k, but an example of this type is not currently known.

7. Newton Methods

So far, we have dealt with methods that use first-order (gradient or subgradient) information about the objective function. We have shown that such algorithms can yield sequences of iterates that converge at linear or sublinear rates. We turn our attention in this chapter to methods that exploit second-derivative (Hessian) information. The canonical method here is Newton's method, named after Isaac Newton, who proposed a version of the method for polynomial equations in around 1670.

For many functions, including many that arise in data analysis, second-order information is not difficult to compute, in the sense that the functions that we deal with are simple (usually compositions of elementary functions). In comparing with first-order methods, there is a tradeoff. Second-order methods typically have local superlinear or quadratic convergence rates: Once the iterates reach a neighborhood of a solution at which second-order sufficient conditions are satisfied, convergence is rapid. Moreover, their global convergence properties are attractive. With appropriate enhancements, they can provably avoid convergence to saddle points. But the costs of calculating and handling the second-order information and of computing the step is higher. Whether this tradeoff makes them appealing depends on the specifics of the application and on whether the second-derivative computations are able to take advantage of structure in the objective function.

We start by sketching the basic Newton's method for the unconstrained smooth optimization problem $\min f(x)$, and prove local convergence to a minimizer x^* that satisfies second-order sufficient conditions. Subsection 7.2 discusses performance of Newton's method on convex functions, where the use of Newton search directions in the line search framework (4.0.1) can yield global convergence. Modifications of Newton's method for nonconvex functions are discussed in Subsection 7.3. Subsection 7.4 discusses algorithms for smooth nonconvex functions that use gradient and Hessian information but guarantee convergence to points that approximately satisfy second-order necessary conditions. Some variants of these methods are related closely to the trust-region methods discussed in Subsection 7.3, but the motivation and mechanics are somewhat different.

7.1. Basic Newton's Method Consider the problem

$$\min f(x), \tag{7.1.1}$$

where $f : \mathbb{R}^n \to \mathbb{R}$ is a Lipschitz twice continuously differentiable function, where the Hessian has Lipschitz constant M, that is,

$$\|\nabla^2 f(x') - \nabla^2 f(x'')\| \leqslant M\|x' - x''\|, \tag{7.1.2}$$

where $\|\cdot\|$ denotes the Euclidean vector norm and its induced matrix norm. Newton's method generates a sequence of iterates $\{x^k\}_{k=0,1,2,\dots}$.

A second-order Taylor series approximation to f around the current iterate x^k is

$$f(x^k + p) \approx f(x^k) + \nabla f(x^k)^T p + \frac{1}{2} p^T \nabla^2 f(x^k) p. \tag{7.1.3}$$

When $\nabla^2 f(x^k)$ is positive definite, the minimizer p^k of the right-hand side is unique; it is

$$p^k = -\nabla^2 f(x^k)^{-1} \nabla f(x^k). \tag{7.1.4}$$

This is the Newton step. In its most basic form, then, Newton's method is defined by the following iteration:

$$x^{k+1} = x^k - \nabla^2 f(x^k)^{-1} \nabla f(x^k). \tag{7.1.5}$$

We have the following local convergence result in the neighborhood of a point x^* satisfying second-order sufficient conditions.

Theorem 7.1.6. *Consider the problem* (7.1.1) *with* f *twice Lipschitz continuously differentiable with Lipschitz constant* M *defined in* (7.1.2). *Suppose that the second-order sufficient conditions are satisfied for the problem* (7.1.1) *at the point* x^*, *that is,* $\nabla f(x^*) = 0$ *and* $\nabla^2 f(x^*) \succeq \gamma I$ *for some* $\gamma > 0$. *Then if* $\|x^0 - x^*\| \leqslant \frac{\gamma}{2M}$, *the sequence defined by* (7.1.5) *converges to* x^* *at a quadratic rate, with*

$$\|x^{k+1} - x^*\| \leqslant \frac{M}{\gamma} \|x^k - x^*\|^2, \quad k = 0, 1, 2, \ldots. \tag{7.1.7}$$

Proof. From (7.1.4) and (7.1.5), and using $\nabla f(x^*) = 0$, we have

$$\begin{aligned} x^{k+1} - x^* &= x^k - x^* - \nabla^2 f(x^k)^{-1} \nabla f(x^k) \\ &= \nabla^2 f(x^k)^{-1} [\nabla^2 f(x^k)(x^k - x^*) - (\nabla f(x^k) - \nabla f(x^*))]. \end{aligned}$$

so that

$$\|x^{k+1} - x^*\| \leqslant \|\nabla^2 f(x^k)^{-1}\| \|\nabla^2 f(x^k)(x^k - x^*) - (\nabla f(x^k) - \nabla f(x^*))\|. \tag{7.1.8}$$

By using Taylor's theorem (see (3.3.4) with $x = x^k$ and $p = x^* - x^k$), we have

$$\nabla f(x^k) - \nabla f(x^*) = \int_0^1 \nabla^2 f(x^k + t(x^* - x^k))(x^k - x^*)\, dt.$$

By using this result along with the Lipschitz condition (7.1.2), we have

$$\begin{aligned} &\|\nabla^2 f(x^k)(x^k - x^*) - (\nabla f(x^k) - \nabla f(x^*))\| \\ &= \left\| \int_0^1 [\nabla^2 f(x^k) - \nabla^2 f(x^k + t(x^* - x^k))](x^k - x^*)\, dt \right\| \\ &\leqslant \int_0^1 \|\nabla^2 f(x^k) - \nabla^2 f(x^k + t(x^* - x^k))\| \|x^k - x^*\|\, dt \\ &\leqslant \left(\int_0^1 Mt\, dt \right) \|x^k - x^*\|^2 = \tfrac{1}{2} M \|x^k - x^*\|^2. \end{aligned} \tag{7.1.9}$$

From the Weilandt-Hoffman inequality[28] and (7.1.2), we have that

$$|\lambda_{\min}(\nabla^2 f(x^k)) - \lambda_{\min}(\nabla^2 f(x^*))| \leqslant \|\nabla^2 f(x^k) - \nabla^2 f(x^*)\| \leqslant M \|x^k - x^*\|,$$

where $\lambda_{\min}(\cdot)$ denotes the smallest eigenvalue of a symmetric matrix. Thus for

$$\|x^k - x^*\| \leqslant \frac{\gamma}{2M}, \tag{7.1.10}$$

we have

$$\lambda_{\min}(\nabla^2 f(x^k)) \geqslant \lambda_{\min}(\nabla^2 f(x^*)) - M\|x^k - x^*\| \geqslant \gamma - M\frac{\gamma}{2M} \geqslant \frac{\gamma}{2},$$

so that $\|\nabla^2 f(x^k)^{-1}\| \leqslant 2/\gamma$. By substituting this result together with (7.1.9) into (7.1.8), we obtain

$$\|x^{k+1} - x^*\| \leqslant \frac{2}{\gamma}\frac{M}{2}\|x^k - x^*\|^2 = \frac{M}{\gamma}\|x^k - x^*\|^2,$$

verifying the local quadratic convergence rate. By applying (7.1.10) again, we have

$$\|x^{k+1} - x^*\| \leqslant \left(\frac{M}{\gamma}\|x^k - x^*\|\right)\|x^k - x^*\| \leqslant \frac{1}{2}\|x^k - x^*\|,$$

so, by arguing inductively, we see that the sequence converges to x^* provided that x^0 satisfies (7.1.10), as claimed. □

Of course, we do not need to explicitly identify a starting point x^0 in the stated region of convergence. Any sequence that approaches to x^* will eventually enter this region, and thereafter the quadratic convergence guarantees apply.

We have established that Newton's method converges rapidly once the iterates enter the neighborhood of a point x^* satisfying second-order sufficient optimality conditions. But what happens when we start far from such a point?

7.2. Newton's Method for Convex Functions When the function f is convex as well as smooth, we can devise variants of Newton's method for which global convergence and complexity results (in particular, results based on those of Section 4.5) can be proved in addition to local quadratic convergence.

When f is strongly convex with modulus γ and satisfies Lipschitz continuity of the gradient (3.3.6), the Hessian $\nabla^2 f(x^k)$ is positive definite for all k, with all eigenvalues in the interval $[\gamma, L]$. Thus, the Newton direction (7.1.4) is well defined at all iterates x^k, and is a descent direction satisfying the condition (4.5.1) with $\eta = \gamma/L$. To verify this claim, note first

$$\|p^k\| \leqslant \|\nabla^2 f(x^k)^{-1}\|\|\nabla f(x^k)\| \leqslant \frac{1}{\gamma}\|\nabla f(x^k)\|.$$

Then

$$\begin{aligned}(p^k)^T \nabla f(x^k) &= -\nabla f(x^k)^T \nabla^2 f(x^k)^{-1} \nabla f(x^k) \\ &\leqslant -\frac{1}{L}\|\nabla f(x^k)\|^2 \\ &\leqslant -\frac{\gamma}{L}\|\nabla f(x^k)\|\|p^k\|.\end{aligned}$$

We can use the Newton direction in the line-search framework of Subsection 4.5 to obtain a method for which $x^k \to x^*$, where x^* is the (unique) global minimizer of f. (This claim follows from the property (4.5.6) together with the fact that x^* is the only point for which $\nabla f(x^*) = 0$.) We can even obtain a complexity result — and $O(1/\sqrt{T})$ bound on $\min_{0 \leqslant k \leqslant T-1} \|\nabla f(x^k)\|$ — from Theorem 4.5.3.

These global convergence properties are enhanced by the local quadratic convergence property of Theorem 7.1.6 if we modify the line-search framework by accepting the step length $\alpha_k = 1$ in (4.0.1) whenever it satisfies the weak Wolfe conditions (4.5.2). (It can be shown, by again using arguments based on Taylor's theorem (Theorem 3.3.1), that these conditions *will* be satisfied by $\alpha_k = 1$ for all x^k sufficiently close to the minimizer x^*.)

Consider now the case in which f is convex and satisfies condition (3.3.6) but is not strongly convex. Here, the Hessian $\nabla^2 f(x^k)$ may be singular for some k, so the direction (7.1.4) may not be well defined. However, by adding any positive number $\lambda_k > 0$ to the diagonal, we can ensure that the modified Newton direction defined by

$$p^k = -[\nabla^2 f(x^k) + \lambda_k I]^{-1} \nabla f(x^k), \tag{7.2.1}$$

is well defined and is a descent direction for f. For any $\eta \in (0, 1)$ in (4.5.1), we have by choosing λ_k large enough that $\lambda_k/(L + \lambda_k) \geqslant \eta$ that the condition (4.5.1) is satisfied too, so we can use the resulting direction p^k in the line-search framework of Subsection 4.5, to obtain a method that convergence to a solution x^* of (1.0.1), when one exists.

If, in addition, the minimizer x^* is unique and satisfies a second-order sufficient condition (so that $\nabla^2 f(x^*)$ is positive definite), then $\nabla^2 f(x^k)$ will be positive definite too for k sufficiently large. Thus, provided that η is sufficiently small, the *unmodified* Newton direction (with $\lambda_k = 0$ in (7.2.1)) will satisfy the condition (4.5.1). If we use (7.2.1) in the line-search framework of Section 4.5, but set $\lambda_k = 0$ where possible, and accept $\alpha_k = 1$ as the step length whenever it satisfies (4.5.2), we can obtain local quadratic convergence to x^*, in addition to the global convergence and complexity promised by Theorem 4.5.3.

7.3. Newton Methods for Nonconvex Functions For smooth nonconvex f, the Hessian $\nabla^2 f(x^k)$ may be indefinite for some k. The Newton direction (7.1.4) may not exist (when $\nabla^2 f(x^k)$ is singular) or it may not be a descent direction (when $\nabla^2 f(x^k)$ has negative eigenvalues). However, we can still define a modified Newton direction as in (7.2.1), which will be a descent direction for λ_k sufficiently large, and thus can be used in the line-search framework of Section 4.5. For a given η in (4.5.1), a sufficient condition for p^k from (7.2.1) to satisfy (4.5.1) is that

$$\frac{\lambda_k + \lambda_{\min}(\nabla^2 f(x^k))}{\lambda_k + L} \geqslant \eta,$$

where $\lambda_{\min}(\nabla^2 f(x^k))$ is the minimum eigenvalue of the Hessian, which may be negative. The line-search framework of Section 4.5 can then be applied to ensure that $\nabla f(x^k) \to 0$.

Once again, if the iterates $\{x^k\}$ enter the neighborhood of a local solution x^* for which $\nabla^2 f(x^*)$ is positive definite, some enhancements of the strategy for choosing λ_k and the step length α_k can recover the local quadratic convergence of Theorem 7.1.6.

Formula (7.2.1) is not the only way to modify the Newton direction to ensure descent in a line-search framework. Other approaches are outlined in [36, Chapter 3]. One such technique is to modify the Cholesky factorization of $\nabla^2(f^k)$ by adding positive elements to the diagonal only as needed to allow the factorization to proceed (that is, to avoid taking the square root of a negative number), then using the modified factorization in place of $\nabla^2 f(x^k)$ in the calculation of the Newton step p^k. Another technique is to compute an eigenvalue decomposition $\nabla^2 f(x^k) = Q_k \Lambda_k Q_k^T$ (where Q_k is orthogonal and Λ_k is the diagonal matrix containing the eigenvalues), then define $\tilde{\Lambda}_k$ to be a modified version of Λ^k in which all the diagonals are positive. Then, following (7.1.4), p^k can be defined as

$$p^k := -Q_k \tilde{\Lambda}_k^{-1} Q_k^T \nabla f(x^k).$$

When an appropriate strategy is used to define $\tilde{\Lambda}_k$, we can ensure satisfaction of the descent condition (4.5.1) for some $\eta > 0$. As above, the line-search framework of Section 4.5 can be used to obtain an algorithm that generates a sequence $\{x^k\}$ such that $\nabla f(x^k) \to 0$. We noted earlier that this condition ensures that all accumulation points $\hat{x}$ are stationary points, that is, they satisfy $\nabla f(\hat{x}) = 0$.

Stronger guarantees can be obtained from a *trust-region* version of Newton's method, which ensures convergence to a point satisfying second-order necessary conditions, that is, $\nabla^2 f(\hat{x}) \succeq 0$ in addition to $\nabla f(\hat{x}) = 0$. The trust-region approach was developed in the late 1970s and early 1980s, and has become popular again recently because of this appealing global convergence behavior. A trust-region Newton method also recovers quadratic convergence to solutions x^* satisfying second-order-sufficient conditions, without any special modifications. (The trust-region Newton approach is closely related to cubic regularization [26, 35], which we discuss in the next section.)

We now outline the trust-region approach. (Further details can be found in [36, Chapter 4].) The subproblem to be solved at each iteration is

$$\min_d f(x^k) + \nabla f(x^k)^T d + \frac{1}{2} d^T \nabla^2 f(x^k) d \quad \text{subject to } \|d\|_2 \leqslant \Delta_k. \tag{7.3.1}$$

The objective is a second-order Taylor-series approximation while Δ_k is the *radius* of the trust region — the region within which we trust the second-order model to capture the true behavior of f. Somewhat surprisingly, the problem (7.3.1) is not too difficult to solve, even when the Hessian $\nabla^2 f(x^k)$ is indefinite. In fact, the solution d^k of (7.3.1) satisfies the linear system

$$[\nabla^2 f(x^k) + \lambda I] d^k = -\nabla f(x^k), \quad \text{for some } \lambda \geqslant 0, \tag{7.3.2}$$

where λ is chosen such that $\nabla^2 f(x^k) + \lambda I$ is positive semidefinite and $\lambda > 0$ only if $\|d^k\| = \Delta_k$ (see [31]). Solving (7.3.1) thus reduces to a search for the appropriate value of the scalar λ_k, for which specialized methods have been devised.

For large-scale problems, it may be too expensive to solve (7.3.1) near-exactly, since the process may require several factorizations of an $n \times n$ matrix (namely, the coefficient matrix in (7.3.2), for different values of λ). A popular approach for finding approximate solutions of (7.3.1), which can be used when $\nabla^2 f(x^k)$ is positive definite, is the *dogleg* method. In this method the curved path traced out by solutions of (7.3.2) for values of λ in the interval $[0, \infty)$ is approximated by simpler path consisting of two line segments. The first segment joins 0 to the point d_C^k that minimizes the objective in (7.3.1) along the direction $-\nabla f(x^k)$, while the second segment joins d_C^k to the pure Newton step defined in (7.1.4). The approximate solution is taken to be the point at which this "dogleg" path crosses the boundary of the trust region $\|d\| \leqslant \Delta_k$. If the dogleg path lies entirely inside the trust region, we take d^k to be the pure Newton step. See [36, Section 4.1].

Having discussed the trust-region subproblem (7.3.1), let us outline how it can be used as the basis for a complete algorithm. A crucial role is played by the ratio between the *amount of decrease in* f *predicted by the quadratic objective in* (7.3.1) and the *actual decrease in* f, *namely,* $f(x^k) - f(x^k + d^k)$. Ideally, this ratio would be close to 1. If it is at least greater than a small tolerance (say, 10^{-4}) we accept the step and proceed to the next iteration. Otherwise, we conclude that the trust-region radius Δ_k is too large, so we do not take the step, shrink the trust region, and re-solve (7.3.1) to obtain a new step. Additionally, when the actual-to-predicted ratio is close to 1, we conclude that a larger trust region may hasten progress, so we increase Δ for the next iteration, provided that the bound $\|d^k\| \leqslant \Delta_k$ really is active at the solution of (7.3.1).

Unlike a basic line-search method, the trust-region Newton method can "escape" from a saddle point. Suppose we have $\nabla f(x^k) = 0$ and $\nabla^2 f(x^k)$ indefinite with some strictly negative eigenvalues. Then, the solution d^k to (7.3.1) will be nonzero, and the algorithm will step away from the saddle point, in the direction of most negative curvature for $\nabla^2 f(x^k)$. Another appealing feature of the trust-region Newton approach is that when the sequence $\{x^k\}$ approaches a point x^* satisfying second-order sufficient conditions, the trust region bound becomes inactive, and the method takes pure Newton steps (7.1.4) for all sufficiently large k so the local quadratic convergence that characterizes Newton's method.

The basic difference between line-search and trust-region methods can be summarized as follows. Line-search methods first choose a direction p^k, then decide how far to move along that direction. Trust-region methods do the opposite: They choose the distance Δ_k first, then find the direction that makes the best progress for this step length.

7.4. A Cubic Regularization Approach Trust-region Newton methods have the significant advantage of guaranteeing that any accumulation points will satisfy second-order necessary conditions. A related approach based on *cubic regularization* has similar properties, plus some additional complexity guarantees. Cubic regularization requires the Hessian to be Lipschitz continuous, as in (7.1.2). It follows that the following cubic function yields a *global upper* bound for f:

$$T_M(z; x) := f(x) + \nabla f(x)^T (z - x) + \frac{1}{2}(z - x)^T \nabla^2 f(x)(z - x) + \frac{M}{6}\|z - x\|^3. \tag{7.4.1}$$

Specifically, we have for any x that

$$f(z) \leqslant T_M(z;x), \quad \text{for all } z.$$

The basic cubic regularization algorithm starting from x^0 proceeds as follows:

$$x^{k+1} = \arg\min_z T_M(z;x^k), \quad k = 0,1,2,\ldots. \tag{7.4.2}$$

The complexity properties of this approach were analyzed in [35], with variants being studied in [26] and [12, 13]. Rather than present the theory for the method based on (7.4.2), we describe an elementary algorithm that makes use of the expansion (7.4.1) as well as the steepest-descent theory of Subsection 4.1. Our algorithm aims to identify a point that *approximately* satisfies second-order necessary conditions, that is,

$$\|\nabla f(x)\| \leqslant \epsilon_g, \quad \lambda_{\min}(\nabla^2 f(x)) \geqslant -\epsilon_H, \tag{7.4.3}$$

where ϵ_g and ϵ_H are two small constants. In addition to Lipschitz continuity of the Hessian (7.1.2), we assume Lipschitz continuity of the gradient with constant L (see (3.3.6)), and also that the objective f is lower-bounded by some number $\bar{f}$.

Our algorithm takes steps of two types: a steepest-descent step, as in Subsection 4.1, or a step in a negative curvature direction for $\nabla^2 f$. Iteration k proceeds as follows:

(i) If $\|\nabla f(x^k)\| > \epsilon_g$, take the steepest descent step (4.1.1).

(ii) Otherwise, if $\lambda_{\min}(\nabla^2 f(x^k)) < -\epsilon_H$, choose p^k to be the eigenvector corresponding to the most negative eigenvalue of $\nabla^2 f(x^k)$. Choose the size and sign of p^k such that $\|p^k\| = 1$ and $(p^k)^T \nabla f(x^k) \leqslant 0$, and set

$$x^{k+1} = x^k + \alpha_k p^k, \quad \text{where } \alpha_k = \frac{2\epsilon_H}{M}. \tag{7.4.4}$$

If neither of these conditions hold, then x^k satisfies the approximate second-order necessary conditions (7.4.3), so we terminate.

For the steepest-descent step (i), we have from (4.1.3) that

$$f(x^{k+1}) \leqslant f(x^k) - \frac{1}{2L}\|\nabla f(x^k)\|^2 \leqslant f(x^k) - \frac{\epsilon_g^2}{2L}. \tag{7.4.5}$$

For a step of type (ii), we have from (7.4.1) that

$$\begin{aligned} f(x^{k+1}) &\leqslant f(x^k) + \alpha_k \nabla f(x^k)^T p^k + \frac{1}{2}\alpha_k^2 (p^k)^T \nabla^2 f(x^k) p^k + \frac{1}{6} M \alpha_k^3 \|p^k\|^3 \\ &\leqslant f(x^k) - \frac{1}{2}\left(\frac{2\epsilon_H}{M}\right)^2 \epsilon_H + \frac{1}{6} M \left(\frac{2\epsilon_H}{M}\right)^3 \\ &= f(x^k) - \frac{2}{3}\frac{\epsilon_H^3}{M^2}. \end{aligned} \tag{7.4.6}$$

By aggregating (7.4.5) and (7.4.6), we have that at each x^k for which the condition (7.4.3) does *not* hold, we attain a decrease in the objective of at least

$$\min\left(\frac{\epsilon_g^2}{2L}, \frac{2}{3}\frac{\epsilon_H^3}{M^2}\right).$$

Using the lower bound $\bar{f}$ on the objective f, we see that the number of iterations K required must satisfy the condition

$$K \min\left(\frac{\epsilon_g^2}{2L}, \frac{2}{3}\frac{\epsilon_H^3}{M^2}\right) \leqslant f(x^0) - \bar{f},$$

from which we conclude that

$$K \leqslant \max\left(2L\epsilon_g^{-2}, \frac{3}{2}M^2\epsilon_H^{-3}\right)\left(f(x^0) - \bar{f}\right).$$

We also observe that that the maximum number of iterates required to identify a point at which only the approximate stationarity condition $\|\nabla f(x^k)\| \leqslant \epsilon_g$ holds is $2L\epsilon_g^{-2}(f(x^0) - \bar{f})$. (We can just omit the second-order part of the algorithm.) Note too that it is easy to devise *approximate* versions of this algorithm with similar complexity. For example, the negative curvature direction p^k in step (ii) above can be replaced by an approximation to the direction of most negative curvature, obtained by the Lanczos iteration with random initialization.

In algorithms that make more complete use of the cubic model (7.4.1), the term ϵ_g^{-2} in the complexity expression becomes $\epsilon_g^{-3/2}$, and the constants are different. The subproblems (7.4.1) are more complicated to solve than those in the simple scheme above. Active research is going on into other algorithms that achieve complexities similar to those of the cubic regularization approach. A variety of methods that make use of Newton-type steps, approximate negative curvature directions, accelerated gradient methods, random perturbations, randomized Lanczos and conjugate gradient methods, and other algorithmic elements have been proposed.

8. Conclusions

We have outlined various algorithmic tools from optimization that are useful for solving problems in data analysis and machine learning, and presented their basic theoretical properties. The intersection of optimization and machine learning is a fruitful and very popular area of current research. All the major machine learning conferences have a large contingent of optimization papers, and there is a great deal of interest in developing algorithmic tools to meet new challenges and in understanding their properties. The edited volume [41] contains a snapshot of the state of the art circa 2010, but this is a fast-moving field and there have been many developments since then.

Acknowledgments

I thank Ching-pei Lee for a close reading and many helpful suggestions, and David Hong and an anonymous referee for detailed, excellent comments.

References

[1] L. Balzano, R. Nowak, and B. Recht, *Online identification and tracking of subspaces from highly incomplete information*, 48th Annual Allerton Conference on Communication, Control, and Computing, 2010, pp. 704–711. ←57

[2] L. Balzano and S. J. Wright, *Local convergence of an algorithm for subspace identification from partial data*, Found. Comput. Math. **15** (2015), no. 5, 1279–1314, DOI 10.1007/s10208-014-9227-7. MR3394711 ←58

[3] A. Beck and M. Teboulle, *A fast iterative shrinkage-thresholding algorithm for linear inverse problems*, SIAM J. Imaging Sci. **2** (2009), no. 1, 183–202, DOI 10.1137/080716542. MR2486527 ←80, 83

[4] B. E. Boser, I. M. Guyon, and V. N. Vapnik, *A training algorithm for optimal margin classifiers*, Proceedings of the Fifth Annual Workshop on Computational Learning Theory, 1992, pp. 144–152. ←60

[5] S. Boyd, N. Parikh, E. Chu, B. Peleato, and J. Eckstein, *Distributed optimization and statistical learning via the alternating direction methods of multipliers*, Foundations and Trends in Machine Learning **3** (2011), no. 1, 1–122. ←51

[6] S. Boyd and L. Vandenberghe, *Convex optimization*, Cambridge University Press, Cambridge, 2004. MR2061575 ←51

[7] S. Bubeck, *Convex optimization: Algorithms and complexity*, Foundations and Trends in Machine Learning **8** (2015), no. 3–4, 231–357. ←83, 84

[8] S. Bubeck, Y. T. Lee, and M. Singh, *A geometric alternative to Nesterov's accelerated gradient descent*, Technical Report arXiv:1506.08187, Microsoft Research, 2015. ←83

[9] S. Burer and R. D. C. Monteiro, *A nonlinear programming algorithm for solving semidefinite programs via low-rank factorization*, Math. Program. **95** (2003), no. 2, Ser. B, 329–357, DOI 10.1007/s10107-002-0352-8. Computational semidefinite and second order cone programming: the state of the art. MR1976484 ←55

[10] E. Candès and B. Recht, *Exact matrix completion via convex optimization*, Foundations of Computational Mathematics **9** (2009), 717–772. ←55

[11] E. J. Candès, X. Li, Y. Ma, and J. Wright, *Robust principal component analysis?*, J. ACM **58** (2011), no. 3, Art. 11, 37, DOI 10.1145/1970392.1970395. MR2811000 ←57

[12] C. Cartis, N. I. M. Gould, and P. L. Toint, *Adaptive cubic regularisation methods for unconstrained optimization. Part I: motivation, convergence and numerical results*, Math. Program. **127** (2011), no. 2, Ser. A, 245–295, DOI 10.1007/s10107-009-0286-5. MR2776701 ←94

[13] C. Cartis, N. I. M. Gould, and P. L. Toint, *Adaptive cubic regularisation methods for unconstrained optimization. Part II: worst-case function- and derivative-evaluation complexity*, Math. Program. **130** (2011), no. 2, Ser. A, 295–319, DOI 10.1007/s10107-009-0337-y. MR2855872 ←94

[14] V. Chandrasekaran, S. Sanghavi, P. A. Parrilo, and A. S. Willsky, *Rank-sparsity incoherence for matrix decomposition*, SIAM J. Optim. **21** (2011), no. 2, 572–596, DOI 10.1137/090761793. MR2817479 ←57

[15] Y. Chen and M. J. Wainwright, *Fast low-rank estimation by projected gradent descent: General statistical and algorithmic guarantees*, UC-Berkeley, 2015, arXiv:1509.03025. ←57

[16] C. Cortes and V. N. Vapnik, *Support-vector networks*, Machine Learning **20** (1995), 273–297. ←60

[17] A. d'Aspremont, O. Banerjee, and L. El Ghaoui, *First-order methods for sparse covariance selection*, SIAM J. Matrix Anal. Appl. **30** (2008), no. 1, 56–66, DOI 10.1137/060670985. MR2399568 ←56

[18] A. d'Aspremont, L. El Ghaoui, M. I. Jordan, and G. R. G. Lanckriet, *A direct formulation for sparse PCA using semidefinite programming*, SIAM Rev. **49** (2007), no. 3, 434–448, DOI 10.1137/050645506. MR2353806 ←56

[19] T. Dasu and T. Johnson, *Exploratory data mining and data cleaning*, Wiley Series in Probability and Statistics, Wiley-Interscience [John Wiley & Sons], Hoboken, NJ, 2003. MR1979601 ←52

[20] P. Drineas and M. W. Mahoney, *Lectures on randomized numerical linear algebra*, The Mathematics of Data, IAS/Park City Math. Ser., vol. 25, Amer. Math. Soc., Providence, RI, 2018. ←54

[21] D. Drusvyatskiy, M. Fazel, and S. Roy, *An optimal first order method based on optimal quadratic averaging*, SIAM J. Optim. **28** (2018), no. 1, 251–271, DOI 10.1137/16M1072528. MR3757113 ←83

[22] J. C. Duchi, *Introductory lectures on stochastic optimization*, The Mathematics of Data, IAS/Park City Math. Ser., vol. 25, Amer. Math. Soc., Providence, RI, 2018. ←51, 64, 71

[23] J. Eckstein and D. P. Bertsekas, *On the Douglas-Rachford splitting method and the proximal point algorithm for maximal monotone operators*, Math. Programming **55** (1992), no. 3, Ser. A, 293–318, DOI 10.1007/BF01581204. MR1168183 ←51

[24] M. Frank and P. Wolfe, *An algorithm for quadratic programming*, Naval Res. Logist. Quart. **3** (1956), 95–110, DOI 10.1002/nav.3800030109. MR0089102 ←76

[25] J. Friedman, T. Hastie, and R. Tibshirani, *Sparse inverse covariance estimation with the graphical lasso*, Biostatistics **9** (2008), no. 3, 432–441. ←56

[26] A. Griewank, *The modification of Newton's method for unconstrained optimization by bounding cubic terms*, Technical Report NA/12, DAMTP, Cambridge University, 1981. ←92, 94

[27] M. R. Hestenes and E. Stiefel, *Methods of conjugate gradients for solving linear systems*, J. Research Nat. Bur. Standards **49** (1952), 409–436 (1953). MR0060307 ←81

[28] A. J. Hoffman and H. W. Wielandt, *The variation of the spectrum of a normal matrix*, Duke Math. J. **20** (1953), 37–39. MR0052379 ←89

[29] J. D Lee, M. Simchowitz, M. I Jordan, and B. Recht, *Gradient descent only converges to minimizers*, Conference on learning theory, 2016, pp. 1246–1257. ←75

[30] D. C. Liu and J. Nocedal, *On the limited memory BFGS method for large scale optimization*, Math. Programming **45** (1989), no. 3, (Ser. B), 503–528, DOI 10.1007/BF01589116. MR1038245 ←51, 82

[31] J. J. Moré and D. C. Sorensen, *Computing a trust region step*, SIAM J. Sci. Statist. Comput. **4** (1983), no. 3, 553–572, DOI 10.1137/0904038. MR723110 ←92

[32] A. S. Nemirovsky and D. B. Yudin, *Problem complexity and method efficiency in optimization*, John Wiley & Sons, Inc., New York, 1983. Translated from the Russian and with a preface by E. R. Dawson; Wiley-Interscience Series in Discrete Mathematics. MR702836 ←87

[33] Yu. E. Nesterov, *A method for solving the convex programming problem with convergence rate* $O(1/k^2)$ (Russian), Dokl. Akad. Nauk SSSR **269** (1983), no. 3, 543–547. MR701288 ←80

[34] Y. Nesterov, *Introductory lectures on convex optimization*, Applied Optimization, vol. 87, Kluwer Academic Publishers, Boston, MA, 2004. A basic course. MR2142598 ←80, 84

[35] Y. Nesterov and B. T. Polyak, *Cubic regularization of Newton method and its global performance*, Math. Program. **108** (2006), no. 1, Ser. A, 177–205, DOI 10.1007/s10107-006-0706-8. MR2229459 ←92, 94

[36] J. Nocedal and S. J. Wright, *Numerical Optimization*, Second, Springer, New York, 2006. ←51, 74, 82, 92, 93

[37] B. T. Poljak, *Some methods of speeding up the convergence of iterative methods* (Russian), Ž. Vyčisl. Mat. i Mat. Fiz. **4** (1964), 791–803. MR0169403 ←80

[38] B. T. Polyak, *Introduction to optimization*, Translations Series in Mathematics and Engineering, Optimization Software, Inc., Publications Division, New York, 1987. Translated from the Russian; With a foreword by Dimitri P. Bertsekas. MR1099605 ←80, 81

[39] B. Recht, M. Fazel, and P. A. Parrilo, *Guaranteed minimum-rank solutions of linear matrix equations via nuclear norm minimization*, SIAM Rev. **52** (2010), no. 3, 471–501, DOI 10.1137/070697835. MR2680543 ←55

[40] R. T. Rockafellar, *Convex analysis*, Princeton Mathematical Series, No. 28, Princeton University Press, Princeton, N.J., 1970. MR0274683 ←65

[41] S. Sra, S. Nowozin, and S. J. Wright (eds.), *Optimization for machine learning*, NIPS Workshop Series, MIT Press, 2011. ←95

[42] R. Tibshirani, *Regression shrinkage and selection via the lasso*, J. Roy. Statist. Soc. Ser. B **58** (1996), no. 1, 267–288. MR1379242 ←54

[43] M. J. Todd, *Semidefinite optimization*, Acta Numer. **10** (2001), 515–560, DOI 10.1017/S0962492901000071. MR2009698 ←51

[44] B. A. Turlach, W. N. Venables, and S. J. Wright, *Simultaneous variable selection*, Technometrics **47** (2005), no. 3, 349–363, DOI 10.1198/004017005000000139. MR2164706 ←57

[45] L. Vandenberghe and S. Boyd, *Semidefinite programming*, SIAM Rev. **38** (1996), no. 1, 49–95, DOI 10.1137/1038003. MR1379041 ←51

[46] S. J. Wright, *Primal-dual interior-point methods*, Society for Industrial and Applied Mathematics (SIAM), Philadelphia, PA, 1997. MR1422257 ←51

[47] S. J. Wright, *Coordinate descent algorithms*, Math. Program. **151** (2015), no. 1, Ser. B, 3–34, DOI 10.1007/s10107-015-0892-3. MR3347548 ←51

[48] X. Yi, D. Park, Y. Chen, and C. Caramanis, *Fast algorithms for robust PCA via gradient descent*, Advances in Neural Information Processing Systems 29, 2016, pp. 4152–4160. ←57

Computer Sciences Department, University of Wisconsin-Madison, Madison, WI 53706
Email address: swright@cs.wisc.edu

IAS/Park City Mathematics Series
Volume 25, Pages 99–185
https://doi.org/10.1090/pcms/025/00831

Introductory Lectures on Stochastic Optimization

John C. Duchi

Contents

2010 *Mathematics Subject Classification.* Primary 65Kxx; Secondary 90C15, 62C20.
Key words and phrases. Convexity, stochastic optimization, subgradients, mirror descent, minimax optimal.

1. Introduction

In this set of four lectures, we study the basic analytical tools and algorithms necessary for the solution of stochastic convex optimization problems, as well as for providing optimality guarantees associated with the methods. As we proceed through the lectures, we will be more exact about the precise problem formulations, providing a number of examples, but roughly, by a stochastic optimization problem we mean a numerical optimization problem that arises from observing data from some (random) data-generating process. We focus almost exclusively on first-order methods for the solution of these types of problems, as they have proven quite successful in the large scale problems that have driven many advances throughout the 2000s.

Our main goal in these lectures, as in the lectures by S. Wright [61] in this volume, is to develop methods for the solution of optimization problems arising in large-scale data analysis. Our route will be somewhat circuitous, as we will first build the necessary convex analytic and other background (see Lecture 2), but broadly, the problems we wish to solve are the problems arising in stochastic convex optimization. In these problems, we have samples S coming from a sample space $\mathcal{S}$, drawn from a distribution P, and we have some *decision vector* $x \in \mathbb{R}^n$ that we wish to choose to minimize the expected loss

$$f(x) := \mathbb{E}_P[F(x; S)] = \int_{\mathcal{S}} F(x; s) dP(s), \tag{1.0.1}$$

where F is convex in its first argument.

The methods we consider for minimizing problem (1.0.1) are typically simple methods that are slower to converge than more advanced methods—such as Newton or other second-order methods—for deterministic problems, but have the advantage that they are robust to noise in the optimization problem itself. Consequently, it is often relatively straightforward to derive generalization bounds for these procedures: if they produce an estimate $\widehat{x}$ exhibiting good performance on some sample $S_1, \ldots, S_m$ drawn from P, then they are likely to exhibit good performance (on average) for future data, that is, to have small objective $f(\widehat{x})$; see Lecture 3, and especially Theorem 3.4.13. It is of course often advantageous to take advantage of problem structure and geometric aspects of the problem, broadly defined, which is the goal of mirror descent and related methods, which we discuss in Lecture 4.

The last part of our lectures is perhaps the most unusual for material on optimization, which is to investigate optimality guarantees for stochastic optimization problems. In Lecture 5, we study the sample complexity of solving problems of the form (1.0.1). More precisely, we measure the performance of an optimization procedure given samples $S_1, \ldots, S_m$ drawn independently from the population distribution P, denoted by $\widehat{x} = \widehat{x}(S_{1:m})$, in a uniform sense: for a class of objective functions $\mathcal{F}$, a procedure's performance is its expected error—or risk—for the worst member of the class $\mathcal{F}$. We provide lower bounds on this maximum risk, showing that the first-order procedures we have developed satisfy certain notions of optimality.

We briefly outline the coming lectures. The first lecture provides definitions and the convex analytic tools necessary for the development of our algorithms and other ideas, developing separation properties of convex sets as well as other properties of convex functions from basic principles. The second two lectures investigate subgradient methods and their application to certain stochastic optimization problems, demonstrating a number of convergence results. The second lecture focuses on standard subgradient-type methods, while the third investigates more advanced material on mirror descent and adaptive methods, which require more care but can yield substantial practical performance benefits. The final lecture investigates optimality guarantees for the various methods we study, demonstrating two standard techniques for proving lower bounds on the ability of any algorithm to solve stochastic optimization problems.

1.1. Scope, limitations, and other references The lectures assume some limited familiarity with convex functions and convex optimization problems and their formulation, which will help appreciation of the techniques herein. All that is truly essential is a level of mathematical maturity that includes some real analysis, linear algebra, and introductory probability. In terms of real analysis, a typical undergraduate course, such as one based on Marsden and Hoffman's *Elementary Real Analysis* [37] or Rudin's *Principles of Mathematical Analysis* [50], are sufficient. Readers should not consider these lectures in any way a comprehensive view of convex analysis or stochastic optimization. These subjects are well-established, and there are numerous references.

Our lectures begin with convex analysis, whose study Rockafellar, influenced by Fenchel, launched in his 1970 book *Convex Analysis* [49]. We develop the basic ideas necessary for our treatment of first-order (gradient-based) methods for optimization, which includes separating and supporting hyperplane theorems, but we provide essentially no treatment of the important concepts of Lagrangian and Fenchel duality, support functions, or saddle point theory more broadly. For these and other important ideas, I have found the books of Rockafellar [49], Hiriart-Urruty and Lemaréchal [27,28], Bertsekas [8], and Boyd and Vandenberghe [12] illuminating.

Convex optimization itself is a huge topic, with thousands of papers and numerous books on the subject. Because of our focus on solution methods for large-scale problems arising out of data collection, we are somewhat constrained in our views. Boyd and Vandenberghe [12] provide an excellent treatment of the possibilities of modeling engineering and scientific problems as convex optimization problems, as well as some important numerical methods. Polyak [47] provides a treatment of stochastic and non-stochastic methods for optimization from which ours borrows substantially. Nocedal and Wright [46] and Bertsekas [9] also describe more advanced methods for the solution of optimization problems, focusing on non-stochastic optimization problems for which there are many sophisticated methods.

Because of our goal to solve problems of the form (1.0.1), we develop first-order methods that are robust to many types of noise from sampling. There are other approaches to dealing with data uncertainty, notably robust optimization [6], where researchers study and develop tractable (polynomial-time-solvable) formulations for a variety of data-based problems in engineering and the sciences. The book of Shapiro et al. [54] provides a more comprehensive picture of stochastic modeling problems and optimization algorithms than we have been able to in our lectures, as stochastic optimization is by itself a major field. Several recent surveys on online learning and online convex optimization provide complementary treatments to ours [26, 52].

The last lecture traces its roots to seminal work in information-based complexity by Nemirovski and Yudin in the early 1980s [41], who investigate the limits of "optimal" algorithms, where optimality is defined in a worst-case sense according to an oracle model of algorithms given access to function, gradient, or other types of local information about the problem at hand. Issues of optimal estimation in statistics are as old as the field itself, and the minimax formulation we use is originally due to Wald in the late 1930s [59, 60]. We prove our results using information theoretic tools, which have broader applications across statistics, and that have been developed by many authors [31, 33, 62, 63].

1.2. Notation We use mostly standard notation throughout these notes, but for completeness, we collect it here. We let $\mathbb{R}$ denote the typical field of real numbers, with $\mathbb{R}^n$ having its usual meaning as n-dimensional Euclidean space. Given vectors x and y, we let $\langle x, y \rangle$ denote the inner product between x and y. Given a norm $\|\cdot\|$, its dual norm $\|\cdot\|_*$ is defined as

$$\|z\|_* := \sup\{\langle z, x \rangle \mid \|x\| \leqslant 1\}.$$

Hölder's inequality (see Exercise B.2.4) shows that the ℓ_p and ℓ_q norms, defined by

$$\|x\|_p = \left(\sum_{j=1}^{n} |x_j|^p\right)^{\frac{1}{p}}$$

(and as the limit $\|x\|_\infty = \max_j |x_j|$) are dual to one another, where $1/p + 1/q = 1$ and $p, q \in [1, \infty]$. Throughout, we will assume that $\|x\|_2 = \sqrt{\langle x, x \rangle}$ is the norm defined by the inner product $\langle \cdot, \cdot \rangle$.

We also require notation related to sets. For a sequence of vectors $v_1, v_2, v_3, \ldots$, we let (v_n) denote the entire sequence. Given sets A and B, we let $A \subset B$ denote that A is a subset (possibly equal to) B, and $A \subsetneq B$ that A is a strict subset of B. The notation $\operatorname{cl} A$ denotes the closure of A, while $\operatorname{int} A$ denotes the interior of the set A. For a function f, the set $\operatorname{dom} f$ is its domain. If $f : \mathbb{R}^n \to \mathbb{R} \cup \{+\infty\}$ is convex, we let $\operatorname{dom} f := \{x \in \mathbb{R}^n \mid f(x) < +\infty\}$.

2. Basic Convex Analysis

Lecture Summary: In this lecture, we will outline several standard facts from convex analysis, the study of the mathematical properties of convex functions and sets. For the most part, our analysis and results will all be with the aim of setting the necessary background for understanding first-order convex optimization methods, though some of the results we state will be quite general.

2.1. Introduction and Definitions This set of lecture notes considers convex optimization problems, numerical optimization problems of the form

$$\begin{aligned} &\text{minimize } f(x) \\ &\text{subject to } x \in C, \end{aligned} \tag{2.1.1}$$

where f is a convex function and C is a convex set. While we will consider tools to solve these types of optimization problems presently, this first lecture is concerned most with the analytic tools and background that underlies solution methods for these problems.

The starting point for any study of convex functions is the definition and study of convex sets, which are intimately related to convex functions. To that end, we recall that a set $C \subset \mathbb{R}^n$ is *convex* if for all $x, y \in C$,

$$\lambda x + (1-\lambda)y \in C \text{ for } \lambda \in [0,1].$$

See Figure 2.1.2.

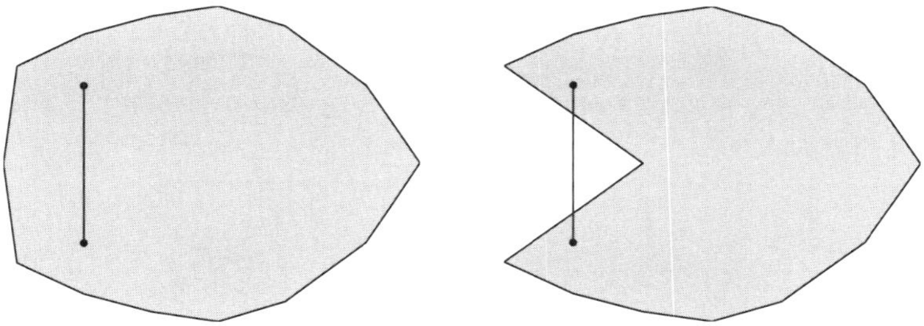

(a) (b)

FIGURE 2.1.2. (a) A convex set (b) A non-convex set.

A convex function is similarly defined: a function $f : \mathbb{R}^n \to (-\infty, \infty]$ is *convex* if for all $x, y \in \operatorname{dom} f := \{x \in \mathbb{R}^n \mid f(x) < +\infty\}$

$$f(\lambda x + (1-\lambda)y) \leqslant \lambda f(x) + (1-\lambda)f(y) \text{ for } \lambda \in [0,1].$$

The epigraph of a function is defined as

$$\operatorname{epi} f := \{(x,t) : f(x) \leqslant t\},$$

and by inspection, a function is convex if and only if its epigraph is a convex set. A convex function f is closed if its epigraph is a closed set; continuous convex functions are always closed. We will assume throughout that any convex function we deal with is closed. See Figure 2.1.3 for graphical representations of these ideas, which make clear that the epigraph is indeed a convex set.

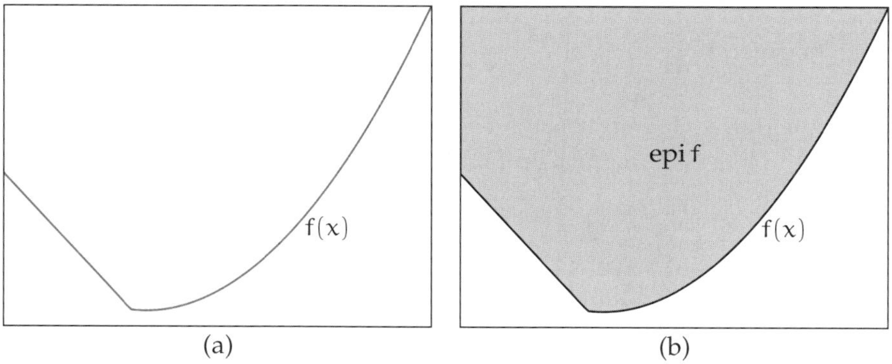

FIGURE 2.1.3. (a) The convex function $f(x) = \max\{x^2, -2x - .2\}$ and (b) its epigraph, which is a convex set.

One may ask why, precisely, we focus on convex functions. In short, as Rockafellar [49] notes, convex optimization problems are the clearest dividing line between numerical problems that are efficiently solvable, often by iterative methods, and numerical problems for which we have no hope. We give one simple result in this direction first:

Observation. *Let* $f : \mathbb{R}^n \to \mathbb{R}$ *be convex and* x *be a local minimum of* f *(respectively a local minimum over a convex set* C*). Then* x *is a global minimum of* f *(resp. a global minimum of* f *over* C*).*

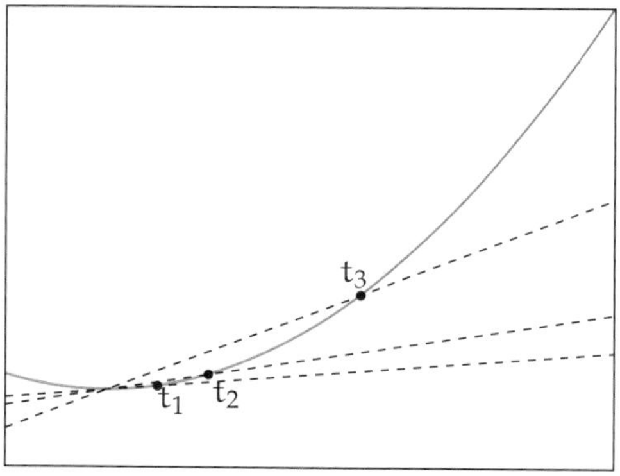

FIGURE 2.1.4. The slopes $\frac{f(x+t)-f(x)}{t}$ increase, with $t_1 < t_2 < t_3$.

To see this, note that if x is a local minimum then for any $y \in C$, we have for small enough $t > 0$ that

$$f(x) \leqslant f(x + t(y - x)) \text{ or } 0 \leqslant \frac{f(x + t(y - x)) - f(x)}{t}.$$

We now use the *criterion of increasing slopes*, that is, for any convex function f the function

$$t \mapsto \frac{f(x + tu) - f(x)}{t} \tag{2.1.5}$$

is *increasing* in $t > 0$. (See Fig. 2.1.4.) Indeed, let $0 \leqslant t_1 \leqslant t_2$. Then

$$\begin{aligned}
\frac{f(x + t_1 u) - f(x)}{t_1} &= \frac{t_2}{t_1} \frac{f(x + t_2(t_1/t_2)u) - f(x)}{t_2} \\
&= \frac{t_2}{t_1} \frac{f((1 - t_1/t_2)x + (t_1/t_2)(x + t_2 u)) - f(x)}{t_2} \\
&\leqslant \frac{t_2}{t_1} \frac{(1 - t_1/t_2)f(x) + (t_1/t_2)f(x + t_2 u) - f(x)}{t_2} \\
&= \frac{f(x + t_2 u) - f(x)}{t_2}.
\end{aligned}$$

In particular, because $0 \leqslant f(x + t(y - x))$ for small enough $t > 0$, we see that for all $t > 0$ we have

$$0 \leqslant \frac{f(x + t(y - x)) - f(x)}{t} \text{ or } f(x) \leqslant \inf_{t \geqslant 0} f(x + t(y - x)) \leqslant f(y)$$

for all $y \in C$.

Most of the results herein apply in general Hilbert (complete inner product) spaces, and many of our proofs will not require anything particular about finite dimensional spaces, but for simplicity we use $\mathbb{R}^n$ as the underlying space on which all functions and sets are defined.[1] While we present all proofs in the chapter, we try to provide geometric intuition that will aid a working knowledge of the results, which we believe is the most important.

2.2. Properties of Convex Sets Convex sets enjoy a number of very nice properties that allow efficient and elegant descriptions of the sets, as well as providing a number of nice properties concerning their separation from one another. To that end, in this section, we give several fundamental properties on separating and supporting hyperplanes for convex sets. The results here begin by showing that there is a unique (Euclidean) projection to any convex set C, then use this fact to show that whenever a point is not contained in a set, it can be separated from the set by a hyperplane. This result can be extended to show separation of convex sets from one another and that points in the boundary of a convex set have a hyperplane tangent to the convex set running through them. We leverage these results in the sequel by making connections of supporting hyperplanes to

[1] The generality of Hilbert, or even Banach, spaces in convex analysis is seldom needed. Readers familiar with arguments in these spaces will, however, note that the proofs can generally be extended to infinite dimensional spaces in reasonably straightforward ways.

epigraphs and gradients, results that in turn find many applications in the design of optimization algorithms as well as optimality certificates.

A few basic properties We list a few simple properties that convex sets have, which are evident from their definitions. First, if C_α are convex sets for each $\alpha \in \mathcal{A}$, where $\mathcal{A}$ is an arbitrary index set, then the intersection

$$C = \bigcap_{\alpha \in \mathcal{A}} C_\alpha$$

is also convex. Additionally, convex sets are closed under scalar multiplication: if $\alpha \in \mathbb{R}$ and C is convex, then

$$\alpha C := \{\alpha x : x \in C\}$$

is evidently convex. The Minkowski sum of two convex sets is defined by

$$C_1 + C_2 := \{x_1 + x_2 : x_1 \in C_1, x_2 \in C_2\},$$

and is also convex. To see this, note that if $x_i, y_i \in C_i$, then

$$\lambda(x_1 + x_2) + (1-\lambda)(y_1 + y_2) = \underbrace{\lambda x_1 + (1-\lambda)y_1}_{\in C_1} + \underbrace{\lambda x_2 + (1-\lambda)y_2}_{\in C_2} \in C_1 + C_2.$$

In particular, convex sets are closed under all linear combination: if $\alpha \in \mathbb{R}^m$, then $C = \sum_{i=1}^m \alpha_i C_i$ is also convex.

We also define the convex hull of a set of points $x_1, \ldots, x_m \in \mathbb{R}^n$ by

$$\text{Conv}\{x_1, \ldots, x_m\} = \left\{ \sum_{i=1}^m \lambda_i x_i : \lambda_i \geqslant 0, \sum_{i=1}^m \lambda_i = 1 \right\}.$$

This set is clearly a convex set.

Projections We now turn to a discussion of orthogonal projection onto a convex set, which will allow us to develop a number of separation properties and alternate characterizations of convex sets. See Figure 2.2.1 for a geometric view of projection.

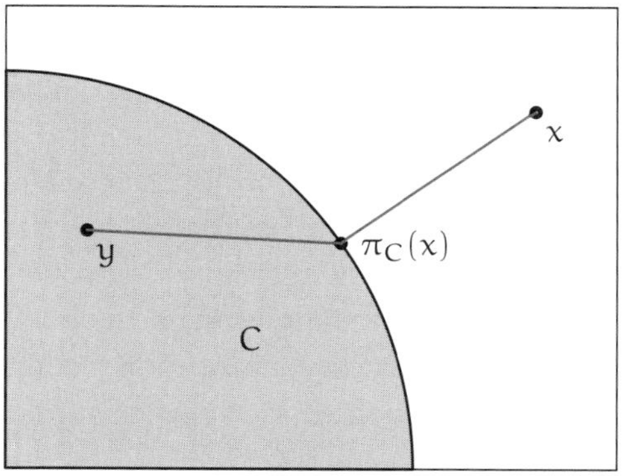

FIGURE 2.2.1. Projection of the point x onto the set C (with projection $\pi_C(x)$), exhibiting $\langle x - \pi_C(x), y - \pi_C(x) \rangle \leqslant 0$.

We begin by stating a classical result about the projection of zero onto a convex set.

Theorem 2.2.2 (Projection of zero). *Let C be a closed convex set not containing the origin 0. Then there is a unique point $x_C \in C$ such that $\|x_C\|_2 = \inf_{x \in C} \|x\|_2$. Moreover, $\|x_C\|_2 = \inf_{x \in C} \|x\|_2$ if and only if*

$$\langle x_C, y - x_C \rangle \geqslant 0 \tag{2.2.3}$$

for all $y \in C$.

Proof. The key to the proof is the following parallelogram identity, which holds in any inner product space: for any x, y,

$$\frac{1}{2}\|x - y\|_2^2 + \frac{1}{2}\|x + y\|_2^2 = \|x\|_2^2 + \|y\|_2^2. \tag{2.2.4}$$

Define $M := \inf_{x \in C} \|x\|_2$. Now, let $(x_n) \subset C$ be a sequence of points in C such that $\|x_n\|_2 \to M$ as $n \to \infty$. By the parallelogram identity (2.2.4), for any $n, m \in \mathbb{N}$, we have

$$\frac{1}{2}\|x_n - x_m\|_2^2 = \|x_n\|_2^2 + \|x_m\|_2^2 - \frac{1}{2}\|x_n + x_m\|_2^2.$$

Fix $\epsilon > 0$, and choose $N \in \mathbb{N}$ such that $n \geqslant N$ implies that $\|x_n\|_2^2 \leqslant M^2 + \epsilon$. Then for any $m, n \geqslant N$, we have

$$\frac{1}{2}\|x_n - x_m\|_2^2 \leqslant 2M^2 + 2\epsilon - \frac{1}{2}\|x_n + x_m\|_2^2. \tag{2.2.5}$$

Now we use the convexity of the set C. We have $\frac{1}{2}x_n + \frac{1}{2}x_m \in C$ for any n, m, which implies

$$\frac{1}{2}\|x_n + x_m\|_2^2 = 2\left\|\frac{1}{2}x_n + \frac{1}{2}x_m\right\|_2^2 \geqslant 2M^2$$

by definition of M. Using the above inequality in the bound (2.2.5), we see that

$$\frac{1}{2}\|x_n - x_m\|_2^2 \leqslant 2M^2 + 2\epsilon - 2M^2 = 2\epsilon.$$

In particular, $\|x_n - x_m\|_2 \leqslant 2\sqrt{\epsilon}$; since ϵ was arbitrary, (x_n) forms a Cauchy sequence and so must converge to a point x_C. The continuity of the norm $\|\cdot\|_2$ implies that $\|x_C\|_2 = \inf_{x \in C} \|x\|_2$, and the fact that C is closed implies that the point x_C lies in C.

Now we show the inequality (2.2.3) holds if and only if x_C is the projection of the origin 0 onto C. Suppose that inequality (2.2.3) holds. Then

$$\|x_C\|_2^2 = \langle x_C, x_C \rangle \leqslant \langle x_C, y \rangle \leqslant \|x_C\|_2 \|y\|_2,$$

the last inequality following from the Cauchy-Schwartz inequality. Dividing each side by $\|x_C\|_2$ implies that $\|x_C\|_2 \leqslant \|y\|_2$ for all $y \in C$. For the converse, let x_C minimize $\|x\|_2$ over C. Then for any $t \in [0, 1]$ and any $y \in C$, we have

$$\begin{aligned}\|x_C\|_2^2 &\leqslant \|(1-t)x_C + ty\|_2^2 \\ &= \|x_C + t(y - x_C)\|_2^2 = \|x_C\|_2^2 + 2t\langle x_C, y - x_C \rangle + t^2\|y - x_C\|_2^2.\end{aligned}$$

Subtracting $\|x_C\|_2^2$ and $t^2\|y-x_C\|_2^2$ from both sides of the above inequality, we have

$$-t^2\|y-x_C\|_2^2 \leqslant 2t\langle x_C, y-x_C\rangle.$$

Dividing both sides of the above inequality by 2t, we have

$$-\frac{t}{2}\|y-x_C\|_2^2 \leqslant \langle x_C, y-x_C\rangle$$

for all $t \in (0,1]$. Letting $t \downarrow 0$ gives the desired inequality. □

With this theorem in place, a simple shift gives a characterization of more general projections onto convex sets.

Corollary 2.2.6 (Projection onto convex sets). *Let C be a closed convex set and let $x \in \mathbb{R}^n$. Then there is a unique point $\pi_C(x)$, called the* projection of x onto C, *such that $\|x-\pi_C(x)\|_2 = \inf_{y\in C}\|x-y\|_2$, that is, $\pi_C(x) = \operatorname{argmin}_{y\in C}\|y-x\|_2^2$. The projection is characterized by the inequality*

$$\langle \pi_C(x)-x, y-\pi_C(x)\rangle \geqslant 0 \tag{2.2.7}$$

for all $y \in C$.

Proof. When $x \in C$, the statement is clear. For $x \notin C$, the corollary simply follows by considering the set $C' = C - x$, then using Theorem 2.2.2 applied to the recentered set. □

Corollary 2.2.8 (Non-expansive projections). *Projections onto convex sets are non-expansive, in particular,*

$$\|\pi_C(x)-y\|_2 \leqslant \|x-y\|_2$$

for any $x \in \mathbb{R}^n$ and $y \in C$.

Proof. When $x \in C$, the inequality is clear, so assume that $x \notin C$. Now use inequality (2.2.7) from the previous corollary. By adding and subtracting y in the inner product, we have

$$\begin{aligned}0 &\leqslant \langle \pi_C(x)-x, y-\pi_C(x)\rangle\\ &= \langle \pi_C(x)-y+y-x, y-\pi_C(x)\rangle\\ &= -\|\pi_C(x)-y\|_2^2 + \langle y-x, y-\pi_C(x)\rangle\end{aligned}$$

We rearrange the above and then use the Cauchy-Schwartz or Hölder's inequality, which gives

$$\|\pi_C(x)-y\|_2^2 \leqslant \langle y-x, y-\pi_C(x)\rangle \leqslant \|y-x\|_2\|y-\pi_C(x)\|_2.$$

Now divide both sides by $\|\pi_C(x)-y\|_2$. □

Separation Properties Projections are important not just because of their existence, but because they also guarantee, first, that convex sets can be described by halfplanes that contain them and, second, that any two convex sets can be separated by hyperplanes. Moreover, the separation can be strict if one of the sets is compact.

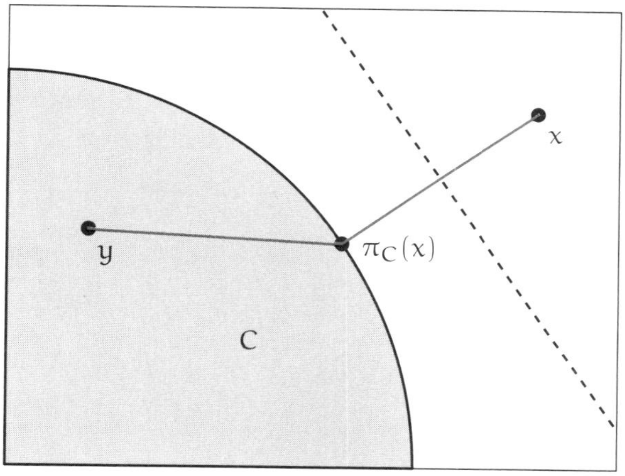

FIGURE 2.2.9. Separation of the point x from the set C by the vector $v = x - \pi_C(x)$.

Proposition 2.2.10 (Strict separation of points). *Let C be a closed convex set. Given any point $x \notin C$, there is a vector v such that*

$$\langle v, x\rangle > \sup_{y \in C} \langle v, y\rangle \tag{2.2.11}$$

Moreover, we can take the vector $v = x - \pi_C(x)$, and $\langle v, x\rangle \geqslant \sup_{y \in C} \langle v, y\rangle + \|v\|_2^2$. See Figure 2.2.9.

Proof. Indeed, since $x \notin C$, we have $x - \pi_C(x) \neq 0$. By setting $v = x - \pi_C(x)$, we have from the characterization (2.2.7) that

$$0 \geqslant \langle v, y - \pi_C(x)\rangle = \langle v, y - x + x - \pi_C(x)\rangle = \langle v, y - x + v\rangle = \langle v, y - x\rangle + \|v\|_2^2.$$

In particular, we see that $\langle v, x\rangle \geqslant \langle v, y\rangle + \|v\|^2$ for all $y \in C$. □

Proposition 2.2.12 (Strict separation of convex sets). *Let C_1, C_2 be closed convex sets, with C_2 compact. Then there is a vector v such that*

$$\inf_{x \in C_1} \langle v, x\rangle > \sup_{x \in C_2} \langle v, x\rangle.$$

Proof. The set $C = C_1 - C_2$ is convex and closed.[2] Moreover, we have $0 \notin C$, so that there is a vector v such that $0 < \inf_{z \in C} \langle v, z\rangle$ by Proposition 2.2.10. Thus we have

$$0 < \inf_{z \in C_1 - C_2} \langle v, z\rangle = \inf_{x \in C_1} \langle v, x\rangle - \sup_{x \in C_2} \langle v, x\rangle,$$

which is our desired result. □

[2]If C_1 is closed and C_2 is compact, then $C_1 + C_2$ is closed. Indeed, let $z_n = x_n + y_n$ be a convergent sequence of points (say $z_n \to z$) with $z_n \in C_1 + C_2$. We claim that $z \in C_1 + C_2$. Indeed, passing to a subsequence if necessary, we may assume $y_n \to y$. Passing to this subsequence, we have $x_n = z_n - y_n \to z - y$, so that x_n is convergent and necessarily converges to a point $x \in C_1$.

We can also investigate the existence of hyperplanes that support the convex set C, meaning that they touch only its boundary and never enter its interior. Such hyperplanes—and the halfspaces associated with them—provide alternatative descriptions of convex sets and functions. See Figure 2.2.13.

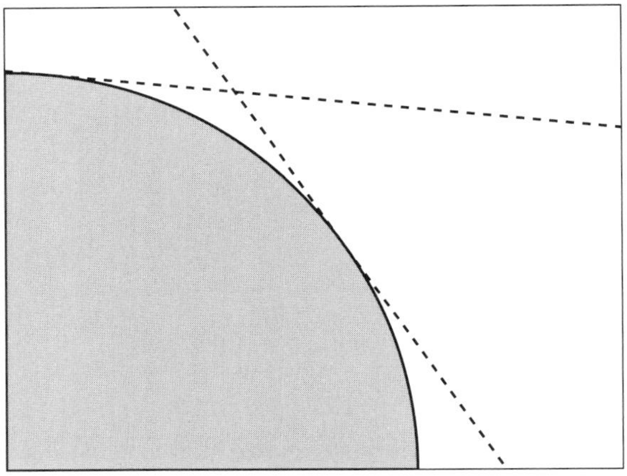

FIGURE 2.2.13. Supporting hyperplanes to a convex set.

Theorem 2.2.14 (Supporting hyperplanes). *Let C be a closed convex set and $x \in \operatorname{bd} C$, the boundary of C. Then there exists a vector $v \neq 0$ supporting C at x, that is,*

$$\langle v, x \rangle \geqslant \langle v, y \rangle \text{ for all } y \in C. \tag{2.2.15}$$

Proof. Let (x_n) be a sequence of points approaching x from outside C, that is, $x_n \notin C$ for any n, but $x_n \to x$. For each n, we can take $s_n = x_n - \pi_C(x_n)$ and define $v_n = s_n / \|s_n\|_2$. Then (v_n) is a sequence satisfying $\langle v_n, x \rangle > \langle v_n, y \rangle$ for all $y \in C$, and since $\|v_n\|_2 = 1$, the sequence (v_n) belongs to the compact set $\{v : \|v\|_2 \leqslant 1\}$.[3] Passing to a subsequence if necessary, it is clear that there is a vector v such that $v_n \to v$, and we have $\langle v, x \rangle \geqslant \langle v, y \rangle$ for all $y \in C$. □

Theorem 2.2.16 (Halfspace intersections). *Let $C \subsetneq \mathbb{R}^n$ be a closed convex set. Then C is the intersection of all the spaces containing it; moreover,*

$$C = \bigcap_{x \in \operatorname{bd} C} H_x \tag{2.2.17}$$

where H_x denotes the intersection of the halfspaces contained in hyperplanes supporting C at x.

Proof. It is clear that $C \subseteq \bigcap_{x \in \operatorname{bd} C} H_x$. Indeed, let $h_x \neq 0$ be a hyperplane supporting to C at $x \in \operatorname{bd} C$ and consider $H_x = \{y : \langle h_x, x \rangle \geqslant \langle h_x, y \rangle\}$. By Theorem 2.2.14 we see that $H_x \supseteq C$.

[3] In a general Hilbert space, this set is actually weakly compact by Alaoglu's theorem. However, in a weakly compact set, any sequence has a weakly convergent subsequence, that is, there exists a subsequence $n(m)$ and vector v such that $\langle v_{n(m)}, y \rangle \to \langle v, y \rangle$ for all y.

Now we show the other inclusion: $\bigcap_{x \in \mathrm{bd}\, C} H_x \subseteq C$. Suppose for the sake of contradiction that $z \in \bigcap_{x \in \mathrm{bd}\, C} H_x$ satisfies $z \notin C$. We will construct a hyperplane supporting C that separates z from C, which will be a contradiction to our supposition. Since C is closed, the projection of $\pi_C(z)$ of z onto C satisfies $\langle z - \pi_C(z), z \rangle > \sup_{y \in C} \langle z - \pi_C(z), y \rangle$ by Proposition 2.2.10. In particular, defining $v_z = z - \pi_C(z)$, the hyperplane $\{y : \langle v_z, y \rangle = \langle v_z, \pi_C(z) \rangle\}$ is supporting to C at the point $\pi_C(z)$ (Corollary 2.2.6) and the halfspace $\{y : \langle v_z, y \rangle \leqslant \langle v_z, \pi_C(z) \rangle\}$ does not contain z but does contain C. This contradicts the assumption that $z \in \bigcap_{x \in \mathrm{bd}\, C} H_x$. □

As a not too immediate consequence of Theorem 2.2.16 we obtain the following characterization of a convex function as the supremum of all affine functions that minorize the function (that is, affine functions that are everywhere less than or equal to the original function). This is intuitive: if f is a closed convex function, meaning that epi f is closed, then epi f is the intersection of all the halfspaces containing it. The challenge is showing that we may restrict this intersection to non-vertical halfspaces. See Figure 2.2.18.

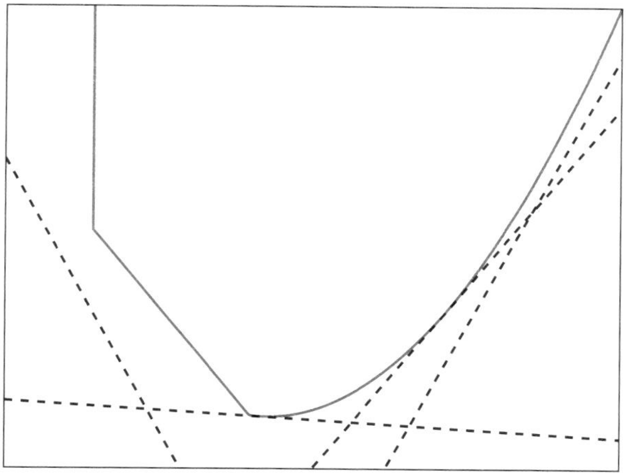

FIGURE 2.2.18. The function f (solid blue line) and affine underestimators (dotted lines).

Corollary 2.2.19. *Let* f *be a closed convex function that is not identically* $-\infty$. *Then*

$$f(x) = \sup_{v \in \mathbb{R}^n, b \in \mathbb{R}} \{\langle v, x \rangle + b : f(y) \geqslant b + \langle v, y \rangle \text{ for all } y \in \mathbb{R}^n\}.$$

Proof. First, we note that epi f is closed by definition. Moreover, we know that we can write

$$\mathrm{epi}\, f = \cap\{H : H \supset \mathrm{epi}\, f\},$$

where H denotes a halfspace. More specifically, we may index each halfspace by $(v, a, c) \in \mathbb{R}^n \times \mathbb{R} \times \mathbb{R}$, and we have $H_{v,a,c} = \{(x, t) \in \mathbb{R}^n \times \mathbb{R} : \langle v, x \rangle + at \leqslant c\}$. Now, because $H \supset \mathrm{epi}\, f$, we must be able to take $t \to \infty$ so that $a \leqslant 0$. If $a < 0$, we may divide by $|a|$ and assume without loss of generality that $a = -1$, while otherwise $a = 0$. So if we let

$$\mathcal{H}_1 := \left\{(v, c) : H_{v,-1,c} \supset \operatorname{epi} f\right\} \text{ and } \mathcal{H}_0 := \{(v, c) : H_{v,0,c} \supset \operatorname{epi} f\}.$$

then

$$\operatorname{epi} f = \bigcap_{(v,c)\in\mathcal{H}_1} H_{v,-1,c} \cap \bigcap_{(v,c)\in\mathcal{H}_0} H_{v,0,c}.$$

We would like to show that $\operatorname{epi} f = \cap_{(v,c)\in\mathcal{H}_1} H_{v,-1,c}$, as the set $H_{v,0,c}$ is a vertical hyperplane separating the domain of f, dom f, from the rest of the space.

To that end, we show that if $(v_1, c_1) \in \mathcal{H}_1$ and $(v_0, c_0) \in \mathcal{H}_0$, then

$$H := \bigcap_{\lambda\geqslant 0} H_{v_1+\lambda v_0,-1,c_1+\lambda c_1} = H_{v_1,-1,c_1} \cap H_{v_0,0,c_0}.$$

Indeed, suppose that $(x, t) \in H_{v_1,-1,c_1} \cap H_{v_0,0,c_0}$. Then

$$\langle v_1, x\rangle - t \leqslant c_1 \text{ and } \lambda \langle v_0, x\rangle \leqslant \lambda c_0 \text{ for all } \lambda \geqslant 0.$$

Summing these, we have

$$\langle v_1 + \lambda v_0, x\rangle - t \leqslant c_1 + \lambda c_0 \text{ for all } \lambda \geqslant 0, \tag{2.2.20}$$

or $(x, t) \in H$. Conversely, if $(x, t) \in H$ then inequality (2.2.20) holds, so that taking $\lambda \to \infty$ we have $\langle v_0, x\rangle \leqslant c_0$, while taking $\lambda = 0$ we have $\langle v_1, x\rangle - t \leqslant c_1$.

Noting that $H \in \{H_{v,-1,c} : (v, c) \in \mathcal{H}_1\}$, we see that

$$\operatorname{epi} f = \bigcap_{(v,c)\in\mathcal{H}_1} H_{v,-1,c} = \{(x, t) \in \mathbb{R}^n \times \mathbb{R} : \langle v, x\rangle - t \leqslant c \text{ for all } (v, c) \in \mathcal{H}_1\}.$$

This is equivalent to the claim in the corollary. □

2.3. Continuity and Local Differentiability of Convex Functions Here we discuss several important results concerning convex functions in finite dimensions. We will see that assuming that a function f is convex is quite strong. In fact, we will see the (intuitive if one pictures a convex function) facts that f is continuous, has a directional derivative everywhere, and in fact is locally Lipschitz. We prove the first two results on continuity in Appendix A.1, as they are not fully necessary for our development.

We begin with the fact that if f is defined on a compact domain, then f has an upper bound. The first step in this direction is to argue that this holds for ℓ_1 balls, which can be proved by a simple argument with the definition of convexity.

Lemma 2.3.1. *Let f be convex and defined on* $B_1 = \{x \in \mathbb{R}^n : \|x\|_1 \leqslant 1\}$, *the* ℓ_1 *ball in* n *dimensions. Then there exist* $-\infty < m \leqslant M < \infty$ *such that* $m \leqslant f(x) \leqslant M$ *for all* $x \in B_1$.

We provide a proof of this lemma, as well as the coming theorem, in Appendix A.1, as they are not central to our development, relying on a few results in the sequel. The theorem makes use of the above lemma to show that on compact domains, convex functions are Lipschitz continuous. The proof of the theorem begins by showing that if a convex function is bounded in some set, then it is Lipschitz continuous in the set, then using Lemma 2.3.1 we can show that on compact sets f is indeed bounded.

Theorem 2.3.2. *Let* f *be convex and defined on a set* C *with non-empty interior. Let* $B \subseteq \operatorname{int} C$ *be compact. Then there is a constant* L *such that* $|f(x) - f(y)| \leqslant L \|x - y\|$ *on* B*, that is,* f *is* L*-Lipschitz continuous on* B.

The last result, which we make strong use of in the next section, concerns the existence of directional derivatives for convex functions.

Definition 2.3.3. The *directional derivative* of a function f at a point x in the direction u is

$$f'(x; u) := \lim_{\alpha \downarrow 0} \frac{1}{\alpha} [f(x + \alpha u) - f(x)].$$

This definition makes sense for convex f by our earlier arguments that convex functions have increasing slopes (recall expression (2.1.5)). To see that the above definition makes sense, we restrict our attention to $x \in \operatorname{int} \operatorname{dom} f$, so that we can approach x from all directions. By taking $u = y - x$ for any $y \in \operatorname{dom} f$,

$$f(x + \alpha(y - x)) = f((1 - \alpha)x + \alpha y) \leqslant (1 - \alpha) f(x) + \alpha f(y)$$

so that

$$\frac{1}{\alpha} [f(x + \alpha(y - x)) - f(x)] \leqslant \frac{1}{\alpha} [\alpha f(y) - \alpha f(x)] = f(y) - f(x) = f(x + u) - f(x).$$

We also know from Theorem 2.3.2 that f is locally Lipschitz, so for small enough α there exists some L such that $f(x + \alpha u) \geqslant f(x) - L\alpha \|u\|$, and thus for which $f'(x; u) \geqslant -L \|u\|$. Further, an argument by convexity (the criterion (2.1.5) of increasing slopes) shows that the function

$$\alpha \mapsto \frac{1}{\alpha} [f(x + \alpha u) - f(x)]$$

is increasing, so we can replace the limit in the definition of $f'(x; u)$ with an infimum over $\alpha > 0$, that is, $f'(x; u) = \inf_{\alpha > 0} \frac{1}{\alpha} [f(x + \alpha u) - f(x)]$. Noting that if x is on the boundary of $\operatorname{dom} f$ and $x + \alpha u \notin \operatorname{dom} f$ for any $\alpha > 0$, then we must have $f'(x; u) = +\infty$, we have proved the following theorem.

Theorem 2.3.4. *For convex* f*, at any point* $x \in \operatorname{dom} f$ *and for any* u*, the directional derivative* $f'(x; u)$ *exists and is*

$$f'(x; u) = \lim_{\alpha \downarrow 0} \frac{1}{\alpha} [f(x + \alpha u) - f(x)] = \inf_{\alpha > 0} \frac{1}{\alpha} [f(x + \alpha u) - f(x)].$$

If $x \in \operatorname{int} \operatorname{dom} f$*, there exists a constant* $L < \infty$ *such that* $|f'(x; u)| \leqslant L \|u\|$ *for any* $u \in \mathbb{R}^n$*. If* f *is Lipschitz continuous with respect to the norm* $\|\cdot\|$*, we can take* L *to be the Lipschitz constant of* f.

Lastly, we state a well-known condition that is equivalent to convexity. This is inuitive: if a function is bowl-shaped, it should have positive second derivatives.

Theorem 2.3.5. *Let* $f : \mathbb{R}^n \to \mathbb{R}$ *be twice continuously differentiable. Then* f *is convex if and only if* $\nabla^2 f(x) \succeq 0$ *for all* x*, that is,* $\nabla^2 f(x)$ *is positive semidefinite.*

Proof. We may essentially reduce the argument to one-dimensional problems, because if f is twice continuously differentiable, then for each $v \in \mathbb{R}^n$ we may define $h_v : \mathbb{R} \to \mathbb{R}$ by

$$h_v(t) = f(x + tv),$$

and f is convex if and only if h_v is convex for each v (because convexity is a property only of lines, by definition). Moreover, we have

$$h_v''(0) = v^\top \nabla^2 f(x) v,$$

and $\nabla^2 f(x) \succeq 0$ if and only if $h_v''(0) \geqslant 0$ for all v.

Thus, with no loss of generality, we assume $n = 1$ and show that f is convex if and only if $f''(x) \geqslant 0$. First, suppose that $f''(x) \geqslant 0$ for all x. Then using that

$$f(y) = f(x) + f'(x)(y - x) + \frac{1}{2}(y - x)^2 f''(\widetilde{x})$$

for some $\widetilde{x}$ between x and y, we have that $f(y) \geqslant f(x) + f'(x)(y - x)$ for all x, y. Let $\lambda \in [0, 1]$. Then we have

$$f(y) \geqslant f(\lambda x + (1-\lambda)y) + \lambda f'(\lambda x + (1-\lambda)y)(y - x) \text{ and}$$
$$f(x) \geqslant f(\lambda x + (1-\lambda)y) + (1-\lambda) f'(\lambda x + (1-\lambda)y)(x - y).$$

Multiplying the first equation by $1 - \lambda$ and the second by λ, then adding, we obtain

$$(1-\lambda)f(y) + \lambda f(x) \geqslant (1-\lambda)f(\lambda x + (1-\lambda)y) + \lambda f(\lambda x + (1-\lambda)y) = f(\lambda x + (1-\lambda)y),$$

that is, f is convex.

For the converse, let $\delta > 0$ and define $x_1 = x + \delta > x > x - \delta = x_0$. Then we have $x_1 - x_0 = 2\delta$, and

$$f(x_1) = f(x) + f'(x)\delta + 2\delta^2 f''(\widetilde{x}_1) \text{ and } f(x_0) = f(x) - f'(x)\delta + 2\delta^2 f''(\widetilde{x}_0)$$

for $\widetilde{x}_1, \widetilde{x}_0 \in [x - \delta, x + \delta]$. Defining $c_\delta = f(x_1) + f(x_0) - 2f(x) \geqslant 0$ (the last inequality by convexity), these equations show that

$$c_\delta = 2\delta^2 [f''(\widetilde{x}_1) + f''(\widetilde{x}_0)].$$

By continuity, we have $f''(\widetilde{x}_i) \to f''(x)$ as $\delta \to 0$, and as $c_\delta / 2\delta^2 \geqslant 0$ for all $\delta > 0$, we must have

$$2f''(x) = \limsup_{\delta \to 0} \{f''(\widetilde{x}_1) + f''(\widetilde{x}_0)\} = \limsup_{\delta \to 0} \frac{c_\delta}{2\delta^2} \geqslant 0.$$

This gives the result. □

2.4. Subgradients and Optimality Conditions The subgradient set of a function f at a point $x \in \operatorname{dom} f$ is defined as follows:

$$\partial f(x) := \{g : f(y) \geqslant f(x) + \langle g, y - x \rangle \text{ for all } y\}. \tag{2.4.1}$$

Intuitively, since a function is convex if and only if epi f is convex, the subgradient set ∂f should be non-empty and consist of supporting hyperplanes to epi f. That is, f should always have global linear underestimators of itself. When a function f is convex, the subgradient generalizes the derivative of f (which is a global linear underestimator of f when f is differentiable), and is also intimately related to optimality conditions for convex minimization.

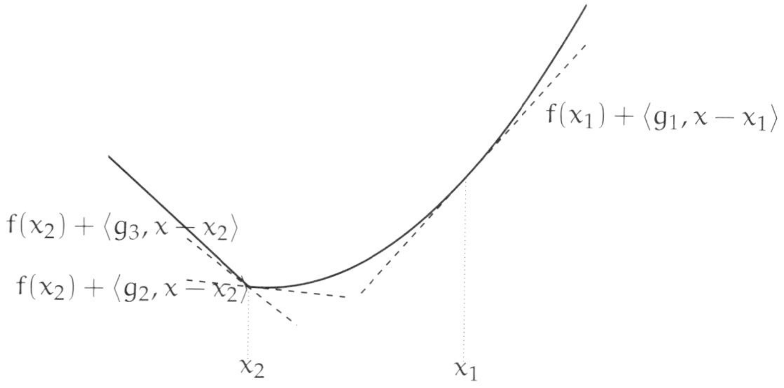

FIGURE 2.4.2. Subgradients of a convex function. At the point x_1, the subgradient g_1 is the gradient. At the point x_2, there are multiple subgradients, because the function is non-differentiable. We show the linear functions given by $g_2, g_3 \in \partial f(x_2)$.

Existence and characterizations of subgradients Our first theorem guarantees that the subdifferential set is non-empty.

Theorem 2.4.3. *Let* $x \in \operatorname{int} \operatorname{dom} f$. *Then* $\partial f(x)$ *is nonempty, closed convex, and compact.*

Proof. The fact that $\partial f(x)$ is closed and convex is straightforward. Indeed, all we need to see this is to recognize that

$$\partial f(x) = \bigcap_z \{g : f(z) \geqslant f(x) + \langle g, z - x \rangle\}$$

which is an intersection of half-spaces, which are all closed and convex.

Now we need to show that $\partial f(x) \neq \emptyset$. This will essentially follow from the following fact: the set $\operatorname{epi} f$ has a non-zero supporting hyperplane at the point $(x, f(x))$. Indeed, from Theorem 2.2.14, we know that there exist a vector v and scalar b, not identically zero, such that

$$\langle v, x \rangle + bf(x) \geqslant \langle v, y \rangle + bt$$

for all $(y, t) \in \operatorname{epi} f$ (that is, y and t such that $f(y) \leqslant t$). Rearranging slightly, we have

$$\langle v, x - y \rangle \geqslant b(t - f(x))$$

and setting $y = x$ shows that $b \leqslant 0$. This is close to what we desire, since if $b < 0$ we set $t = f(y)$ and see that

$$-bf(y) \geqslant -bf(x) + \langle v, y - x \rangle \text{ or } f(y) \geqslant f(x) - \left\langle \frac{v}{b}, y - x \right\rangle$$

for all y, by dividing both sides by $-b$. In particular, $-v/b$ is a subgradient. Thus, suppose for the sake of contradiction that $b = 0$. Then we have $\langle v, x - y \rangle \geqslant 0$ for all $y \in \operatorname{dom} f$, but we assumed that $x \in \operatorname{int} \operatorname{dom} f$, so for small enough $\epsilon > 0$, we can set $y = x + \epsilon v$. This would imply that $\langle v, x - y \rangle = -\epsilon \langle v, v \rangle = 0$, i.e. $v = 0$, contradicting the fact that at least one of v and b must be non-zero.

For the compactness of $\partial f(x)$, we use Lemma 2.3.1, which implies that f is bounded in an ℓ_1-ball around of x. As $x \in \operatorname{int} \operatorname{dom} f$ by assumption, there is some $\epsilon > 0$ such that $x + \epsilon B \subset \operatorname{int} \operatorname{dom} f$ for the ℓ_1-ball $B = \{v : \|v\|_1 \leqslant 1\}$. Lemma 2.3.1 implies that $\sup_{v \in B} f(x + \epsilon v) = M < \infty$ for some M, so we have $M \geqslant f(x + \epsilon v) \geqslant f(x) + \epsilon \langle g, v \rangle$ for all $v \in B$ and $g \in \partial f(x)$, or $\|g\|_\infty \leqslant (M - f(x))/\epsilon$. Thus $\partial f(x)$ is closed and bounded, hence compact. □

The next two results require a few auxiliary results related to the directional derivative of a convex function. The reason for this is that both require connecting the local properties of the convex function f with the sub-differential $\partial f(x)$, which is difficult in general since $\partial f(x)$ can consist of multiple vectors. However, by looking at directional derivatives, we can accomplish what we desire. The connection between a directional derivative and the subdifferential is contained in the next two lemmas.

Lemma 2.4.4. *An equivalent characterization of the subdifferential $\partial f(x)$ of f at x is*

$$\partial f(x) = \left\{ g : \langle g, u \rangle \leqslant f'(x; u) \text{ for all } u \right\}. \tag{2.4.5}$$

Proof. Denote by $S = \{g : \langle g, u \rangle \leqslant f'(x; u)\}$, the set on the right hand side of the equality (2.4.5), and let $g \in S$. By the increasing slopes condition, we have

$$\langle g, u \rangle \leqslant f'(x; u) \leqslant \frac{f(x + \alpha u) - f(x)}{\alpha}$$

for all u and $\alpha > 0$; in particular, by taking $\alpha = 1$ and $u = y - x$, we have the standard subgradient inequality that $f(x) + \langle g, y - x \rangle \leqslant f(y)$. So if $g \in S$, then $g \in \partial f(x)$. Conversely, for any $g \in \partial f(x)$, the definition of a subgradient implies that

$$f(x + \alpha u) \geqslant f(x) + \langle g, x + \alpha u - x \rangle = f(x) + \alpha \langle g, u \rangle .$$

Subtracting $f(x)$ from both sides and dividing by α gives that

$$\frac{1}{\alpha} [f(x + \alpha u) - f(x)] \geqslant \sup_{g \in \partial f(x)} \langle g, u \rangle$$

for all $\alpha > 0$; in particular, $g \in S$. □

The representation (2.4.5) gives another proof that $\partial f(x)$ is compact, as claimed in Theorem 2.4.3. Because we know that $f'(x; u)$ is finite for all u as $x \in \operatorname{int} \operatorname{dom} f$, any $g \in \partial f(x)$ satisfies

$$\|g\|_2 = \sup_{u : \|u\|_2 \leqslant 1} \langle g, u \rangle \leqslant \sup_{u : \|u\|_2 \leqslant 1} f'(x; u) < \infty.$$

Lemma 2.4.6. *Let* f *be closed convex and* $\partial f(x) \neq \emptyset$. *Then*

$$f'(x;u) = \sup_{g \in \partial f(x)} \langle g, u \rangle . \tag{2.4.7}$$

Proof. Certainly, Lemma 2.4.4 shows that $f'(x;u) \geqslant \sup_{g \in \partial f(x)} \langle g, u \rangle$. We must show the other direction. To that end, note that viewed as a function of u, $f'(x;u)$ is convex and positively homogeneous, meaning that $f'(x;tu) = tf'(x;u)$ for $t \geqslant 0$. Thus, we can always write (by Corollary 2.2.19)

$$f'(x;u) = \sup \left\{ \langle v, u \rangle + b : f'(x;w) \geqslant b + \langle v, w \rangle \text{ for all } w \in \mathbb{R}^n \right\}.$$

Using the positive homogeneity, we have $f'(x;0) = 0$ and thus we must have $b = 0$, so that $u \mapsto f'(x;u)$ is characterized as the supremum of linear functions:

$$f'(x;u) = \sup \left\{ \langle v, u \rangle : f'(x;w) \geqslant \langle v, w \rangle \text{ for all } w \in \mathbb{R}^n \right\}.$$

But the set $\{v : \langle v, w \rangle \leqslant f'(x;w) \text{ for all } w\}$ is simply $\partial f(x)$ by Lemma 2.4.4. □

A relatively straightforward calculation using Lemma 2.4.4, which we give in the next proposition, shows that the subgradient is simply the gradient of differentiable convex functions. Note that as a consequence of this, we have the first-order inequality that $f(y) \geqslant f(x) + \langle \nabla f(x), y - x \rangle$ for any differentiable convex function.

Proposition 2.4.8. *Let* f *be convex and differentiable at a point* x. *Then* $\partial f(x) = \{\nabla f(x)\}$.

Proof. If f is differentiable at a point x, then the chain rule implies that

$$f'(x;u) = \langle \nabla f(x), u \rangle \geqslant \langle g, u \rangle$$

for any $g \in \partial f(x)$, the inequality following from Lemma 2.4.4. By replacing u with $-u$, we have $f'(x;-u) = -\langle \nabla f(x), u \rangle \geqslant -\langle g, u \rangle$ as well, or $\langle g, u \rangle = \langle \nabla f(x), u \rangle$ for all u. Letting u vary in (for example) the set $\{u : \|u\|_2 \leqslant 1\}$ gives the result. □

Lastly, we have the following consequence of the previous lemmas, which relates the norms of subgradients $g \in \partial f(x)$ to the Lipschitzian properties of f. Recall that a function f is L-Lipschitz with respect to the norm $\|\cdot\|$ over a set C if

$$|f(x) - f(y)| \leqslant L \|x - y\|$$

for all $x, y \in C$. Then the following proposition is an immediate consequence of Lemma 2.4.6.

Proposition 2.4.9. *Suppose that* f *is* L*-Lipschitz with respect to the norm* $\|\cdot\|$ *over a set* C, *where* $C \subset \operatorname{int} \operatorname{dom} f$. *Then*

$$\sup\{\|g\|_* : g \in \partial f(x), x \in C\} \leqslant L.$$

Examples We can provide a number of examples of subgradients. A general rule of thumb is that, if it is possible to compute the function, it is possible to compute its subgradients. As a first example, we consider

$$f(x) = |x|.$$

Then by inspection, we have

$$\partial f(x) = \begin{cases} -1 & \text{if } x < 0 \\ [-1, 1] & \text{if } x = 0 \\ 1 & \text{if } x > 0. \end{cases}$$

A more complex example is given by any vector norm $\|\cdot\|$. In this case, we use the fact that the dual norm is defined by

$$\|y\|_* := \sup_{x : \|x\| \leqslant 1} \langle x, y \rangle .$$

Moreover, we have that $\|x\| = \sup_{y : \|y\|_* \leqslant 1} \langle y, x \rangle$. Fixing $x \in \mathbb{R}^n$, we thus see that if $\|g\|_* \leqslant 1$ and $\langle g, x \rangle = \|x\|$, then

$$\|x\| + \langle g, y - x \rangle = \|x\| - \|x\| + \langle g, y \rangle \leqslant \sup_{v : \|v\|_* \leqslant 1} \langle v, y \rangle = \|y\| .$$

It is possible to show a converse—we leave this as an exercise for the interested reader—and we claim that

$$\partial \|x\| = \{g \in \mathbb{R}^n : \|g\|_* \leqslant 1, \langle g, x \rangle = \|x\|\}.$$

For a more concrete example, we have

$$\partial \|x\|_2 = \begin{cases} x / \|x\|_2 & \text{if } x \neq 0 \\ \{u : \|u\|_2 \leqslant 1\} & \text{if } x = 0. \end{cases}$$

Optimality properties Subgradients also allow us to characterize solutions to convex optimization problems, giving similar characterizations as those we provided for projections.

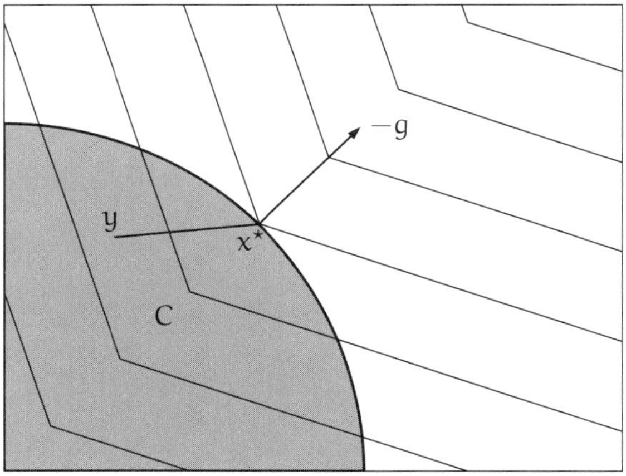

FIGURE 2.4.10. The point $x^\star$ minimizes f over C (the shown level curves) if and only if for some $g \in \partial f(x^\star)$, $\langle g, y - x^\star \rangle \geqslant 0$ for all $y \in C$. Note that not all subgradients satisfy this inequality.

The next theorem, containing necessary and sufficient conditions for a point x to minimize a convex function f, generalizes the standard first-order optimality conditions for differentiable f (e.g., Section 4.2.3 in [12]). The intuition for Theorem 2.4.11 is that there is a vector g in the subgradient set $\partial f(x)$ such that $-g$ is a supporting hyperplane to the feasible set C at the point x. That is, the directions of decrease of the function f lie outside the optimization set C. Figure 2.4.10 shows this behavior.

Theorem 2.4.11. *Let f be convex. The point $x \in \operatorname{int} \operatorname{dom} f$ minimizes f over a convex set C if and only if there exists a subgradient $g \in \partial f(x)$ such that simultaneously for all $y \in C$,*

$$\langle g, y - x \rangle \geqslant 0. \tag{2.4.12}$$

Proof. One direction of the theorem is easy. Indeed, pick $y \in C$. Then certainly there exists $g \in \partial f(x)$ for which $\langle g, y - x \rangle \geqslant 0$. Then by definition,

$$f(y) \geqslant f(x) + \langle g, y - x \rangle \geqslant f(x).$$

This holds for any $y \in C$, so x is clearly optimal.

For the converse, suppose that x minimizes f over C. Then for any $y \in C$ and any $t \geqslant 0$ such that $x + t(y - x) \in C$, we have

$$f(x + t(y - x)) \geqslant f(x) \text{ or } 0 \leqslant \frac{f(x + t(y - x)) - f(x)}{t}.$$

Taking the limit as $t \to 0$, we have $f'(x; y - x) \geqslant 0$ for all $y \in C$. Now, let us suppose for the sake of contradiction that there exists a y such that for all $g \in \partial f(x)$, we have $\langle g, y - x \rangle < 0$. Because

$$\partial f(x) = \{g : \langle g, u \rangle \leqslant f'(x; u) \text{ for all } u \in \mathbb{R}^n\}$$

by Lemma 2.4.6, and $\partial f(x)$ is compact, we have that $\sup_{g \in \partial f(x)} \langle g, y - x \rangle$ is attained, which would imply

$$f'(x; y - x) < 0.$$

This is a contradiction. □

2.5. Calculus rules with subgradients We present a number of calculus rules that show how subgradients are, essentially, similar to derivatives, with a few exceptions (see also Ch. VII of [27]). When we develop methods for optimization problems based on subgradients, these basic calculus rules will prove useful.

Scaling. If we let $h(x) = \alpha f(x)$ for some $\alpha \geqslant 0$, then $\partial h(x) = \alpha \partial f(x)$.

Finite sums. Suppose that $f_1, \ldots, f_m$ are convex functions and let $f = \sum_{i=1}^m f_i$. Then

$$\partial f(x) = \sum_{i=1}^m \partial f_i(x),$$

where the addition is Minkowski addition. To see that $\sum_{i=1}^m \partial f_i(x) \subset \partial f(x)$, let $g_i \in \partial f_i(x)$ for each i. Clearly, $f(y) = \sum_{i=1}^m f_i(y) \geqslant \sum_{i=1}^m f_i(x) + \langle g_i, y - x \rangle$, so that $\sum_{i=1}^{m} g_i \in \partial f(x)$. The converse is somewhat more technical and is a special case of the results to come.

Integrals. More generally, we can extend this summation result to integrals, assuming the integrals exist. These calculations are essential for our development of stochastic optimization schemes based on stochastic (sub)gradient information in the coming lectures. Indeed, for each $s \in \mathcal{S}$, where $\mathcal{S}$ is some set, let f_s be convex. Let μ be a positive measure on the set $\mathcal{S}$, and define the convex function $f(x) = \int f_s(x) d\mu(s)$. In the notation of the introduction (Eq. (1.0.1)) and the problems coming in Section 3.4, we take μ to be a probability distribution on a set $\mathcal{S}$, and if $F(\cdot; s)$ is convex in its first argument for all $s \in \mathcal{S}$, then we may take

$$f(x) = \mathbb{E}[F(x; S)]$$

and satisfy the conditions above. We shall see many such examples in the sequel.

Then if we let $g_s(x) \in \partial f_s(x)$ for each $s \in \mathcal{S}$, we have (assuming the integral exists and that the selections $g_s(x)$ are appropriately measurable)

$$\int g_s(x) d\mu(s) \in \partial f(x). \tag{2.5.1}$$

To see the inclusion, note that for any y we have

$$\begin{aligned}\left\langle \int g_s(x) d\mu(s), y - x \right\rangle &= \int \langle g_s(x), y - x \rangle \, d\mu(s) \\ &\leqslant \int (f_s(y) - f_s(x)) d\mu(s) = f(y) - f(x),\end{aligned}$$

so the inclusion (2.5.1) holds. Eliding a few technical details, one generally obtains the equality

$$\partial f(x) = \left\{ \int g_s(x) d\mu(s) : g_s(x) \in \partial f_s(x) \text{ for each } s \in \mathcal{S} \right\}.$$

Returning to our running example of stochastic optimization, if we have a collection of functions $F : \mathbb{R}^n \times \mathcal{S} \to \mathbb{R}$, where for each $s \in \mathcal{S}$ the function $F(\cdot; s)$ is convex, then $f(x) = \mathbb{E}[f(x; S)]$ is convex when we take expectations over S, and taking

$$g(x; s) \in \partial F(x; s)$$

gives a *stochastic gradient* with the property that $\mathbb{E}[g(x; S)] \in \partial f(x)$. For more on these calculations and conditions, see the classic paper of Bertsekas [7], which addresses the measurability issues.

Affine transformations. Let $f : \mathbb{R}^m \to \mathbb{R}$ be convex and $A \in \mathbb{R}^{m \times n}$ and $b \in \mathbb{R}^m$. Then $h : \mathbb{R}^n \to \mathbb{R}$ defined by $h(x) = f(Ax + b)$ is convex and has subdifferential

$$\partial h(x) = A^T \partial f(Ax + b).$$

Indeed, let $g \in \partial f(Ax + b)$, so that

$$h(y) = f(Ay + b) \geqslant f(Ax + b) + \langle g, (Ay + b) - (Ax + b) \rangle = h(x) + \left\langle A^T g, y - x \right\rangle,$$

giving the result.

Finite maxima. Let $f(x) = \max_{i \leqslant m} f_i(x)$ where f_i, $i = 1, \ldots, m$ are convex functions. Then we have

$$\operatorname{epi} f = \bigcap_{i \leqslant m} \operatorname{epi} f_i,$$

which is convex, and f is convex. Now, let i be any index such that $f_i(x) = f(x)$, and let $g_i \in \partial f_i(x)$. Then we have for any $y \in \mathbb{R}^n$ that

$$f(y) \geqslant f_i(y) \geqslant f_i(x) + \langle g_i, y - x \rangle = f(x) + \langle g_i, y - x \rangle .$$

So $g_i \in \partial f(x)$. More generally, we have the result that

$$\partial f(x) = \operatorname{Conv}\{\partial f_i(x) : f_i(x) = f(x)\}, \tag{2.5.2}$$

that is, the subgradient set of f is the convex hull of the subgradients of *active* functions at x, that is, those attaining the maximum. If there is only a single unique active function f_i, then $\partial f(x) = \partial f_i(x)$. See Figure 2.5.3 for a graphical representation.

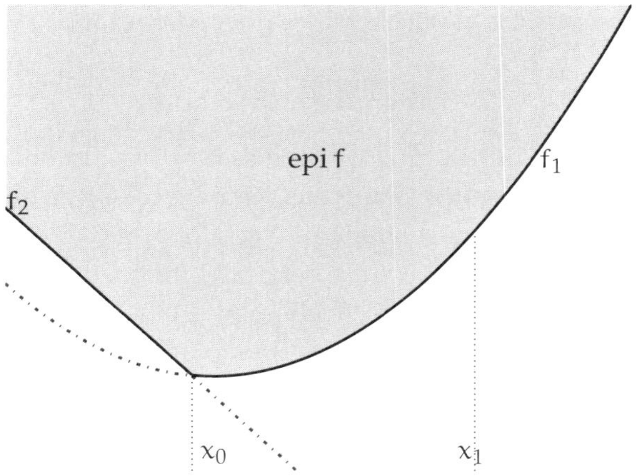

FIGURE 2.5.3. Subgradients of finite maxima. The function $f(x) = \max\{f_1(x), f_2(x)\}$ where $f_1(x) = x^2$ and $f_2(x) = -2x - \frac{1}{5}$, and f is differentiable everywhere except at $x_0 = -1 + \sqrt{4/5}$.

Uncountable maxima (supremum). Lastly, consider $f(x) = \sup_{\alpha \in \mathcal{S}} f_\alpha(x)$, where $\mathcal{A}$ is an arbitrary index set and f_α is convex for each α. First, let us assume that the supremum is attained at some $\alpha \in \mathcal{A}$. Then, identically to the above, we have that $\partial f_\alpha(x) \subset \partial f(x)$. More generally, we have

$$\partial f(x) \supset \operatorname{Conv}\{\partial f_\alpha(x) : f_\alpha(x) = f(x)\}.$$

Achieving equality in the preceding definition requires a number of conditions, and if the supremum is not attained, the function f *may* not be subdifferentiable.

Notes and further reading The study of convex analysis and optimization originates, essentially, with Rockafellar's 1970 book *Convex Analysis* [49]. Because of the limited focus of these lecture notes, we have only barely touched on many topics in convex analysis, developing only those we need. Two omissions are perhaps the most glaring: except tangentially, we have provided no discussion of conjugate functions and conjugacy, and we have not discussed Lagrangian duality, both of which are central to any study of convex analysis and optimization.

A number of books provide coverage of convex analysis in finite and infinite dimensional spaces and make excellent further reading. For broad coverage of convex optimization problems, theory, and algorithms, Boyd and Vandenberghe [12] is an excellent reference, also providing coverage of basic convex duality theory and conjugate functions. For deeper forays into convex analysis, personal favorites of mine include the books of Hiriart-Urruty and Lemaréchal [27, 28], as well as the shorter volume [29], and Bertsekas [8] also provides an elegant geometric picture of convex analysis and optimization. Our approach here follows Hiriart-Urruty and Lemaréchal's most closely. For a treatment of the issues of separation, convexity, duality, and optimization in infinite dimensional spaces, an excellent reference is the classic book by Luenberger [36].

3. Subgradient Methods

Lecture Summary: In this lecture, we discuss first order methods for the minimization of convex functions. We focus almost exclusively on subgradient-based methods, which are essentially universally applicable for convex optimization problems, because they rely very little on the structure of the problem being solved. This leads to effective but slow algorithms in classical optimization problems. In large scale problems arising out of machine learning and statistical tasks, however, subgradient methods enjoy a number of (theoretical) optimality properties and have excellent practical performance.

3.1. Introduction In this lecture, we explore a basic subgradient method, and a few variants thereof, for solving general convex optimization problems. Throughout, we will attack the problem

$$\underset{x}{\text{minimize}}\ f(x) \quad \text{subject to } x \in C \tag{3.1.1}$$

where $f : \mathbb{R}^n \to \mathbb{R}$ is convex (though it may take on the value $+\infty$ for $x \notin \operatorname{dom} f$) and C is a closed convex set. Certainly in this generality, finding a universally good method for solving the problem (3.1.1) is hopeless, though we will see that the subgradient method does essentially apply in this generality.

Convex programming methodologies developed in the last fifty years or so have given powerful methods for solving optimization problems. The performance of many methods for solving convex optimization problems is measured by the amount of time or the number of iterations required of them to give an ϵ-optimal solution to the problem (3.1.1), roughly, how long it takes to find some $\widehat{x}$ such that $f(\widehat{x}) - f(x^\star) \leqslant \epsilon$ and $\operatorname{dist}(\widehat{x}, C) \leqslant \epsilon$ for an optimal $x^\star \in C$. Essentially any problem for which we can compute subgradients efficiently can be solved to accuracy ϵ in time polynomial in the dimension n of the problem and $\log \frac{1}{\epsilon}$ by the ellipsoid method (cf. [41, 45]). Moreover, for somewhat better structured (but still quite general) convex problems, interior point and second order methods [12, 45] are practically and theoretically quite efficient, sometimes requiring only $\mathcal{O}(\log \log \frac{1}{\epsilon})$ iterations to achieve optimization error ϵ. (See the lectures by S. Wright in this volume.) These methods use the Newton method as a basic solver, along with specialized representations of the constraint set C, and can be quite powerful.

However, for large scale problems, the time complexity of standard interior point and Newton methods can be prohibitive. Indeed, for problems in n-dimensions—that is, when $x \in \mathbb{R}^n$—interior point methods scale at best as $\mathcal{O}(n^3)$, and can be much worse. When n is large (where today, large may mean $n \approx 10^9$), this becomes highly non-trivial. In such large scale problems and problems arising from any type of data-collection process, it is reasonable to expect that our representation of problem data is inexact at best. In statistical machine learning problems, for example, this is often the case; generally, many applications do not require accuracy higher than, say $\epsilon = 10^{-2}$ or 10^{-3}, in which case faster but less exact methods become attractive.

It is with this motivation that we approach the problem (3.1.1) in this lecture, showing classical subgradient algorithms. These algorithms have the advantage that their per-iteration costs are low—$\mathcal{O}(n)$ or smaller for n-dimensional problems—but they achieve low accuracy solutions to (3.1.1) very quickly. Moreover, depending on problem structure, they can sometimes achieve convergence rates that are *independent* of problem dimension. More precisely, and as we will see later, the methods we study will guarantee convergence to an ϵ-optimal solution to problem (3.1.1) in $O(1/\epsilon^2)$ iterations, while methods that achieve better dependence on ϵ require at least $n \log \frac{1}{\epsilon}$ iterations.

3.2. The gradient and subgradient methods We begin by focusing on the unconstrained case, that is, when the set C in problem (3.1.1) is $C = \mathbb{R}^n$. That is, we wish to solve

$$\underset{x \in \mathbb{R}^n}{\text{minimize}} \; f(x).$$

We first review the gradient descent method, using it as motivation for what follows. In the gradient descent method, we minimize the objective (3.1.1) by iteratively updating

$$x_{k+1} = x_k - \alpha_k \nabla f(x_k), \tag{3.2.1}$$

where $\alpha_k > 0$ is a positive sequence of stepsizes. The original motivations for this choice of update come from the fact that $x^\star$ minimizes a convex f if and only if $0 = \nabla f(x^\star)$; we believe a more compelling justification comes from the idea of modeling the convex function being minimized. Indeed, the update (3.2.1) is equivalent to

$$x_{k+1} = \operatorname*{argmin}_x \left\{ f(x_k) + \langle \nabla f(x_k), x - x_k \rangle + \frac{1}{2\alpha_k} \|x - x_k\|_2^2 \right\}. \tag{3.2.2}$$

The interpretation is that the linear functional $x \mapsto \{f(x_k) + \langle \nabla f(x_k), x - x_k \rangle\}$ is the best linear approximation to the function f at the point x_k, and we would like to make progress minimizing x. So we minimize this linear approximation, but to make sure that it has fidelity to the function f, we add a quadratic $\|x - x_k\|_2^2$ to penalize moving too far from x_k, which would invalidate the linear approximation. See Figure 3.2.3.

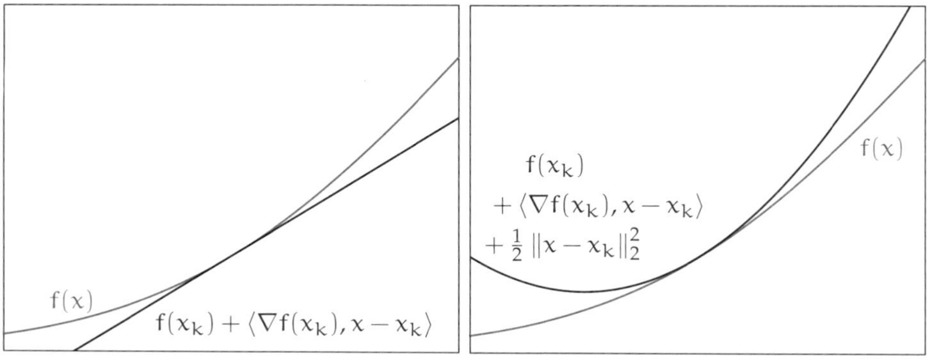

FIGURE 3.2.3. Left: linear approximation (in black) to the function $f(x) = \log(1 + e^x)$ (in blue) at the point $x_k = 0$. Right: linear plus quadratic upper bound for the function $f(x) = \log(1 + e^x)$ at the point $x_k = 0$. This is the upper-bound and approximation of the gradient method (3.2.2) with the choice $\alpha_k = 1$.

Assuming that f is continuously differentiable (often, one assumes the gradient $\nabla f(x)$ is Lipschitz), then gradient descent is a descent method if the stepsize $\alpha_k > 0$ is small enough—it monotonically decreases the objective $f(x_k)$. We spend no more time on the convergence of gradient-based methods, except to say that the choice of the stepsize α_k is often extremely important, and there is a body of research on carefully choosing directions as well as stepsize lengths; Nesterov [44] provides an excellent treatment of many of the basic issues.

Subgradient algorithms The subgradient method is a variant of the method of (3.2.1) in which, instead of using the gradient, we use a subgradient. The method can be written simply: for $k = 1, 2, \ldots$, we iterate

i. Choose any subgradient

$$g_k \in \partial f(x_k).$$

ii. Take the subgradient step

$$x_{k+1} = x_k - \alpha_k g_k. \tag{3.2.4}$$

Unfortunately, the subgradient method is not, in general, a descent method. For a simple example, take the function $f(x) = |x|$, and let $x_1 = 0$. Then except for the choice $g = 0$, all subgradients $g \in \partial f(0) = [-1, 1]$ are *ascent* directions. This is not just an artifact of 0 being optimal for f; in higher dimensions, this behavior is common. Consider, for example, $f(x) = \|x\|_1$ and let $x = e_1 \in \mathbb{R}^n$, the first standard basis vector. Then $\partial f(x) = e_1 + \sum_{i=2}^n t_i e_i$, where $t_i \in [-1, 1]$. Any vector $g = e_1 + \sum_{i=2}^n t_i e_i$ with $\sum_{i=2}^n |t_i| > 1$ is an ascent direction for f, meaning that $f(x - \alpha g) > f(x)$ for all $\alpha > 0$. If we were to pick a uniformly random $g \in \partial f(e_1)$, for example, then the probability that g is a descent direction is exponentially small in the dimension n.

In general, the characterization of the subgradient set $\partial f(x)$ as in Lemma 2.4.4, as $\{g : f'(x; u) \geqslant \langle g, u \rangle \text{ for all } u\}$ where $f'(x; u) = \lim_{t \to 0} \frac{f(x+tu) - f(x)}{t}$ is the directional derivative, and the fact that $f'(x; u) = \sup_{g \in \partial f(x)} \langle g, u \rangle$ guarantees that

$$\operatorname*{argmin}_{g \in \partial f(x)} \{\|g\|_2^2\}$$

is a descent direction, but we do not prove this here. Indeed, finding such a descent direction would require explicitly calculating the entire subgradient set $\partial f(x)$, which for a number of functions is non-trivial and breaks the simplicity of the subgradient method (3.2.4), which works with *any* subgradient.

It is the case, however, that so long as the point x does not minimize $f(x)$, then subgradients descend on a related quantity: the distance of x to *any* optimal point. Indeed, let $g \in \partial f(x)$, and let $x^\star \in \operatorname{argmin} f(x)$ (we assume such a point exists), which need not be unique. Then we have for any α that

$$\frac{1}{2} \|x - \alpha g - x^\star\|_2^2 = \frac{1}{2} \|x - x^\star\|_2^2 - \alpha \langle g, x - x^\star \rangle + \frac{\alpha^2}{2} \|g\|_2^2.$$

The key is that for small enough $\alpha > 0$, the quantity on the right is strictly smaller than $\frac{1}{2} \|x - x^\star\|_2^2$, as we now show. We use the defining inequality of the subgradient, that is, that $f(y) \geqslant f(x) + \langle g, y - x \rangle$ for all y, including $x^\star$. This gives $-\langle g, x - x^\star \rangle = \langle g, x^\star - x \rangle \leqslant f(x^\star) - f(x)$, and thus

$$\text{(3.2.5)} \qquad \frac{1}{2} \|x - \alpha g - x^\star\|_2^2 \leqslant \frac{1}{2} \|x - x^\star\|_2^2 - \alpha (f(x) - f(x^\star)) + \frac{\alpha^2}{2} \|g\|_2^2.$$

From inequality (3.2.5), we see immediately that, no matter our choice $g \in \partial f(x)$, we have

$$0 < \alpha < \frac{2(f(x) - f(x^\star))}{\|g\|_2^2} \text{ implies } \|x - \alpha g - x^\star\|_2^2 < \|x - x^\star\|_2^2.$$

Summarizing, by noting that $f(x) - f(x^\star) > 0$, we have

Observation 3.2.6. *If* $0 \notin \partial f(x)$, *then for any* $x^\star \in \operatorname{argmin}_x f(x)$ *and any* $g \in \partial f(x)$, *there is a stepsize* $\alpha > 0$ *such that* $\|x - \alpha g - x^\star\|_2^2 < \|x - x^\star\|_2^2$.

This observation is the key to the analysis of subgradient methods.

Convergence guarantees Unsurprisingly, given the simplicity of the subgradient method, the analysis of convergence for the method is also quite simple. We begin by stating a general result on the convergence of subgradient methods; we provide a number of variants in the sequel. We make a few simplifying assumptions in stating our result, several of which are not completely necessary, but which considerably simplify the analysis. We enumerate them here:

i. There is at least one (not necessarily unique) point $x^\star \in \operatorname{argmin}_x f(x)$ with $f(x^\star) = \inf_x f(x) > -\infty$.
ii. The subgradients are bounded: for all x and all $g \in \partial f(x)$, we have the subgradient bound $\|g\|_2 \leqslant M < \infty$ (independently of x).

Theorem 3.2.7. *Let $\alpha_k \geqslant 0$ be any sequence of stepsizes for which the assumptions above hold. Let x_k be generated by the subgradient iteration* (3.2.4). *Then for all $K \geqslant 1$,*

$$\sum_{k=1}^{K} \alpha_k [f(x_k) - f(x^\star)] \leqslant \frac{1}{2} \|x_1 - x^\star\|_2^2 + \frac{1}{2} \sum_{k=1}^{K} \alpha_k^2 M^2.$$

Proof. The entire proof essentially amounts to writing down explicitly the distance $\|x_{k+1} - x^\star\|_2^2$ and expanding the square, which we do. By applying inequality (3.2.5), we have

$$\begin{aligned}\frac{1}{2} \|x_{k+1} - x^\star\|_2^2 &= \frac{1}{2} \|x_k - \alpha_k g_k - x^\star\|_2^2 \\ &\overset{(3.2.5)}{\leqslant} \frac{1}{2} \|x_k - x^\star\|_2^2 - \alpha_k (f(x_k) - f(x^\star)) + \frac{\alpha_k^2}{2} \|g_k\|_2^2.\end{aligned}$$

Rearranging this inequality and using that $\|g_k\|_2^2 \leqslant M^2$, we obtain

$$\begin{aligned}\alpha_k [f(x_k) - f(x^\star)] &\leqslant \frac{1}{2} \|x_k - x^\star\|_2^2 - \frac{1}{2} \|x_{k+1} - x^\star\|_2^2 + \frac{\alpha_k^2}{2} \|g_k\|_2^2 \\ &\leqslant \frac{1}{2} \|x_k - x^\star\|_2^2 - \frac{1}{2} \|x_{k+1} - x^\star\|_2^2 + \frac{\alpha_k^2}{2} M^2.\end{aligned}$$

By summing the preceding expression from $k = 1$ to $k = K$ and canceling the alternating $\pm \|x_k - x^\star\|_2^2$ terms, we obtain the theorem. □

Theorem 3.2.7 is the starting point from which we may derive a number of useful consquences. First, we use convexity to obtain the following immediate corollary (we assume that $\alpha_k > 0$ in the corollary).

Corollary 3.2.8. *Let $A_k = \sum_{i=1}^{k} \alpha_i$ and define $\overline{x}_K = \frac{1}{A_K} \sum_{k=1}^{K} \alpha_k x_k$. Then*

$$f(\overline{x}_K) - f(x^\star) \leqslant \frac{\|x_1 - x^\star\|_2^2 + \sum_{k=1}^{K} \alpha_k^2 M^2}{2 \sum_{k=1}^{K} \alpha_k}.$$

Proof. Noting that $A_K^{-1} \sum_{k=1}^{K} \alpha_k = 1$, we see by convexity that

$$f(\overline{x}_K) - f(x^\star) \leqslant \frac{1}{\sum_{k=1}^{K} \alpha_k} \sum_{k=1}^{K} \alpha_k f(x_k) - f(x^\star) = A_K^{-1} \left[\sum_{k=1}^{K} \alpha_k (f(x_k) - f(x^\star)) \right].$$

Applying Theorem 3.2.7 gives the result. □

Corollary 3.2.8 allows us to give a number of basic convergence guarantees based on our stepsize choices. For example, we see that whenever we have

$$\alpha_k \to 0 \text{ and } \sum_{k=1}^{\infty} \alpha_k = \infty,$$

then $\sum_{k=1}^K \alpha_k^2 / \sum_{k=1}^K \alpha_k \to 0$ and so

$$f(\overline{x}_K) - f(x^\star) \to 0 \text{ as } K \to \infty.$$

Moreover, we can give specific stepsize choices to optimize the bound. For example, let us assume for simplicity that $R^2 = \|x_1 - x^\star\|_2^2$ is our distance (radius) to optimality. Then choosing a fixed stepsize $\alpha_k = \alpha$, we have

$$f(\overline{x}_K) - f(x^\star) \leqslant \frac{R^2}{2K\alpha} + \frac{\alpha M^2}{2}. \tag{3.2.9}$$

Optimizing this bound by taking $\alpha = \frac{R}{M\sqrt{K}}$ gives

$$f(\overline{x}_K) - f(x^\star) \leqslant \frac{RM}{\sqrt{K}}.$$

Given that subgradient descent methods are not descent methods, it often makes sense, instead of tracking the (weighted) average of the points or using the final point, to use the best point observed thus far. Naturally, if we let

$$x_k^{\text{best}} = \operatorname*{argmin}_{x_i : i \leqslant k} f(x_i)$$

and define $f_k^{\text{best}} = f(x_k^{\text{best}})$, then we have the same convergence guarantees that

$$f(x_k^{\text{best}}) - f(x^\star) \leqslant \frac{R^2 + \sum_{k=1}^K \alpha_k^2 M^2}{2\sum_{k=1}^K \alpha_k}.$$

A number of more careful stepsize choices are possible, and we refer to the notes at the end of this lecture for more on these choices and applications outside of those we consider, as our focus is naturally circumscribed.

Example We now present an example that has applications in robust statistics and other data fitting scenarios. As a motivating scenario, suppose we have a sequence of vectors $a_i \in \mathbb{R}^n$ and target responses $b_i \in \mathbb{R}$, and we would like to predict b_i via the inner product $\langle a_i, x\rangle$ for some vector x. If there are outliers or other data corruptions in the targets b_i, a natural objective for this task, given the data matrix $A = [a_1 \ \cdots \ a_m]^\top \in \mathbb{R}^{m \times n}$ and vector $b \in \mathbb{R}^m$, is the absolute error

$$f(x) = \frac{1}{m}\|Ax - b\|_1 = \frac{1}{m}\sum_{i=1}^m |\langle a_i, x\rangle - b_i|. \tag{3.2.10}$$

We perform subgradient descent on this objective, which has subgradient

$$g(x) = \frac{1}{m}A^T \operatorname{sign}(Ax - b) = \frac{1}{m}\sum_{i=1}^m a_i \operatorname{sign}(\langle a_i, x\rangle - b_i) \in \partial f(x)$$

at the point x, for $K = 4000$ iterations with a fixed stepsize $\alpha_k \equiv \alpha$ for all k.

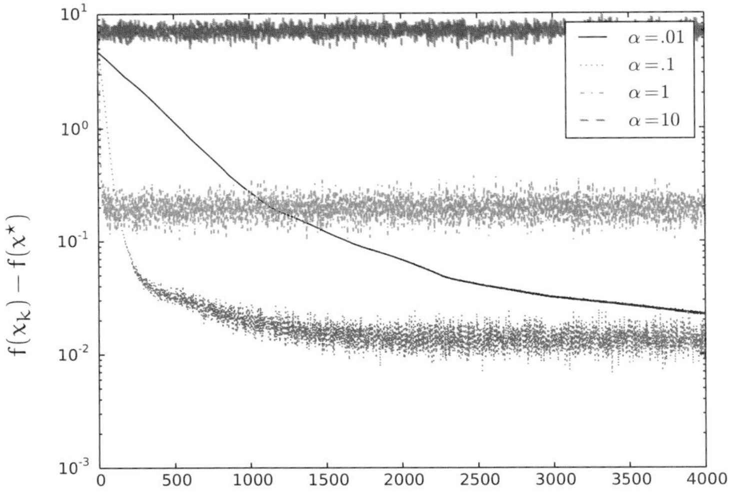

FIGURE 3.2.11. Subgradient method applied to the robust regression problem (3.2.10) with fixed stepsizes.

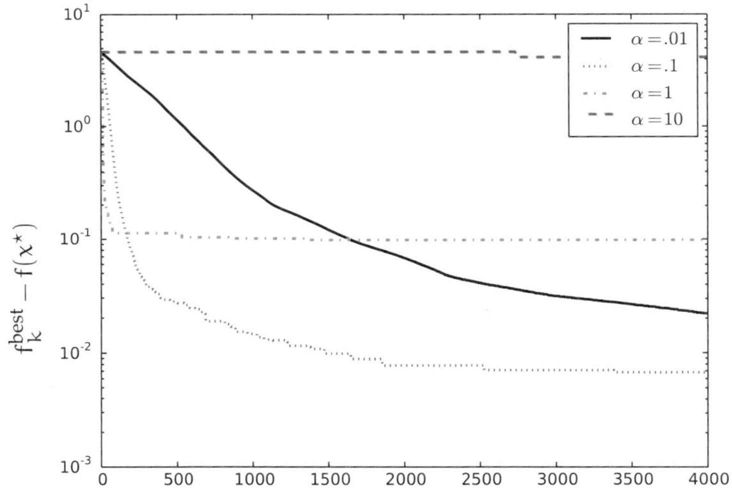

FIGURE 3.2.12. Subgradient method applied to the robust regression problem (3.2.10) with fixed stepsizes, showing performance of the best iterate $f_k^{\rm best} - f(x^\star)$.

We give the results in Figures 3.2.11 and 3.2.12, which exhibit much of the typical behavior of subgradient methods. From the plots, we see roughly a few phases of behavior: the method with stepsize $\alpha = 1$ makes progress very quickly initially, but then enters its "jamming" phase, where it essentially makes no more progress. (The largest stepsize, $\alpha = 10$, simply jams immediately.) The accuracy of the methods with different stepsizes varies greatly, as well—the smaller the stepsize, the better the (final) performance of the iterates x_k, but initial progress is much slower.

3.3. Projected subgradient methods We often wish to solve problems not over $\mathbb{R}^n$ but over some constrained set, for example, in the Lasso [57] and in compressed sensing applications [20] one minimizes an objective such as $\|Ax - b\|_2^2$ subject to $\|x\|_1 \leqslant R$ for some constant $R < \infty$. Recalling the problem (3.1.1), we more generally wish to solve the problem

$$\text{minimize } f(x) \text{ subject to } x \in C \subset \mathbb{R}^n,$$

where C is a closed convex set, not necessarily $\mathbb{R}^n$. The projected subgradient method is close to the subgradient method, except that we replace the iteration with

$$x_{k+1} = \pi_C(x_k - \alpha_k g_k) \tag{3.3.1}$$

where

$$\pi_C(x) = \operatorname*{argmin}_{y \in C}\{\|x - y\|_2\}$$

denotes the (Euclidean) projection onto C. As in the gradient case (3.2.2), we can reformulate the update as making a linear approximation, with quadratic damping, to f and minimizing this approximation: by algebraic manipulation, the update (3.3.1) is equivalent to

$$x_{k+1} = \operatorname*{argmin}_{x \in C}\left\{f(x_k) + \langle g_k, x - x_k\rangle + \frac{1}{2\alpha_k}\|x - x_k\|_2^2\right\}. \tag{3.3.2}$$

Figure 3.3.3 shows an example of the iterations of the projected gradient method applied to minimizing $f(x) = \|Ax - b\|_2^2$ subject to the ℓ_1-constraint $\|x\|_1 \leqslant 1$. Note that the method iterates between moving outside the ℓ_1-ball toward the minimum of f (the level curves) and projecting back onto the ℓ_1-ball.

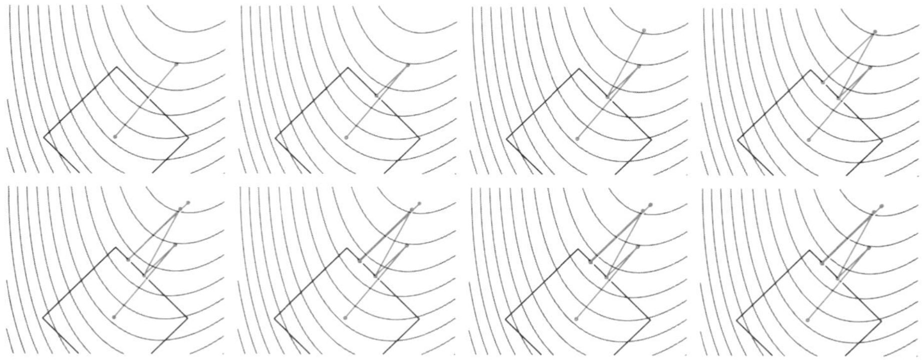

FIGURE 3.3.3. Example execution of the projected gradient method (3.3.1), on minimizing $f(x) = \frac{1}{2}\|Ax - b\|_2^2$ subject to $\|x\|_1 \leqslant 1$.

It is very important in the projected subgradient method that the projection mapping π_C be efficiently computable—the method is effective essentially only in problems where this is true. Often this is the case, but some care is necessary if the objective f is simple but the set C is complex. Then projecting onto the set C may be as complex as solving the original optimization problem (3.1.1). For example, a general linear programming problem is described by

$$\underset{x}{\text{minimize}}\ \langle c, x\rangle \text{ subject to } Ax = b,\ Cx \preceq d.$$

Then computing the projection onto the set $\{x : Ax = b, Cx \preceq d\}$ is at least as difficult as solving the original problem.

Examples of projections As noted above, it is important that projections π_C be efficiently calculable, and often a method's effectiveness is governed by how quickly one can compute the projection onto the constraint set C. With that in mind, we now provide two examples exhibiting convex sets C onto which projection is reasonably straightforward and for which we can write explicit, concrete projected subgradient updates.

Example 3.3.4: Suppose that C is an affine set, given as $C = \{x \in \mathbb{R}^n : Ax = b\}$ for $A \in \mathbb{R}^{m \times n}$, $m \leqslant n$, where A is full rank. (So that A is a short and fat matrix and $AA^T \succ 0$.) Then the projection of x onto C is

$$\pi_C(x) = (I - A^T(AA^T)^{-1}A)x + A^T(AA^T)^{-1}b,$$

and if we begin the iterates from a point $x_k \in C$, i.e. with $Ax_k = b$, then

$$x_{k+1} = \pi_C(x_k - \alpha_k g_k) = x_k - \alpha_k(I - A^T(AA^T)^{-1}A)g_k,$$

that is, we project g_k onto the nullspace of A and iterate. ◊

Example 3.3.5 (Some norm balls)**:** Consider updates when $C = \{x : \|x\|_p \leqslant 1\}$ for $p \in \{1, 2, \infty\}$, each reasonably simple, though the projections are no longer affine. First, for $p = \infty$, we consider each coordinate $j = 1, 2, \ldots, n$ in turn, giving

$$[\pi_C(x)]_j = \min\{1, \max\{x_j, -1\}\},$$

that is, we truncate the coordinates of x to be in the range $[-1, 1]$. For $p = 2$, we have a similarly simple to describe update:

$$\pi_C(x) = \begin{cases} x & \text{if } \|x\|_2 \leqslant 1 \\ x/\|x\|_2 & \text{otherwise.} \end{cases}$$

When $p = 1$, that is, $C = \{x : \|x\|_1 \leqslant 1\}$, the update is somewhat more complex. If $\|x\|_1 \leqslant 1$, then $\pi_C(x) = x$. Otherwise, we find the (unique) $t \geqslant 0$ such that

$$\sum_{j=1}^{n} \left[|x_j| - t\right]_+ = 1,$$

and then set the coordinates j via $[\pi_C(x)]_j = \text{sign}(x_j)\left[|x_j| - t\right]_+$. There are numerous efficient algorithms for finding this t (e.g. [14, 23]). ◊

Convergence results We prove the convergence of the projected subgradient using an argument similar to our proof of convergence for the classic (unconstrained) subgradient method. We assume that the set C is contained in the interior of the domain of the function f, which (as noted in the lecture on convex analysis) guarantees that f is Lipschitz continuous and subdifferentiable, so that there exists $M < \infty$ with $\|g\|_2 \leqslant M$ for all $g \in \partial f$. We make the following assumptions in the next theorem.

i. The set $C \subset \mathbb{R}^n$ is compact and convex, and $\|x - x^\star\|_2 \leqslant R < \infty$ for all $x \in C$.

ii. There exists $M < \infty$ such that $\|g\|_2 \leqslant M$ for all $g \in \partial f(x)$ and $x \in C$.

We make the compactness assumption to allow for a slightly different result than Theorem 3.2.7.

Theorem 3.3.6. *Let x_k be generated by the projected subgradient iteration* (3.3.1), *where the stepsizes $\alpha_k > 0$ are non-increasing. Then*

$$\sum_{k=1}^{K} [f(x_k) - f(x^\star)] \leqslant \frac{R^2}{2\alpha_K} + \frac{1}{2} \sum_{k=1}^{K} \alpha_k M^2.$$

Proof. The starting point of the proof is the same basic inequality as we have been using, that is, the distance $\|x_{k+1} - x^\star\|_2^2$. In this case, we note that projections can never increase distances to points $x^\star \in C$, so that

$$\|x_{k+1} - x^\star\|_2^2 = \|\pi_C(x_k - \alpha_k g_k) - x^\star\|_2^2 \leqslant \|x_k - \alpha_k g_k - x^\star\|_2^2.$$

Now, as in our earlier derivation, we apply inequality (3.2.5) to obtain

$$\frac{1}{2}\|x_{k+1} - x^\star\|_2^2 \leqslant \frac{1}{2}\|x_k - x^\star\|_2^2 - \alpha_k[f(x_k) - f(x^\star)] + \frac{\alpha_k^2}{2}\|g_k\|_2^2.$$

Rearranging this slightly by dividing by α_k, we find that

$$f(x_k) - f(x^\star) \leqslant \frac{1}{2\alpha_k}\left[\|x_k - x^\star\|_2^2 - \|x_{k+1} - x^\star\|_2^2\right] + \frac{\alpha_k}{2}\|g_k\|_2^2.$$

Now, using a variant of the telescoping sum in the proof of Theorem 3.2.7 we have

$$\sum_{k=1}^{K} [f(x_k) - f(x^\star)] \leqslant \sum_{k=1}^{K} \left(\frac{1}{2\alpha_k}\left[\|x_k - x^\star\|_2^2 - \|x_{k+1} - x^\star\|_2^2\right] + \frac{\alpha_k}{2}\|g_k\|_2^2 \right). \tag{3.3.7}$$

We rearrange the middle sum in expression (3.3.7), obtaining

$$\begin{aligned}
&\sum_{k=1}^{K} \frac{1}{2\alpha_k}\left[\|x_k - x^\star\|_2^2 - \|x_{k+1} - x^\star\|_2^2\right] \\
&= \sum_{k=2}^{K} \left(\frac{1}{2\alpha_k} - \frac{1}{2\alpha_{k-1}}\right)\|x_k - x^\star\|_2^2 + \frac{1}{2\alpha_1}\|x_1 - x^\star\|_2^2 - \frac{1}{2\alpha_K}\|x_K - x^\star\|_2^2 \\
&\leqslant \sum_{k=2}^{K} \left(\frac{1}{2\alpha_k} - \frac{1}{2\alpha_{k-1}}\right) R^2 + \frac{1}{2\alpha_1} R^2
\end{aligned}$$

because $\alpha_k \leqslant \alpha_{k-1}$. Noting that this last sum telescopes and that $\|g_k\|_2^2 \leqslant M^2$ in inequality (3.3.7) gives the result. □

One application of this result is to use a decreasing stepsize of $\alpha_k = \alpha/\sqrt{k}$. This allows nearly as strong of a convergence rate as in the fixed stepsize case when the number of iterations K is known, but the algorithm provides a guarantee for all iterations k. Here, we have that

$$\sum_{k=1}^{K} \frac{1}{\sqrt{k}} \leqslant \int_0^K t^{-\frac{1}{2}} dt = 2\sqrt{K},$$

and so by taking $\overline{x}_K = \frac{1}{K}\sum_{k=1}^K x_k$ we obtain the following corollary.

Corollary 3.3.8. *In addition to the conditions of the preceding paragraph, let the conditions of Theorem* 3.3.6 *hold. Then*

$$f(\overline{x}_K) - f(x^\star) \leqslant \frac{R^2}{2\alpha\sqrt{K}} + \frac{M^2\alpha}{\sqrt{K}}.$$

So we see that convergence is guaranteed, at the "best" rate $1/\sqrt{K}$, for all iterations. Here, we say "best" because this rate is unimprovable—there are worst case functions for which no method can achieve a rate of convergence faster than $RM/\sqrt{K}$—but in practice, one would hope to attain better behavior by leveraging problem structure.

3.4. Stochastic subgradient methods The real power of subgradient methods, which has become evident in the last ten or fifteen years, is in their applicability to large scale optimization problems. Indeed, while subgradient methods guarantee only slow convergence—requiring $1/\epsilon^2$ iterations to achieve ϵ-accuracy—their simplicity ensures that they are robust to a number of errors. In fact, subgradient methods achieve unimprovable rates of convergence for a number of optimization problems with noise, and they often do so very computationally efficiently.

Stochastic optimization problems The basic building block for stochastic (sub)-gradient methods is the *stochastic (sub)gradient*, often called the stochastic (sub)-gradient oracle. Let $f : \mathbb{R}^n \to \mathbb{R} \cup \{\infty\}$ be a convex function, and fix $x \in \operatorname{dom} f$. (We will typically omit the sub- qualifier in what follows.) Then a random vector g is a stochastic gradient for f at the point x if $\mathbb{E}[g] \in \partial f(x)$, or

$$f(y) \geqslant f(x) + \langle \mathbb{E}[g], y - x\rangle \quad \text{for all } y.$$

Said somewhat more formally, we make the following definition.

Definition 3.4.1. A *stochastic gradient oracle* for the function f is a triple $(g, \mathcal{S}, P)$, where $\mathcal{S}$ is a sample space, P is a probability distribution, and $g : \mathbb{R}^n \times \mathcal{S} \to \mathbb{R}^n$ is a mapping that for each $x \in \operatorname{dom} f$ satisfies

$$\mathbb{E}_P[g(x, S)] = \int g(x, s)dP(s) \in \partial f(x),$$

where $S \in \mathcal{S}$ is a sample drawn from P.

Often, with some abuse of notation, we will use g or $g(x)$ for shorthand of the random vector $g(x, S)$ when this does not cause confusion.

A standard example for these types of problems is *stochastic programming*, where we wish to solve the convex optimization problem

$$\begin{aligned} &\text{minimize } f(x) := \mathbb{E}_P[F(x; S)] \\ &\text{subject to } x \in C. \end{aligned} \tag{3.4.2}$$

Here S is a random variable on the space $\mathcal{S}$ with distribution P (so the expectation $\mathbb{E}_P[F(x; S)]$ is taken according to P), and for each $s \in \mathcal{S}$, the function $x \mapsto F(x; s)$ is convex. Then we immediately see that if we let

$$g(x, s) \in \partial_x F(x; s),$$

then g is a stochastic gradient when we draw $S \sim P$ and set $g = g(x, S)$, as in Lecture 2 (recall expression (2.5.1)). Recalling this calculation, we have

$$f(y) = \mathbb{E}_P[F(y; S)] \geqslant \mathbb{E}_P[F(x; S) + \langle g(x, S), y - x \rangle] = f(x) + \langle \mathbb{E}_P[g(x, S)], y - x \rangle$$

so that $\mathbb{E}_P[g(x, S)]$ is a stochastic subgradient.

To make the setting (3.4.2) more concrete, consider the robust regression problem (3.2.10), which uses

$$f(x) = \frac{1}{m} \|Ax - b\|_1 = \frac{1}{m} \sum_{i=1}^{m} |\langle a_i, x \rangle - b_i|.$$

Then a natural stochastic gradient, which requires time only $\mathcal{O}(n)$ to compute (as opposed to $\mathcal{O}(m \cdot n)$ to compute $Ax - b$), is to uniformly at random draw an index $i \in [m]$, then return

$$g = a_i \operatorname{sign}(\langle a_i, x \rangle - b_i).$$

More generally, given any problem in which one has a large dataset $\{s_1, \ldots, s_m\}$, and we wish to minimize the sum

$$f(x) = \frac{1}{m} \sum_{i=1}^{m} F(x; s_i),$$

then drawing an index $i \in \{1, \ldots, m\}$ uniformly at random and selecting any $g \in \partial_x F(x; s_i)$ is a stochastic gradient. Computing this stochastic gradient requires only the time necessary for computing some element of the subgradient set $\partial_x F(x; s_i)$, while the standard subgradient method applied to these problems is m-times more expensive in each iteration.

More generally, the expectation $\mathbb{E}[F(x; S)]$ is generally intractable to compute, especially if S is a high-dimensional distribution. In statistical and machine learning applications, we may not even know the distribution P, but we can observe samples $S_i \stackrel{\text{iid}}{\sim} P$. In these cases, it may be impossible to even implement the calculation of a subgradient $f'(x) \in \partial f(x)$, but sampling from P is possible, allowing us to compute *stochastic* subgradients.

Stochastic subgradient method With this motivation in place, we can describe the (projected) stochastic subgradient method. Simply, the method iterates as follows:

(1) Compute a stochastic subgradient g_k at the point x_k, where we have $\mathbb{E}[g_k \mid x_k] \in \partial f(x_k)$.

(2) Perform the projected subgradient step

$$x_{k+1} = \pi_C(x_k - \alpha_k g_k).$$

This is essentially identical to the projected gradient method (3.3.1), except that we replace the true subgradient with a stochastic gradient.

In the next section, we analyze the convergence of the procedure, but here we give two examples that exhibit some of the typical behavior of these methods.

Example 3.4.3 (Robust regression): We consider the robust regression problem of equation (3.2.10), solving

$$\text{(3.4.4)} \qquad \underset{x}{\text{minimize}}\ f(x) = \frac{1}{m}\sum_{i=1}^{m} |\langle a_i, x\rangle - b_i| \ \text{subject to}\ \|x\|_2 \leqslant R,$$

using the random sample $g = a_i \operatorname{sign}(\langle a_i, x\rangle - b_i)$ as our stochastic gradient. We generate $A = [a_1 \ \cdots \ a_m]^\top$ by drawing $a_i \stackrel{\text{iid}}{\sim} N(0, I_{n\times n})$ and $b_i = \langle a_i, u\rangle + \varepsilon_i|\varepsilon_i|^3$, where $\varepsilon_i \stackrel{\text{iid}}{\sim} N(0,1)$ and u is a Gaussian random variable with identity covariance. We use $n = 50$, $m = 100$, and $R = 4$ for this experiment.

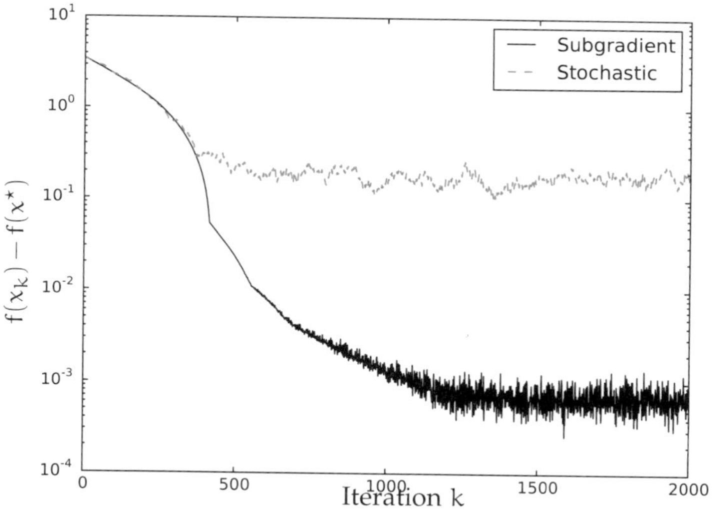

FIGURE 3.4.5. Performance of the stochastic subgradient method and of the non-stochastic subgradient method on problem (3.4.4).

We plot the results of running the stochastic gradient iteration versus standard projected subgradient descent in Figure 3.4.5; both methods run with the fixed stepsize $\alpha = R/M\sqrt{K}$ for $M^2 = \frac{1}{m}\|A\|_{\mathrm{Fr}}^2$, which optimizes the convergence guarantees for the methods. We see in the figure the typical performance of a stochastic gradient method: the initial progress in improving the objective is quite fast, but the method eventually stops making progress once it achieves some low accuracy (in this case, 10^{-1}). In this figure we should make clear, however, that each iteration of the stochastic gradient method requires time $\mathcal{O}(n)$, while each iteration of the (non-noisy) projected gradient method requires times $\mathcal{O}(n \cdot m)$, a factor of approximately 100 times slower. ◇

Example 3.4.6 (Multiclass support vector machine)**:** Our second example is somewhat more complex. We are given a collection of 16×16 grayscale images of handwritten digits $\{0, 1, \ldots, 9\}$, and wish to classify images, represented as vectors $a \in \mathbb{R}^{256}$, as one of the 10 digits. In a general k-class classification problem, we represent the multiclass classifier using the matrix

$$X = [x_1 \; x_2 \; \cdots \; x_k] \in \mathbb{R}^{n \times k},$$

where $k = 10$ for the digit classification problem. Given a data vector $a \in \mathbb{R}^n$, the "score" associated with class l is then $\langle x_l, a \rangle$, and the goal (given image data) is to find a matrix X assigning high scores to the correct image labels. (In machine learning, the typical notation is to use weight vectors $w_1, \ldots, w_k \in \mathbb{R}^n$ instead of $x_1, \ldots, x_k$, but we use X to remain consistent with our optimization focus.) The predicted class for a data vector $a \in \mathbb{R}^n$ is then

$$\operatorname*{argmax}_{l \in [k]} \langle a, x_l \rangle = \operatorname*{argmax}_{l \in [k]} \{[X^T a]_l\}.$$

We represent single training examples as pairs $(a, b) \in \mathbb{R}^n \times \{1, \ldots k\}$, and as a convex surrogate for a misclassification error that the matrix X makes on the pair (a, b), we use the *multiclass hinge* loss function

$$F(X; (a, b)) = \max_{l \neq b} [1 + \langle a, x_l - x_b \rangle]_+$$

where $[t]_+ = \max\{t, 0\}$ denotes the positive part. Then F is convex in X, and for a pair (a, b) we have $F(X; (a, b)) = 0$ if and only if the classifer represented by X has a *large margin*, meaning that

$$\langle a, x_b \rangle \geqslant \langle a, x_l \rangle + 1 \text{ for all } l \neq b.$$

In this example, we have a sample of $N = 7291$ digits $(a_i, b_i) \in \mathbb{R}^n \times \{1, \ldots, k\}$, and we compare the performance of stochastic subgradient descent to standard subgradient descent for solving the problem

$$\text{minimize } f(X) = \frac{1}{N} \sum_{i=1}^{N} F(X; (a_i, b_i)) \text{ subject to } \|X\|_{\mathrm{Fr}} \leqslant R \tag{3.4.7}$$

where $R = 40$. We perform the stochastic gradient descent using the stepsizes $\alpha_k = \alpha_1/\sqrt{k}$, where $\alpha_1 = R/M$ and $M^2 = \frac{1}{N}\sum_{i=1}^{N} \|a_i\|_2^2$ (this is an approximation to the Lipschitz constant of f). For our stochastic gradient oracle, we select an index $i \in \{1, \ldots, N\}$ uniformly at random, then take $g \in \partial_X F(X; (a_i, b_i))$.

For the standard subgradient method, we also perform projected subgradient descent, where we compute subgradients by taking $g_i \in \partial F(X; (a_i, b_i))$ and setting $g = \frac{1}{N}\sum_{i=1}^{N} g_i \in \partial f(X)$. We use an identical stepsize strategy of setting $\alpha_k = \alpha_1/\sqrt{k}$, but use the five stepsizes $\alpha_1 = 10^{-j}R/M$ for $j \in \{-2, -1, \ldots, 2\}$. We plot the results of this experiment in Figure 3.4.8, showing the optimality gap (vertical axis) plotted against the number of matrix-vector products $X^\top a$ computed, normalized by $N = 7291$. The plot makes clear that computing the entire subgradient $\partial f(X)$ is wasteful: the non-stochastic methods' convergence, in terms of iteration count, is potentially faster than that for the stochastic method, but the large ($7291\times$) per-iteration speedup the stochastic method enjoys because of its random sampling yields substantially better performance. Though we do not demonstrate this in the figure, this benefit remains typically true even across a range of stepsize choices, suggesting the benefits of stochastic gradient methods in stochastic programming problems such as problem (3.4.7). $\diamond$

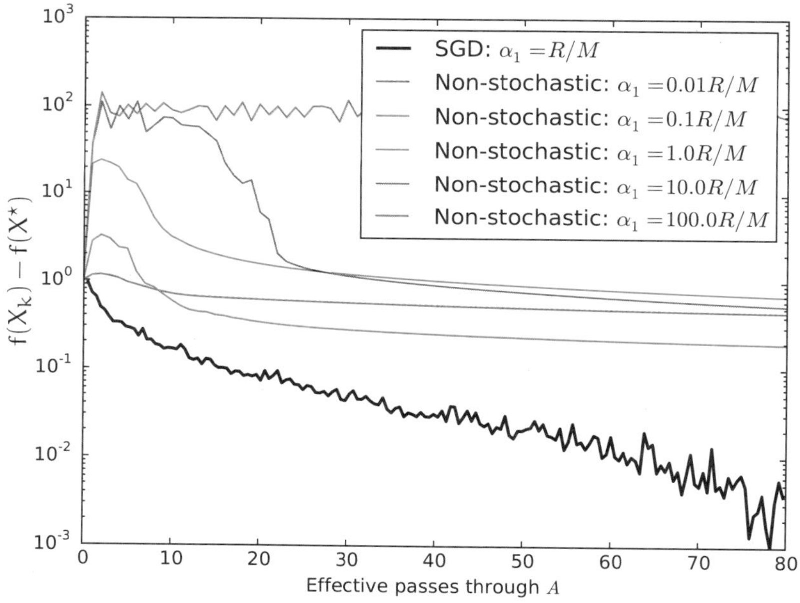

FIGURE 3.4.8. Comparison of stochastic versus non-stochastic methods for the average hinge-loss minimization problem (3.4.7). The horizontal axis is a measure of the time used by each method, represented as the number of times the matrix-vector product $X^\top a_i$ is computed. Stochastic gradient descent vastly outperforms the non-stochastic methods.

Convergence guarantees We now turn to guarantees of convergence for the stochastic subgradient method. As in our analysis of the projected subgradient method, we assume that C is compact and there is some $R < \infty$ such that

$\|x^\star - x\|_2 \leqslant R$ for all $x \in C$, that projections π_C are efficiently computable, and that for all $x \in C$ we have the bound $\mathbb{E}[\|g(x, S)\|_2^2] \leqslant M^2$ for our stochastic oracle g. (The oracle's noise S may depend on the previous iterates, but we always have the unbiased condition $\mathbb{E}[g(x, S)] \in \partial f(x)$.)

Theorem 3.4.9. *Let the conditions of the preceding paragraph hold and let $\alpha_k > 0$ be a non-increasing sequence of stepsizes. Let $\overline{x}_K = \frac{1}{K}\sum_{k=1}^K x_k$. Then*

$$\mathbb{E}[f(\overline{x}_K) - f(x^\star)] \leqslant \frac{R^2}{2K\alpha_K} + \frac{1}{2K}\sum_{k=1}^K \alpha_k M^2.$$

Proof. The analysis is quite similar to our previous analyses, in that we simply expand the error $\|x_{k+1} - x^\star\|_2^2$. Let us define $f'(x) := \mathbb{E}[g(x, S)] \in \partial f(x)$ to be the expected subgradient returned by the stochastic gradient oracle, and then let $\xi_k = g_k - f'(x_k)$ be the error in the kth subgradient. Then

$$\begin{aligned}
\frac{1}{2}\|x_{k+1} - x^\star\|_2^2 &= \frac{1}{2}\|\pi_C(x_k - \alpha_k g_k) - x^\star\|_2^2 \\
&\leqslant \frac{1}{2}\|x_k - \alpha_k g_k - x^\star\|_2^2 \\
&= \frac{1}{2}\|x_k - x^\star\|_2^2 - \alpha_k \langle g_k, x_k - x^\star\rangle + \frac{\alpha_k^2}{2}\|g_k\|_2^2,
\end{aligned}$$

as in the proof of Theorems 3.2.7 and 3.3.6. Now, we can add and subtract a term $\alpha_k \langle f'(x_k), x_k - x^\star\rangle$, which gives

$$\begin{aligned}
&\frac{1}{2}\|x_{k+1} - x^\star\|_2^2 \\
&\leqslant \frac{1}{2}\|x_k - x^\star\|_2^2 - \alpha_k \left\langle f'(x_k), x_k - x^\star\right\rangle + \frac{\alpha_k^2}{2}\|g_k\|_2^2 - \alpha_k \langle \xi_k, x_k - x^\star\rangle \\
&\leqslant \frac{1}{2}\|x_k - x^\star\|_2^2 - \alpha_k [f(x_k) - f(x^\star)] + \frac{\alpha_k^2}{2}\|g_k\|_2^2 - \alpha_k \langle \xi_k, x_k - x^\star\rangle,
\end{aligned}$$

where we have used the standard first-order convexity inequality.

Except for the error term $\langle \xi_k, x_k - x^\star\rangle$, the proof is completely identical to that of Theorem 3.3.6. Indeed, dividing each side of the preceding display by α_k and rearranging, we have

$$f(x_k) - f(x^\star) \leqslant \frac{1}{2\alpha_k}\left(\|x_k - x^\star\|_2^2 - \|x_{k+1} - x^\star\|_2^2\right) + \frac{\alpha_k}{2}\|g_k\|_2^2 - \langle \xi_k, x_k - x^\star\rangle.$$

Summing this inequality, as is done after inequality (3.3.7), yields

$$\sum_{k=1}^K [f(x_k) - f(x^\star)] \leqslant \frac{R^2}{2\alpha_K} + \frac{1}{2}\sum_{k=1}^K \alpha_k \|g_k\|_2^2 - \sum_{k=1}^K \langle \xi_k, x_k - x^\star\rangle. \tag{3.4.10}$$

All our subsequent convergence guarantees follow from this basic inequality.

For this theorem, we need only take expectations, realizing that

$$\mathbb{E}[\langle \xi_k, x_k - x^\star\rangle] = \mathbb{E}\left[\mathbb{E}[\langle g(x_k) - f'(x_k), x_k - x^\star\rangle \mid x_k]\right]$$

$$= \mathbb{E}\Big[\langle \underbrace{\mathbb{E}[g(x_k) \mid x_k]}_{=f'(x_k)} - f'(x_k), x_k - x\rangle\Big] = 0.$$

Thus we obtain

$$\mathbb{E}\left[\sum_{k=1}^{K}(f(x_k) - f(x^\star))\right] \leqslant \frac{R^2}{2\alpha_K} + \frac{1}{2}\sum_{k=1}^{K}\alpha_k M^2$$

once we realize that $\mathbb{E}[\|g_k\|_2^2] \leqslant M^2$, which gives the desired result. □

Theorem 3.4.9 makes it clear that, in expectation, we can achieve the same convergence guarantees as in the non-noisy case. This does not mean that stochastic subgradient methods are always as good as non-stochastic methods, but it does show the robustness of the subgradient method even to substantial noise. So while the subgradient method is *very* slow, its slowness comes with the benefit that it can handle large amounts of noise.

We now provide a few corollaries on the convergence of stochastic gradient descent. For background on probabilistic modes of convergence, see Appendix A.2.

Corollary 3.4.11. *Let the conditions of Theorem* 3.4.9 *hold, and let* $\alpha_k = R/M\sqrt{k}$ *for each* k. *Then*

$$\mathbb{E}[f(\overline{x}_K)] - f(x^\star) \leqslant \frac{3RM}{2\sqrt{K}}$$

for all $K \in \mathbb{N}$.

The proof of the corollary is identical to that of Corollary 3.3.8 for the projected gradient method, once we substitute $\alpha = R/M$ in the bound. We can also obtain convergence in probability of the iterates more generally.

Corollary 3.4.12. *If* α_k *is non-summable but convergent to zero (i.e.* $\sum_{k=1}^{\infty}\alpha_k = \infty$ *and* $\alpha_k \to 0$*), then* $f(\overline{x}_K) - f(x^\star) \overset{p}{\to} 0$ *as* $K \to \infty$. *That is, for all* $\epsilon > 0$ *we have*

$$\limsup_{k\to\infty} \mathbb{P}\left(f(\overline{x}_k) - f(x^\star) \geqslant \epsilon\right) = 0.$$

The above corollaries guarantee convergence of the iterates in expectation and with high probability, but sometimes it is advantageous to give finite sample guarantees of convergence with high probability. We can do this under somewhat stronger conditions on the subgradient noise sequence and using the Azuma-Hoeffding inequality (Theorem A.2.5 in Appendix A.2), which we present now.

Theorem 3.4.13. *In addition to the conditions of Theorem* 3.4.9, *assume that* $\|g\|_2 \leqslant M$ *for all stochastic subgradients* g. *Then for any* $\epsilon > 0$,

$$f(\overline{x}_K) - f(x^\star) \leqslant \frac{R^2}{2K\alpha_K} + \sum_{k=1}^{K}\frac{\alpha_k}{2}M^2 + \frac{RM}{\sqrt{K}}\epsilon$$

with probability at least $1 - e^{-\frac{1}{2}\epsilon^2}$.

Written differently, by taking $\alpha_k = \frac{R}{\sqrt{k}M}$ and setting $\delta = e^{-\frac{1}{2}\epsilon^2}$, we have

$$f(\overline{x}_K) - f(x^\star) \leqslant \frac{3MR}{\sqrt{K}} + \frac{MR\sqrt{2\log\frac{1}{\delta}}}{\sqrt{K}}$$

with probability at least $1-\delta$. That is, we have convergence of $\mathcal{O}(MR/\sqrt{K})$ with high probability.

Before providing the proof proper, we discuss two examples in which the boundedness condition holds. Recall from Lecture 2 that a convex function f is M-Lipschitz if and only if $\|g\|_2 \leqslant M$ for all $g \in \partial f(x)$ and $x \in \mathbb{R}^n$, so Theorem 3.4.13 requires that the random functions $F(\cdot; S)$ are Lipschitz over the domain C. Our robust regression and multiclass support vector machine examples both satisfy the conditions of the theorem so long as the data is bounded. More precisely, for the robust regression problem (3.2.10) with loss $F(x; (a, b)) = |\langle a, x\rangle - b|$, we have $\partial F(x; (a, b)) = a\,\mathrm{sign}(\langle a, x\rangle - b)$ so that the condition $\|g\|_2 \leqslant M$ holds if and only if $\|a\|_2 \leqslant M$. For the multiclass hinge loss problem (3.4.7), with $F(X; (a, b)) = \sum_{l\neq b}[1 + \langle a, x_l - x_b\rangle]_+$, Exercise B.3.1 develops the subgradient calculations, but again, we have the boundedness of $\partial_X F(X; (a, b))$ if and only if $a \in \mathbb{R}^n$ is bounded.

Proof. We begin with the basic inequality of Theorem 3.4.9, inequality (3.4.10). We see that we would like to bound the probability that

$$\sum_{k=1}^{K} \langle \xi_k, x^\star - x_k\rangle$$

is large. First, we note that the iterate x_k is a function of $\xi_1, \ldots, \xi_{k-1}$, and we have the conditional expectation

$$\mathbb{E}[\xi_k \mid \xi_1, \ldots, \xi_{k-1}] = \mathbb{E}[\xi_k \mid x_k] = 0.$$

Moreover, using the boundedness assumption that $\|g\|_2 \leqslant M$, we first obtain $\|\xi_k\|_2 = \|g_k - f'(x_k)\|_2 \leqslant 2M$ and then

$$|\langle \xi_k, x_k - x^\star\rangle| \leqslant \|\xi_k\|_2 \|x_k - x^\star\|_2 \leqslant 2MR.$$

Thus, the sequence $\sum_{k=1}^{K} \langle \xi_k, x_k - x^\star\rangle$ is a bounded difference martingale sequence, and we may apply Azuma's inequality (Theorem A.2.5), which gurantees

$$\mathbb{P}\left(\sum_{k=1}^{K} \langle \xi_k, x^\star - x_k\rangle \geqslant t\right) \leqslant \exp\left(-\frac{t^2}{2KM^2R^2}\right)$$

for all $t \geqslant 0$. Substituting $t = MR\sqrt{K}\epsilon$, we obtain, as desired, that

$$\mathbb{P}\left(\frac{1}{K}\sum_{k=1}^{K} \langle \xi_k, x^\star - x_k\rangle \geqslant \frac{\epsilon MR}{\sqrt{K}}\right) \leqslant \exp\left(-\frac{\epsilon^2}{2}\right). \qquad \square$$

Summarizing the results of this section, we see a number of consequences. First, stochastic gradient methods guarantee that after $\mathcal{O}(1/\epsilon^2)$ iterations, we have error at most $f(x) - f(x^\star) = \mathcal{O}(\epsilon)$. Secondly, this convergence is (at least to the order in ϵ) the same as in the non-noisy case; that is, stochastic gradient methods are robust enough to noise that their convergence is hardly affected by it. In addition to this, they are often applicable in situations in which we cannot even evaluate the objective f, whether for computational reasons or because we do not have access to it, as in statistical problems. This robustness to noise and good performance has led to wide adoption of subgradient-like methods as the *de facto* choice for many large-scale data-based optimization problems. In the coming sections, we give further discussion of the optimality of stochastic gradient methods, showing that—roughly—when we have access only to noisy data, it is impossible to solve (certain) problems to accuracy better than ϵ given $1/\epsilon^2$ data points; thus, using more expensive but accurate optimization methods may have limited benefit (though there may still be some benefit practically!).

Notes and further reading Our treatment in this chapter borrows from a number of resources. The two heaviest are the lecture notes for Stephen Boyd's Stanford's EE364b course [10, 11] and Polyak's *Introduction to Optimization* [47]. Our guarantees of high probability convergence are similar to those originally developed by Cesa-Bianchi et al. [16] in the context of online learning, which Nemirovski et al. [40] more fully develop. More references on subgradient methods include the lecture notes of Nemirovski [43] and Nesterov [44].

A number of extensions of (stochastic) subgradient methods are possible, including to online scenarios in which we observe streaming sequences of functions [25, 64]; our analysis in this section follows closely that of Zinkevich [64]. The classic paper of Polyak and Juditsky [48] shows that stochastic gradient descent methods, coupled with averaging, can achieve asymptotically optimal rates of convergence even to constant factors. Recent work in machine learning by a number of authors [18, 32, 53] has shown how to leverage the structure of optimization problems based on *finite sums*, that is, when $f(x) = \frac{1}{N}\sum_{i=1}^{N} f_i(x)$, to develop methods that achieve convergence rates similar to those of interior point methods but with iteration complexity close to stochastic gradient methods.

4. The Choice of Metric in Subgradient Methods

Lecture Summary: Standard subgradient and projected subgradient methods are inherently Euclidean—they rely on measuring distances using Euclidean norms, and their updates are based on Euclidean steps. In this lecture, we study methods for more carefully choosing the metric, giving rise to mirror descent, also known as non-Euclidean subgradient descent, as well as methods for adapting the updates performed to the problem at hand. By more carefully studying the geometry of the optimization problem being solved, we show how faster convergence guarantees are possible.

4.1. Introduction In the prior lecture, we studied projected subgradient methods for solving the problem (2.1.1) by iteratively updating $x_{k+1} = \pi_C(x_k - \alpha_k g_k)$, where π_C denotes Euclidean projection. The convergence of these methods, as exemplified by Corollaries 3.2.8 and 3.4.11, scales as

$$f(\overline{x}_K) - f(x^\star) \leqslant \frac{MR}{\sqrt{K}} = \mathcal{O}(1)\frac{\text{diam}(C)\text{Lip}(f)}{\sqrt{K}}, \tag{4.1.1}$$

where $R = \sup_{x \in C} \|x - x^\star\|_2$ and M is the Lipschitz constant of f over the set C with respect to the ℓ_2-norm,

$$M = \sup_{x \in C} \sup_{g \in \partial f(x)} \left\{ \|g\|_2 = \left(\sum_{j=1}^n g_j^2 \right)^{\frac{1}{2}} \right\}.$$

The convergence guarantee (4.1.1) reposes on Euclidean measures of scale—the diameter of C and norm of the subgradients g are both measured in ℓ_2-norm. It is thus natural to ask if we can develop methods whose convergence rates depend on other measures of scale of f and C, obtainining better problem-dependent behavior and geometry. With that in mind, in this lecture we derive a number of methods that use either non-Euclidean or adaptive updates to better reflect the geometry of the underlying optimization problem.

4.2. Mirror Descent Methods Our first set of results focuses on mirror descent methods, which modify the basic subgradient update to use a different distance-measuring function rather than the squared ℓ_2-term.

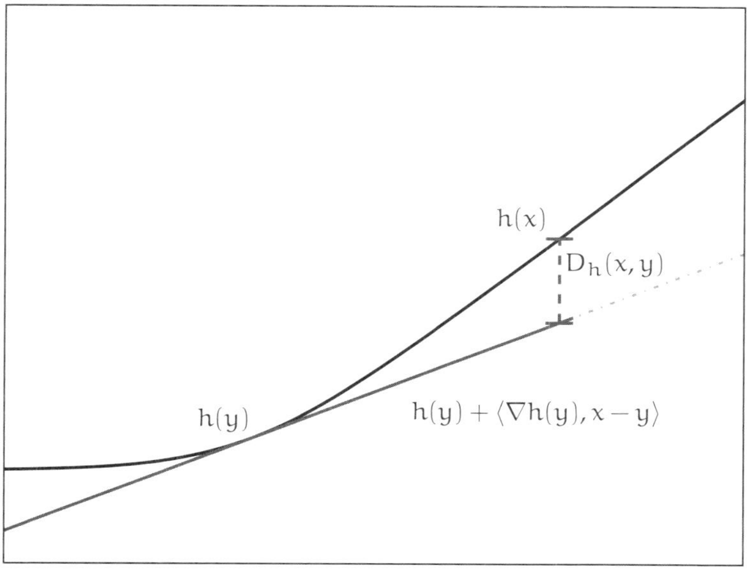

FIGURE 4.2.1. Bregman divergence $D_h(x, y)$. The bottom upper function is $h(x) = \log(1 + e^x)$, the lower (linear) is the linear approximation $x \mapsto h(y) + \langle \nabla h(y), x - y \rangle$ to h at y.

Before presenting these methods, we give a few definitions. Let h be a differentiable convex function, differentiable on C. The *Bregman divergence* D_h associated with h is defined as

$$D_h(x,y) = h(x) - h(y) - \langle \nabla h(y), x - y \rangle . \tag{4.2.2}$$

It is always nonnegative, by the standard first-order inequality for convex functions, and measures the gap between the value $h(x)$ at x and the linear approximation $h(y) + \langle \nabla h(y), x - y\rangle$ for $h(x)$ taken from the point y. See Figure 4.2.1.

As one standard example, if we take $h(x) = \frac{1}{2}\|x\|_2^2$, then we obtain $D_h(x,y) = \frac{1}{2}\|x-y\|_2^2$. A second common example follows by taking the entropy functional $h(x) = \sum_{j=1}^n x_j \log x_j$, restricting x to the probability simplex (i.e. $x \succeq 0$ and $\sum_j x_j = 1$). We then have $D_h(x,y) = \sum_{j=1}^n x_j \log \frac{x_j}{y_j}$, the entropic or Kullback-Leibler divergence.

Because the quantity (4.2.2) is always non-negative and convex in its first argument, it is natural to treat it as a distance-like function in the development of optimization procedures. Indeed, by recalling the updates (3.2.2) and (3.3.2), by analogy we consider the method

i. Compute subgradient $g_k \in \partial f(x_k)$.
ii. Perform update

$$\begin{aligned} x_{k+1} &= \operatorname*{argmin}_{x\in C} \left\{ f(x_k) + \langle g_k, x - x_k\rangle + \frac{1}{\alpha_k} D_h(x, x_k) \right\} \\ &= \operatorname*{argmin}_{x\in C} \left\{ \langle g_k, x\rangle + \frac{1}{\alpha_k} D_h(x, x_k) \right\}. \end{aligned} \tag{4.2.3}$$

This scheme is the *mirror descent method*. Thus, each differentiable convex function h gives a new optimization scheme, where we often attempt to choose h to better match the geometry of the underlying constraint set C.

To this point, we have been vague about the "geometry" of the constraint set, so we attempt to be somewhat more concrete. We say that h is λ-*strongly convex* over C with respect to the norm $\|\cdot\|$ if

$$h(y) \geqslant h(x) + \langle \nabla h(x), y - x\rangle + \frac{\lambda}{2}\|x-y\|^2 \text{ for all } x, y \in C.$$

Importantly, this norm need not be the typical ℓ_2 or Euclidean norm. Then our goal is, roughly, to choose a strongly convex function h so that the diameter of C is small in the norm $\|\cdot\|$ with respect to which h is strongly convex (as we see presently, an analogue of the bound (4.1.1) holds). In the standard updates (3.2.2) and (3.3.2), we use the squared Euclidean norm to trade between making progress on the linear approximation $x \mapsto f(x_k) + \langle g_k, x - x_k\rangle$ and making sure the approximation is reasonable—we *regularize* progress. Thus it is natural to ask that the function h we use provide a similar type of regularization, and consequently, we will require that the function h be 1-strongly convex (usually shortened to the unqualified strongly convex) with respect to some norm $\|\cdot\|$ over the constraint set C in the mirror descent method (4.2.3).[4] Note that strong convexity of h is equivalent to

$$D_h(x,y) \geqslant \frac{1}{2}\|x-y\|^2 \text{ for all } x,y \in C.$$

Examples of mirror descent Before analyzing the method (4.2.3), we present a few examples, showing the updates that are possible as well as verifying that the associated divergence is appropriately strongly convex. One of the nice consequences of allowing different divergence measures D_h, as opposed to only the Euclidean divergence, is that they often yield cleaner or simpler updates.

Example 4.2.4 (Gradient descent is mirror descent): Let $h(x) = \frac{1}{2}\|x\|_2^2$. Then $\nabla h(y) = y$, and

$$D_h(x,y) = \frac{1}{2}\|x\|_2^2 - \frac{1}{2}\|y\|_2^2 - \langle y, x-y\rangle = \frac{1}{2}\|x\|_2^2 + \frac{1}{2}\|y\|_2^2 - \langle x,y\rangle = \frac{1}{2}\|x-y\|_2^2.$$

Thus, substituting into the update (4.2.3), we see the choice $h(x) = \frac{1}{2}\|x\|_2^2$ recovers the standard (stochastic sub)gradient method

$$x_{k+1} = \operatorname*{argmin}_{x \in C}\left\{\langle g_k, x\rangle + \frac{1}{2\alpha_k}\|x-x_k\|_2^2\right\}.$$

It is evident that h is strongly convex with respect to the ℓ_2-norm for any constraint set C. $\diamondsuit$

Example 4.2.5 (Solving problems on the simplex with exponentiated gradient methods): Suppose that our constraint set $C = \{x \in \mathbb{R}_+^n : \langle \mathbf{1}, x\rangle = 1\}$ is the probability simplex in $\mathbb{R}^n$. Then updates with the standard Euclidean distance are somewhat challenging—though there are efficient implementations [14, 23]—and it is natural to ask for a simpler method.

With that in mind, let $h(x) = \sum_{j=1}^n x_j \log x_j$ be the negative entropy, which is convex because it is the sum of convex functions. (The derivatives of $f(t) = t\log t$ are $f'(t) = \log t + 1$ and $f''(t) = 1/t > 0$ for $t \geqslant 0$.) Then we have

$$\begin{aligned} D_h(x,y) &= \sum_{j=1}^n \left[x_j \log x_j - y_j \log y_j - (\log y_j + 1)(x_j - y_j)\right] \\ &= \sum_{j=1}^n x_j \log\frac{x_j}{y_j} + \langle \mathbf{1}, y - x\rangle = D_{kl}(x|y), \end{aligned}$$

the KL-divergence between x and y (when extended to $\mathbb{R}_+^n$, though over C we have $\langle \mathbf{1}, x-y\rangle = 0$). This gives us the form of the update (4.2.3).

Let us consider the update (4.2.3). Simplifying notation, we would like to solve

$$\text{minimize } \langle g, x\rangle + \sum_{j=1}^n x_j \log\frac{x_j}{y_j} \text{ subject to } \langle \mathbf{1}, x\rangle = 1,\ x \succeq 0.$$

[4]This is not strictly a requirement, and sometimes it is analytically convenient to avoid this, but our analysis is simpler when h is strongly convex.

We assume that the $y_j > 0$, though this is not strictly necessary. Though we have not discussed this, we write the Lagrangian for this problem by introducing Lagrange multipliers $\tau \in \mathbb{R}$ for the equality constraint $\langle \mathbf{1}, x\rangle = 1$ and $\lambda \in \mathbb{R}^n_+$ for the inequality $x \succeq 0$.

Then we obtain Lagrangian

$$\mathcal{L}(x, \tau, \lambda) = \langle g, x\rangle + \sum_{j=1}^{n} \left[x_j \log \frac{x_j}{y_j} + \tau x_j - \lambda_j x_j \right] - \tau.$$

Minimizing out x to find the appropriate form for the solution, we take derivatives with respect to x and set them to zero to find

$$0 = \frac{\partial}{\partial x_j} \mathcal{L}(x, \tau, \lambda) = g_j + \log x_j + 1 - \log y_j + \tau - \lambda_j,$$

or

$$x_j(\tau, \lambda) = y_j \exp(-g_j - 1 - \tau + \lambda_j).$$

We may take $\lambda_j = 0$, as the latter expression yields all positive x_j, and to satisfy the constraint that $\sum_j x_j = 1$, we set $\tau = \log(\sum_j y_j e^{-g_j}) - 1$. Thus we have the update

$$x_i = \frac{y_i \exp(-g_i)}{\sum_{j=1}^n y_j \exp(-g_j)}.$$

Rewriting this in terms of the precise update at time k for the mirror descent method, we have for each coordinate i of iterate $k+1$ of the method that

$$x_{k+1,i} = \frac{x_{k,i} \exp(-\alpha_k g_{k,i})}{\sum_{j=1}^n x_{k,j} \exp(-\alpha_k g_{k,j})}. \tag{4.2.6}$$

This is the so-called *exponentiated gradient* update, also known as *entropic mirror descent*.

Later, after stating and proving our main convergence theorems, we will show that the negative entropy is strongly convex with respect to the ℓ_1-norm, meaning that our coming convergence guarantees apply. ◊

Example 4.2.7 (Using ℓ_p-norms): As a final example, we consider using squared ℓ_p-norms for our distance-generating function h. These have nice robustness properties, and are also finite on any compact set (unlike the KL-divergence of Example 4.2.5). Indeed, let $p \in (1, 2]$, and define $h(x) = \frac{1}{2(p-1)} \|x\|_p^2$. We claim without proof that h is strongly convex with respect to the ℓ_p-norm, that is,

$$D_h(x, y) \geqslant \frac{1}{2} \|x - y\|_p^2.$$

(See, for example, the thesis of Shalev-Shwartz [51] and Question B.4.2 in the exercises. This inequality fails for powers other than 2 as well as for $p > 2$.)

We do not address the constrained case here, assuming instead that $C = \mathbb{R}^n$. In this case, we have

$$\nabla h(x) = \frac{1}{p-1} \|x\|_p^{2-p} \left[\operatorname{sign}(x_1)|x_1|^{p-1} \ \cdots \ \operatorname{sign}(x_n)|x_n|^{p-1} \right]^\top.$$

Now, if we define the function $\phi(x) = (p-1)\nabla h(x)$, then a calculation verifies that the function $\varphi : \mathbb{R}^n \to \mathbb{R}^n$ defined coordinate-wise by

$$\phi_j(x) = \|x\|_p^{2-p} \operatorname{sign}(x_j)|x_j|^{p-1} \text{ and } \varphi_j(y) = \|y\|_q^{2-q} \operatorname{sign}(y_j)|y_j|^{q-1},$$

where $\frac{1}{p} + \frac{1}{q} = 1$, satisfies $\varphi(\phi(x)) = x$, that is, $\varphi = \phi^{-1}$ (and similarly $\phi = \varphi^{-1}$). Thus, the mirror descent update (4.2.3) when $C = \mathbb{R}^n$ becomes the somewhat more complex

$$x_{k+1} = \varphi(\phi(x_k) - \alpha_k(p-1)g_k) = (\nabla h)^{-1}(\nabla h(x_k) - \alpha_k g_k). \tag{4.2.8}$$

The second form of the update (4.2.8), that is, that involving the inverse of the gradient mapping $(\nabla h)^{-1}$, holds more generally, that is, for any strictly convex and differentiable h. This is the original form of the mirror descent update (4.2.3), and it justifies the name *mirror* descent, as the gradient is "mirrored" through the distance-generating function h and back again. Nonetheless, we find the modeling perspective of (4.2.3) somewhat easier to explain.

We remark in passing that while constrained updates are somewhat more challenging for this case, a few are efficiently solvable. For example, suppose that $C = \{x \in \mathbb{R}_+^n : \langle \mathbf{1}, x\rangle = 1\}$, the probability simplex. In this case, the update with ℓ_p-norms becomes a problem of solving

$$\underset{x}{\text{minimize}} \ \langle v, x\rangle + \frac{1}{2}\|x\|_p^2 \ \text{ subject to } \langle \mathbf{1}, x\rangle = 1, \ x \succeq 0,$$

where $v = \alpha_k(p-1)g_k - \phi(x_k)$, and φ and ϕ are defined as above. An analysis of the Karush-Kuhn-Tucker conditions for this problem (omitted) yields that the solution to the problem is given by finding the $t^\star \in \mathbb{R}$ such that

$$\sum_{j=1}^n \varphi_j([-v_j + t^\star]_+) = 1 \text{ and setting } x_j = \varphi([-v_j + t^\star]_+).$$

Because φ is increasing in its argument with $\varphi(0) = 0$, this $t^\star$ can be found to accuracy ϵ in time $\mathcal{O}(n \log \frac{1}{\epsilon})$ by binary search. $\diamondsuit$

Convergence guarantees With the mirror descent method described, we now provide an analysis of its convergence behavior. In this case, the analysis is somewhat more complex than that for the subgradient, projected subgradient, and stochastic subgradient methods, as we cannot simply expand the distance $\|x_{k+1} - x^\star\|_2^2$. Thus, we give a variant proof that relies on the optimality conditions for convex optimization problems, as well as a few tricks involving norms and their dual norms. Recall that we assume that the function h is strongly convex with respect to some norm $\|\cdot\|$, and that the associated dual norm $\|\cdot\|_*$ is defined by

$$\|y\|_* := \sup_{x:\|x\|\leqslant 1} \langle y, x\rangle .$$

Theorem 4.2.9. *Let $\alpha_k > 0$ be any sequence of non-increasing stepsizes. Let x_k be generated by the mirror descent iteration* (4.2.3). *If $D_h(x, x^\star) \leqslant R^2$ for all $x \in C$, then* *for all* $K \in \mathbb{N}$

$$\sum_{k=1}^{K} [f(x_k) - f(x^\star)] \leqslant \frac{1}{\alpha_K} R^2 + \sum_{k=1}^{K} \frac{\alpha_k}{2} \|g_k\|_*^2.$$

If $\alpha_k \equiv \alpha$ *is constant, then for all* $K \in \mathbb{N}$

$$\sum_{k=1}^{K} [f(x_k) - f(x^\star)] \leqslant \frac{1}{\alpha} D_h(x^\star, x_1) + \frac{\alpha}{2} \sum_{k=1}^{K} \|g_k\|_*^2.$$

As an immediate consequence of this theorem, we see that if $\overline{x}_K = \frac{1}{K}\sum_{k=1}^{K} x_k$ or $\overline{x}_K = \operatorname{argmin}_{x_k} f(x_k)$ and we have the gradient bound $\|g\|_* \leqslant M$ for all $g \in \partial f(x)$ for $x \in C$, then (say, in the second case) convexity implies

$$f(\overline{x}_K) - f(x^\star) \leqslant \frac{1}{K\alpha} D_h(x^\star, x_1) + \frac{\alpha}{2} M^2. \tag{4.2.10}$$

By comparing with the bound (3.2.9), we see that the mirror descent (non-Euclidean gradient descent) method gives roughly the same type of convergence guarantees as standard subgradient descent. Roughly we expect the following type of behavior with a fixed stepsize: a rate of convergence of roughly $1/\alpha K$ until we are within a radius α of the optimum, after which mirror descent and subgradient descent essentially jam—they just jump back and forth near the optimum.

Proof. We begin by considering the progress made in a single update of x_k, but whereas our previous proofs all began with a Lyapunov function for the distance $\|x_k - x^\star\|_2$, we use function value gaps instead of the distance to optimality. Using the first order convexity inequality—i.e. the definition of a subgradient—we have

$$f(x_k) - f(x^\star) \leqslant \langle g_k, x_k - x^\star \rangle.$$

The idea is to show that replacing x_k with x_{k+1} makes the term $\langle g_k, x_k - x^\star \rangle$ small because of the definition of x_{k+1}, but x_k and x_{k+1} are close together so that this is not much of a difference.

First, we add and subtract $\langle g_k, x_{k+1} \rangle$ to obtain

$$f(x_k) - f(x^\star) \leqslant \langle g_k, x_{k+1} - x^\star \rangle + \langle g_k, x_k - x_{k+1} \rangle. \tag{4.2.11}$$

Now, we use the the first-order necessary and sufficient conditions for optimality of convex optimization problems given by Theorem 2.4.11. Because x_{k+1} solves problem (4.2.3), we have

$$\left\langle g_k + \alpha_k^{-1} (\nabla h(x_{k+1}) - \nabla h(x_k)), x - x_{k+1} \right\rangle \geqslant 0 \text{ for all } x \in C.$$

In particular, this holds for $x = x^\star$, and substituting into (4.2.11) yields

$$f(x_k) - f(x^\star) \leqslant \frac{1}{\alpha_k} \langle \nabla h(x_{k+1}) - \nabla h(x_k), x^\star - x_{k+1} \rangle + \langle g_k, x_k - x_{k+1} \rangle.$$

We now use two tricks: an algebraic identity involving D_h and the Fenchel-Young inequality. By algebraic manipulations, we have that

$$\langle \nabla h(x_{k+1}) - \nabla h(x_k), x^\star - x_{k+1} \rangle = D_h(x^\star, x_k) - D_h(x^\star, x_{k+1}) - D_h(x_{k+1}, x_k).$$

Substituting into the preceding display, we have

$$
\begin{aligned}
&f(x_k) - f(x^\star) \\
&\leqslant \frac{1}{\alpha_k}\left[D_h(x^\star, x_k) - D_h(x^\star, x_{k+1}) - D_h(x_{k+1}, x_k)\right] + \langle g_k, x_k - x_{k+1}\rangle . \qquad (4.2.12)
\end{aligned}
$$

The second insight is that the subtraction of $D_h(x_{k+1}, x_k)$ allows us to cancel some of $\langle g_k, x_k - x_{k+1}\rangle$. To see this, recall the Fenchel-Young inequality, which states that

$$\langle x, y\rangle \leqslant \frac{\eta}{2}\|x\|^2 + \frac{1}{2\eta}\|y\|_*^2$$

for any pair of dual norms $(\|\cdot\|, \|\cdot\|_*)$ and any $\eta > 0$. To see this, note that by definition of the dual norm, we have $\langle x, y\rangle \leqslant \|x\|\,\|y\|_*$, and for any constants $a, b \in \mathbb{R}$ and $\eta > 0$, we have $0 \leqslant \frac{1}{2}(\eta^{\frac{1}{2}}a - \eta^{-\frac{1}{2}}b)^2 = \frac{\eta}{2}a^2 + \frac{1}{2\eta}b^2 - ab$, so that $\|x\|\,\|y\|_* \leqslant \frac{\eta}{2}\|x\|^2 + \frac{1}{2\eta}\|y\|_*^2$. In particular, we have

$$\langle g_k, x_k - x_{k+1}\rangle \leqslant \frac{\alpha_k}{2}\|g_k\|_*^2 + \frac{1}{2\alpha_k}\|x_k - x_{k+1}\|^2 .$$

The strong convexity assumption on h guarantees $D_h(x_k, x_{k+1}) \geqslant \frac{1}{2}\|x_k - x_{k+1}\|^2$, or that

$$-\frac{1}{\alpha_k}D_h(x_{k+1}, x_k) + \langle g_k, x_k - x_{k+1}\rangle \leqslant \frac{\alpha_k}{2}\|g_k\|_*^2 .$$

Substituting this into inequality (4.2.12), we have

$$f(x_k) - f(x^\star) \leqslant \frac{1}{\alpha_k}\left[D_h(x^\star, x_k) - D_h(x^\star, x_{k+1})\right] + \frac{\alpha_k}{2}\|g_k\|_*^2 . \qquad (4.2.13)$$

This inequality should look similar to inequality (3.3.7) in the proof of Theorem 3.3.6 on the projected subgradient method in Lecture 3. Indeed, using that $D_h(x^\star, x_k) \leqslant R^2$ by assumption, an identical derivation to that in Theorem 3.3.6 gives the first result of this theorem. For the second when the stepsize is fixed, note that

$$
\begin{aligned}
\sum_{k=1}^{K}\left[f(x_k) - f(x^\star)\right] &\leqslant \sum_{k=1}^{K}\frac{1}{\alpha}\left[D_h(x^\star, x_k) - D_h(x^\star, x_{k+1})\right] + \sum_{k=1}^{K}\frac{\alpha}{2}\|g_k\|_*^2 \\
&= \frac{1}{\alpha}\left[D_h(x^\star, x_1) - D_h(x^\star, x_{K+1})\right] + \sum_{k=1}^{K}\frac{\alpha}{2}\|g_k\|_*^2 ,
\end{aligned}
$$

which is the second result. □

We briefly provide a few remarks before moving on. As a first remark, all of the preceding analysis carries through in an almost completely identical fashion in the stochastic case. We state the most basic result, as the extension from Section 3.4 is essentially straightforward.

Corollary 4.2.14. *Let the conditions of Theorem 4.2.9 hold, except that instead of receiving a vector $g_k \in \partial f(x_k)$ at iteration k, the vector g_k is a stochastic subgradient satisfying $\mathbb{E}[g_k \mid x_k] \in \partial f(x_k)$. Then for any non-increasing stepsize sequence α_k*

(where α_k may be chosen dependent on $g_1, \ldots, g_k$),

$$\mathbb{E}\left[\sum_{k=1}^{K}(f(x_k)-f(x^\star))\right] \leqslant \mathbb{E}\left[\frac{R^2}{\alpha_K}+\sum_{k=1}^{K}\frac{\alpha_k}{2}\|g_k\|_*^2\right].$$

Proof. We sketch the proof, which is identical to that of Theorem 4.2.9, except that we replace g_k with the vector $f'(x_k)$ satisfying $\mathbb{E}[g_k \mid x_k] = f'(x_k) \in \partial f(x_k)$. Then

$$f(x_k)-f(x^\star) \leqslant \langle f'(x_k), x_k - x^\star\rangle = \langle g_k, x_k - x^\star\rangle + \langle f'(x_k)-g_k, x_k - x^\star\rangle,$$

and an identical derivation yields the following analogue of inequality (4.2.13):

$$\begin{aligned} &f(x_k)-f(x^\star) \\ &\leqslant \frac{1}{\alpha_k}\left[D_h(x^\star, x_k)-D_h(x^\star, x_{k+1})\right]+\frac{\alpha_k}{2}\|g_k\|_*^2+\langle f'(x_k)-g_k, x_k-x^\star\rangle. \end{aligned}$$

This inequality holds regardless of how we choose α_k. Moreover, by iterating expectations, we have

$$\mathbb{E}[\langle f'(x_k)-g_k, x_k-x^\star\rangle] = \mathbb{E}[\langle f'(x_k)-\mathbb{E}[g_k \mid x_k], x_k - x^\star\rangle] = 0,$$

which gives the corollary by a derivation identical to Theorem 4.2.9. □

Thus, if we have the bound $\mathbb{E}[\|g\|_*^2] \leqslant M^2$ for all stochastic subgradients, then taking $\overline{x}_K = \frac{1}{K}\sum_{k=1}^{K} x_k$ and $\alpha_k = R/M\sqrt{k}$, then

$$\mathbb{E}[f(\overline{x}_K)-f(x^\star)] \leqslant \frac{RM}{\sqrt{K}} + \frac{R\max_k \mathbb{E}[\|g_k\|_*^2]}{M}\sum_{k=1}^{K}\frac{1}{2\sqrt{k}} \leqslant 3\frac{RM}{\sqrt{K}} \tag{4.2.15}$$

where we have used that $\mathbb{E}[\|g\|_*^2] \leqslant M^2$ and $\sum_{k=1}^{K} k^{-\frac{1}{2}} \leqslant 2\sqrt{K}$.

In addition, we can provide concrete convergence guarantees for a few methods, revisiting our earlier examples. We begin with Example 4.2.5, exponentiated gradient descent.

Corollary 4.2.16. *Let $C = \{x \in \mathbb{R}_+^n : \langle \mathbf{1}, x\rangle = 1\}$, and take $h(x) = \sum_{j=1}^{n} x_j \log x_j$, the negative entropy. Let $x_1 = \frac{1}{n}\mathbf{1}$, the vector with all entries $1/n$. Then if $\overline{x}_K = \frac{1}{K}\sum_{k=1}^{K} x_k$, the exponentiated gradient method* (4.2.6) *with fixed stepsize α guarantees*

$$f(\overline{x}_K)-f(x^\star) \leqslant \frac{\log n}{K\alpha} + \frac{\alpha}{2K}\sum_{k=1}^{K}\|g_k\|_\infty^2.$$

Proof. To apply Theorem 4.2.9, we must show that the negative entropy h is strongly convex with respect to the ℓ_1-norm, whose dual norm is the ℓ_∞-norm. By a Taylor expansion, we know that for any $x, y \in C$, we have

$$h(x) = h(y) + \langle \nabla h(y), x - y\rangle + \frac{1}{2}(x-y)^\top \nabla^2 h(\widetilde{x})(x-y)$$

for some $\widetilde{x}$ between x and y, that is, $\widetilde{x} = tx + (1-t)y$ for some $t \in [0,1]$. Calculating these quantities, this is equivalent to

$$D_{kl}(x|y) = D_h(x,y) = \frac{1}{2}(x-y)^\top \operatorname{diag}\left(\frac{1}{\widetilde{x}_1}, \ldots, \frac{1}{\widetilde{x}_n}\right)(x-y) = \frac{1}{2}\sum_{j=1}^{n}\frac{(x_j-y_j)^2}{\widetilde{x}_j}.$$

Using the Cauchy-Schwarz inequality and the fact that $\widetilde{x} \in C$, we have

$$\|x - y\|_1 = \sum_{j=1}^n |x_j - y_j| = \sum_{j=1}^n \sqrt{\widetilde{x}_j} \frac{|x_j - y_j|}{\sqrt{\widetilde{x}_j}} \leqslant \underbrace{\left(\sum_{j=1}^n \widetilde{x}_j\right)^{\frac{1}{2}}}_{=1} \left(\sum_{j=1}^n \frac{(x_j - y_j)^2}{\widetilde{x}_j}\right)^{\frac{1}{2}}.$$

That is, we have $D_{kl}(x|y) = D_h(x, y) \geqslant \frac{1}{2}\|x - y\|_1^2$, and h is strongly convex with respect to the ℓ_1-norm over C.

With this strong convexity result in hand, we may apply second result of Theorem 4.2.9, achieving

$$\sum_{k=1}^K [f(x_k) - f(x^\star)] \leqslant \frac{D_{kl}(x^\star|x_1)}{\alpha} + \frac{\alpha}{2}\sum_{k=1}^K \|g_k\|_\infty^2.$$

If $x_1 = \frac{1}{n}\mathbf{1}$, then $D_{kl}(x|x_1) = h(x) + \log n \leqslant \log n$, as $h(x) \leqslant 0$ for $x \in C$. Thus, dividing by K and using that $f(\overline{x}_K) \leqslant \frac{1}{K}\sum_{k=1}^K f(x_k)$ gives the corollary. □

Comparing the guarantee provided by Corollary 4.2.16 to that given by the standard (non-stochastic) projected subgradient method (i.e. using $h(x) = \frac{1}{2}\|x\|_2^2$ as in Theorem 3.3.6) is instructive. In the case of projected subgradient descent, we have $D_h(x^\star, x) = \frac{1}{2}\|x^\star - x\|_2^2 \leqslant 1$ for all $x, x^\star \in C = \{x \in \mathbb{R}_+^n : \langle \mathbf{1}, x\rangle = 1\}$ (and this distance is achieved). But the dual norm to the ℓ_2-norm is ℓ_2, so we measure the size of the gradient terms $\|g_k\|$ in ℓ_2-norm. As $\|g_k\|_\infty \leqslant \|g_k\|_2 \leqslant \sqrt{n}\|g_k\|_\infty$, supposing that $\|g_k\|_\infty \leqslant 1$ for all k, the convergence guarantee $\mathcal{O}(1)\sqrt{\log n / K}$ may be up to $\sqrt{n/\log n}$-times better than that guaranteed by the standard (Euclidean) projected gradient method.

Lastly, we provide a final convergence guarantee for the mirror descent method using ℓ_p-norms, where $p \in (1, 2]$. Using such norms has the benefit that D_h is bounded whenever the set C is compact—distinct from the relative entropy $D_h(x, y) = \sum_j x_j \log \frac{x_j}{y_j}$—and thus providing a nicer guarantee of convergence. Indeed, for $h(x) = \frac{1}{2}\|x\|_p^2$ we always have that

$$D_h(x, y) = \frac{1}{2}\|x\|_p^2 - \frac{1}{2}\|y\|_p^2 - \sum_{j=1}^n \|y\|_p^{2-p}\,\mathrm{sign}(y_j)|y_j|^{p-1}(x_j - y_j)$$

$$(4.2.17) \qquad = \frac{1}{2}\|x\|_p^2 + \frac{1}{2}\|y\|_p^2 \underbrace{- \|y\|_p^{2-p}\sum_{j=1}^n |y_j|^{p-1}\,\mathrm{sign}(y_j)x_j}_{\leqslant \frac{1}{2}\|x\|_p^2 + \frac{1}{2}\|y\|_p^2} \leqslant \|x\|_p^2 + \|y\|_p^2,$$

where the inequality uses that $q(p-1) = p$ and

$$\sum_{j=1}^n \|y\|_p^{2-p}|y_j|^{p-1}|x_j| \leqslant \|y\|_p^{2-p}\left(\sum_{j=1}^n |y_j|^{q(p-1)}\right)^{\frac{1}{q}}\left(\sum_{j=1}^n |x_j|^p\right)^{\frac{1}{p}}$$

$$= \|y\|_p^{2-p}\|y\|_p^{\frac{p}{q}}\|x\|_p = \|y\|_p\|x\|_p \leqslant \frac{1}{2}\|y\|_p^2 + \frac{1}{2}\|x\|_p^2.$$

More generally, $h(x) = \frac{1}{2}\|x - x_0\|_p^2$ gives $D_h(x, y) \leqslant \|x - x_0\|_p^2 + \|y - x_0\|_p^2$. As one example, we obtain the following corollary.

Corollary 4.2.18. *Let* $h(x) = \frac{1}{2(p-1)}\|x\|_p^2$, *where* $p = 1 + \frac{1}{\log(2n)}$, *and assume that* $C \subset \{x \in \mathbb{R}^n : \|x\|_1 \leqslant R_1\}$. *Then*

$$\sum_{k=1}^{K} [f(x_k) - f(x^\star)] \leqslant \frac{2R_1^2 \log(2n)}{\alpha_K} + \frac{e^2}{2}\sum_{k=1}^{K} \alpha_k \|g_k\|_\infty^2.$$

In particular, taking $\alpha_k = R_1\sqrt{\log(2n)/k}/e$ *and* $\overline{x}_K = \frac{1}{K}\sum_{k=1}^K x_k$ *gives*

$$f(\overline{x}_K) - f(x^\star) \leqslant 3e\frac{R_1\sqrt{\log(2n)}}{\sqrt{K}}.$$

Proof. First, we note that $h(x) = \frac{1}{2(p-1)}\|x\|_p^2$ is strongly convex with respect to the ℓ_p-norm, where $1 < p \leqslant 2$. (Recall Example 4.2.7 and see Exercise B.4.2.) Moreover, we know that the dual to the ℓ_p-norm is the conjugate ℓ_q-norm with $1/p + 1/q = 1$, and thus Theorem 4.2.9 implies that

$$\sum_{k=1}^{K} [f(x_k) - f(x^\star)] \leqslant \frac{1}{\alpha_K}\sup_{x \in C} D_h(x, x^\star) + \sum_{k=1}^{K} \frac{\alpha_k}{2}\|g_k\|_q^2.$$

Now, we use that if C is contained in the ℓ_1-ball of radius R_1, then

$$(p-1)D_h(x, y) \leqslant \|x\|_p^2 + \|y\|_p^2 \leqslant \|x\|_1^2 + \|y\|_1^2 \leqslant 2R_1^2.$$

Moreover, because $p = 1 + \frac{1}{\log(2n)}$, we have $q = 1 + \log(2n)$, and

$$\|v\|_q \leqslant \|\mathbf{1}\|_q \|v\|_\infty = n^{\frac{1}{q}}\|v\|_\infty = n^{\frac{1}{\log(2n)}}\|v\|_\infty \leqslant e\|v\|_\infty.$$

Substituting this into the previous display and noting that $1/(p-1) = \log(2n)$ gives the first result. The second follows by first tntegrating $\sum_{k=1}^K k^{-\frac{1}{2}}$ and then using convexity. □

So we see that, in more general cases than the simple simplex constraint afforded by the entropic mirror descent (exponentiated gradient) updates, we have convergence guarantees of order $\sqrt{\log n}/\sqrt{K}$, which may be substantially faster than that guaranteed by the standard projected gradient methods.

A simulated mirror-descent example With our convergence theorems given, we provide a (simulation-based) example of the convergence behavior for an optimization problem for which it is natural to use non-Euclidean norms. We consider a robust regression problem of the following form: we let $A \in \mathbb{R}^{m \times n}$ have entries drawn i.i.d. $N(0, 1)$ with rows $a_1^\top, \ldots, a_m^\top$. We let $b_i = \frac{1}{2}(a_{i,1} + a_{i,2}) + \varepsilon_i$ where $\varepsilon_i \overset{\text{iid}}{\sim} N(0, 10^{-2})$, and $m = 20$ and the dimension $n = 3000$. Then we define

$$f(x) := \|Ax - b\|_1 = \sum_{i=1}^{m} |\langle a_i, x\rangle - b_i|,$$

which has subgradients $A^\top \operatorname{sign}(Ax - b)$. We then minimize f over the simplex $C = \{x \in \mathbb{R}^n_+ : \langle \mathbf{1}, x \rangle = 1\}$; this is the same robust regression problem (3.2.10), except with a particular choice of C.

We compare the subgradient method to exponentiated gradient descent for this problem, noting that the Euclidean projection of a vector $v \in \mathbb{R}^n$ to the set C has coordinates $x_j = [v_j - t]_+$, where $t \in \mathbb{R}$ is chosen so that

$$\sum_{j=1}^{n} x_j = \sum_{j=1}^{n} [v_j - t]_+ = 1.$$

(See the papers [14, 23] for a full derivation of this expression.) We use stepsizes $\alpha_k = \alpha_0/\sqrt{k}$, where the initial stepsize α_0 is chosen to optimize the convergence guarantee for each of the methods (see the coming section). In Figure 4.2.19, we plot the results of performing the projected gradient method versus the exponentiated gradient (entropic mirror decent) method and a method using distance generating functions $h(x) = \frac{1}{2}\|x\|_p^2$ for $p = 1 + 1/\log(2n)$, which can also be shown to be optimal, showing the optimality gap versus iteration count. While all three methods are sensitive to initial stepsize, the mirror descent method (4.2.6) enjoys faster convergence than the standard gradient-based method.

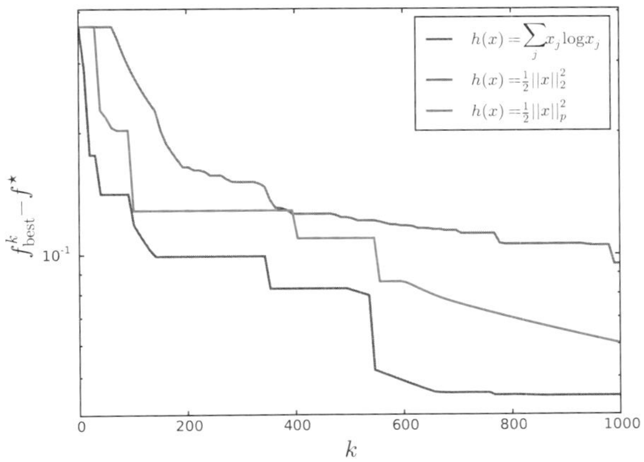

FIGURE 4.2.19. Convergence of mirror descent (entropic gradient method) versus projected gradient method.

4.3. Adaptive stepsizes and metrics In our discussion of mirror descent methods, we assumed we knew enough about the geometry of the problem at hand—or at least the constraint set—to choose an appropriate metric and associated distance-generating function h. In other situations, however, it may be advantageous to *adapt* the metric being used, or at least the stepsizes, to achieve faster convergence guarantees. We begin by describing a simple scheme for choosing stepsizes to optimize bounds on convergence, which means one does not need to know the Lipschitz constants of gradients ahead of time, and then move on to somewhat more involved schemes that use a distance-generating function of the type $h(x) = \frac{1}{2}x^\top Ax$ for some matrix A, which may change depending on information observed during solution of the problem. We leave proofs of the major results in these sections to exercises at the end of the lectures.

Adaptive stepsizes Let us begin by recalling the convergence guarantees for mirror descent in the stochastic case, given by Corollary 4.2.14, which assumes the stepsize α_k used to calculate x_{k+1} is chosen based on the observed gradients $g_1, \ldots, g_k$ (it may be specified ahead of time). In this case, taking $\overline{x}_K = \frac{1}{K}\sum_{k=1}^K x_k$, we have by Corollary 4.2.14 that as long as $D_h(x, x^\star) \leqslant R^2$ for all $x \in C$, then

$$\mathbb{E}[f(\overline{x}_K) - f(x^\star)] \leqslant \mathbb{E}\left[\frac{R^2}{K\alpha_K} + \frac{1}{K}\sum_{k=1}^K \frac{\alpha_k}{2}\|g_k\|_*^2\right]. \tag{4.3.1}$$

Now, if we were to use a fixed stepsize $\alpha_k = \alpha$ for all k, we see that the choice of stepsize minimizing

$$\frac{R^2}{K\alpha} + \frac{\alpha}{2K}\sum_{k=1}^K \|g_k\|_*^2$$

is

$$\alpha^\star = \sqrt{2}R\left(\sum_{k=1}^K \|g_k\|_*^2\right)^{-\frac{1}{2}},$$

which, when substituted into the bound (4.3.1) yields

$$\mathbb{E}[f(\overline{x}_K) - f(x^\star)] \leqslant \sqrt{2}\frac{R}{K}\mathbb{E}\left[\left(\sum_{k=1}^K \|g_k\|_*^2\right)^{\frac{1}{2}}\right]. \tag{4.3.2}$$

While the stepsize choice $\alpha^\star$ and the resulting bound are not strictly possible, as we do not know the magnitudes of the gradients $\|g_k\|_*$ before the procedure executes, in Exercise B.4.1, we prove the following corollary, which uses the "up to now" optimal choice of stepsize α_k.

Corollary 4.3.3. *Let the conditions of Corollary 4.2.14 hold. Let $\alpha_k = R/\sqrt{\sum_{i=1}^k \|g_i\|_*^2}$. Then*

$$\mathbb{E}[f(\overline{x}_K) - f(x^\star)] \leqslant 3\frac{R}{K}\mathbb{E}\left[\left(\sum_{k=1}^K \|g_k\|_*^2\right)^{\frac{1}{2}}\right],$$

where $\overline{x}_K = \frac{1}{K}\sum_{k=1}^K x_k$.

When comparing Corollary 4.3.3 to Corollary 4.2.14, we see by Jensen's inequality that, if $\mathbb{E}[\|g_k\|_*^2] \leqslant M^2$ for all k, then

$$\mathbb{E}\left[\left(\sum_{k=1}^K \|g_k\|_*^2\right)^{\frac{1}{2}}\right] \leqslant \mathbb{E}\left[\sum_{k=1}^K \|g_k\|_*^2\right]^{\frac{1}{2}} \leqslant \sqrt{M^2K} = M\sqrt{K}.$$

Thus, ignoring the $\sqrt{2}$ versus 3 multiplier, the bound of Corollary 4.3.3 is always tighter than that provided by Corollary 4.2.14 and its immediate consequence (4.2.15). We do not explore these particular stepsize choices further, but turn to more sophisticated adaptation strategies.

Variable metric methods and the adaptive gradient method In variable metric methods, the idea is to adjust the metric with which one constructs updates to better reflect local (or non-local) problem structure. The basic framework is very similar to the standard subgradient method (or the mirror descent method), and proceeds as follows.

(i) Receive subgradient $g_k \in \partial f(x_k)$ (or stochastic subgradient g_k satisfying $\mathbb{E}[g_k \mid x_k] \in \partial f(x_k)$).
(ii) Update positive semidefinite matrix $H_k \in \mathbb{R}^{n \times n}$.
(iii) Compute update

$$x_{k+1} = \operatorname*{argmin}_{x \in C} \left\{ \langle g_k, x \rangle + \frac{1}{2} \langle x, H_k x \rangle \right\}. \tag{4.3.4}$$

The method (4.3.4) subsumes a number of standard and less standard optimization methods. If $H_k = \frac{1}{\alpha_k} I_{n \times n}$, a scaled identity matrix, we recover the (stochastic) subgradient method (3.2.4) when $C = \mathbb{R}^n$ (or (3.3.2) generally). If f is twice differentiable and $C = \mathbb{R}^n$, then taking $H_k = \nabla^2 f(x_k)$ to be the Hessian of f at x_k gives the (undamped) Newton method, and using $H_k = \nabla^2 f(x_k)$ even when $C \neq \mathbb{R}^n$ gives a constrained Newton method. More general choices of H_k can even give the ellipsoid method and other classical convex optimization methods [56].

In our case, we specialize the iterations above to focus on diagonal matrices H_k, and we do not assume the function f is smooth (not even differentiable). This, of course, renders unusable standard methods using second order information in the matrix H_k (as it does not exist), but we may still develop useful algorithms. It is possible to consider more general matrices [22], but their additional computational cost generally renders them impractical in large scale and stochastic settings. With that in mind, let us develop a general framework for algorithms and provide their analysis.

We begin with a general convergence guarantee.

Theorem 4.3.5. *Let H_k be a sequence of positive definite matrices, where H_k is a function of $g_1, \ldots, g_k$ (and potentially some additional randomness). Let g_k be (stochastic) subgradients with $\mathbb{E}[g_k \mid x_k] \in \partial f(x_k)$. Then*

$$\begin{aligned}
&\mathbb{E}\left[\sum_{k=1}^{K} (f(x_k) - f(x^\star))\right] \\
&\leqslant \frac{1}{2}\mathbb{E}\left[\sum_{k=2}^{K} \left(\|x_k - x^\star\|_{H_k}^2 - \|x_k - x^\star\|_{H_{k-1}}^2\right) + \|x_1 - x^\star\|_{H_1}^2\right] \\
&+ \frac{1}{2}\mathbb{E}\left[\sum_{k=1}^{K} \|g_k\|_{H_k^{-1}}^2\right].
\end{aligned}$$

Proof. In contrast to mirror descent methods, in this proof we return to our classic Lyapunov-based style of proof for standard subgradient methods, looking at the distance $\|x_k - x^\star\|$. Let $\|x\|_A^2 = \langle x, Ax\rangle$ for any positive semidefinite matrix. We claim that

$$\|x_{k+1} - x^\star\|_{H_k}^2 \leqslant \left\|x_k - H_k^{-1} g_k - x^\star\right\|_{H_k}^2, \tag{4.3.6}$$

the analogue of the fact that projections are non-expansive. This is an immediate consequence of the update (4.3.4): we have that

$$x_{k+1} = \operatorname*{argmin}_{x \in C} \left\{ \left\| x - (x_k - H_k^{-1} g_k) \right\|_{H_k}^2 \right\},$$

which is a Euclidean projection of $x_k - H_k^{-1} g_k$ into C (in the norm $\|\cdot\|_{H_k}$). Then the standard result that projections are non-expansive (Corollary 2.2.8) gives inequality (4.3.6).

Inequality (4.3.6) is the key to our analysis, as previously. Expanding the square on the right side of the inequality, we obtain

$$\begin{aligned}\frac{1}{2}\|x_{k+1} - x^\star\|_{H_k}^2 &\leqslant \frac{1}{2}\left\|x_k - H_k^{-1} g_k - x^\star\right\|_{H_k}^2 \\ &= \frac{1}{2}\|x_k - x^\star\|_{H_k}^2 - \langle g_k, x_k - x^\star\rangle + \frac{1}{2}\|g_k\|_{H_k^{-1}}^2,\end{aligned}$$

and taking expectations we have $\mathbb{E}[\langle g_k, x_k - x^\star\rangle \mid x_k] \geqslant f(x_k) - f(x^\star)$ by convexity and that $\mathbb{E}[g_k \mid x_k] \in \partial f(x_k)$. Thus

$$\frac{1}{2}\mathbb{E}\left[\|x_{k+1} - x^\star\|_{H_k}^2\right] \leqslant \mathbb{E}\left[\frac{1}{2}\|x_k - x^\star\|_{H_k}^2 - [f(x_k) - f(x^\star)] + \frac{1}{2}\|g_k\|_{H_k^{-1}}^2\right].$$

Rearranging, we have

$$\mathbb{E}[f(x_k) - f(x^\star)] \leqslant \mathbb{E}\left[\frac{1}{2}\|x_k - x^\star\|_{H_k}^2 - \frac{1}{2}\|x_{k+1} - x^\star\|_{H_k}^2 + \frac{1}{2}\|g_k\|_{H_k^{-1}}^2\right].$$

Summing this inequality from $k = 1$ to K gives the theorem. □

We may specialize the theorem in a number of ways to develop particular algorithms. One specialization, which is convenient because the computational overhead is fairly small, is to use diagonal matrices H_k. In particular, the AdaGrad method sets

$$H_k = \frac{1}{\alpha}\operatorname{diag}\left(\sum_{i=1}^{k} g_i g_i^T\right)^{\frac{1}{2}}, \tag{4.3.7}$$

where $\alpha > 0$ is a pre-specified constant (stepsize). In this case, the following corollary to Theorem 4.3.5 follows. Exercise B.4.3 sketches the proof of the corollary, which is similar to that of Corollary 4.3.3. In the corollary, recall that the trace of a matrix is $\operatorname{tr}(A) = \sum_{j=1}^{n} A_{jj}$.

Corollary 4.3.8 (AdaGrad convergence). *Let $R_\infty := \sup_{x \in C}\|x - x^\star\|_\infty$ be the ℓ_∞ radius of the set C and let the conditions of Theorem* 4.3.5 *hold. Then with the choice* (4.3.7) *in the variable metric method, we have*

$$\mathbb{E}\left[\sum_{k=1}^{K}(f(x_k) - f(x^\star))\right] \leqslant \frac{1}{2\alpha}R_\infty^2\mathbb{E}[\operatorname{tr}(H_K)] + \alpha\mathbb{E}[\operatorname{tr}(H_K)].$$

Inspecting Corollary 4.3.8 gives a few consequences. First, choosing $\alpha = R_\infty$, we obtain the expected convergence guarantee $\frac{3}{2}R_\infty \mathbb{E}[\mathrm{tr}(H_K)]$. If $\overline{x}_K = \frac{1}{K}\sum_{k=1}^K x_k$ as usual, and let $g_{k,j}$ denote the jth component of the kth gradient observed by the method, then we immediately obtain the convergence guarantee

$$\mathbb{E}\left[f(\overline{x}_K) - f(x^\star)\right] \leqslant \frac{3}{2K} R_\infty \mathbb{E}[\mathrm{tr}(H_K)] = \frac{3}{2K} R_\infty \sum_{j=1}^n \mathbb{E}\left[\left(\sum_{k=1}^K g_{k,j}^2\right)^{\frac{1}{2}}\right]. \tag{4.3.9}$$

In addition to proving the bound (4.3.9), Exercise B.4.3 also shows that when we take $C = \{x \in \mathbb{R}^n : \|x\|_\infty \leqslant 1\}$, the bound (4.3.9) is always better than the bounds (e.g. Corollary 3.4.11) guaranteed by standard stochastic gradient methods. In addition, the bound (4.3.9) is unimprovable—there are stochastic optimization problems for which no algorithm can achieve a faster convergence rate. These types of problems generally involve data in which the gradients g have highly varying components (or components that are often zero, i.e. the gradients g are sparse), as for such problems geometric aspects are quite important.

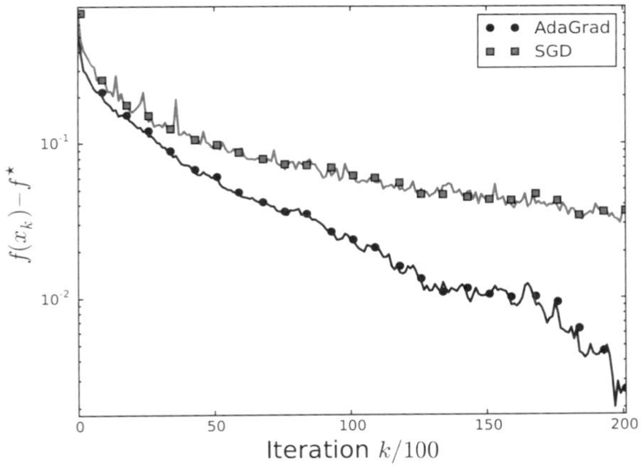

FIGURE 4.3.10. A comparison of the convergence of AdaGrad and SGD on the problem (4.3.11) for the best initial stepsize α for each method.

We now give an example application of the AdaGrad method, showing its performance on a simulated example. We consider solving the problem

$$\text{minimize } f(x) = \frac{1}{m}\sum_{i=1}^m \left[1 - b_i \langle a_i, x\rangle\right]_+ \text{ subject to } \|x\|_\infty \leqslant 1, \tag{4.3.11}$$

where the vectors $a_i \in \{-1, 0, 1\}^n$ with $m = 5000$ and $n = 1000$. This is the objective common to hinge loss (support vector machine) classification problems. For each coordinate $j \in \{1, \ldots, n\}$, we set $a_{i,j} \in \{\pm 1\}$ to have a random sign with probability $1/j$, and $a_{i,j} = 0$ otherwise. Letting $u \in \{-1, 1\}^n$ uniformly at random, we set $b_i = \mathrm{sign}(\langle a_i, u\rangle)$ with probability .95 and $b_i = -\mathrm{sign}(\langle a_i, u\rangle)$ otherwise. For this problem, the coordinates of a_i (and hence subgradients or stochastic subgradients of f) naturally have substantial variability, making it a natural problem for adaptation of the metric H_k.

In Figure 4.3.10, we show the convergence behavior of AdaGrad versus stochastic gradient descent (SGD) on one realization of this problem, where at each iteration we choose a stochastic gradient by selecting $i \in \{1, \ldots, m\}$ uniformly at random, then setting $g_k \in \partial\, [1 - b_i \langle a_i, x_k \rangle]_+$. For SGD, we use stepsizes $\alpha_k = \alpha/\sqrt{k}$, where α is the best stepsize of several choices (based on the eventual convergence of the method), while AdaGrad uses the matrix (4.3.7), with α similarly chosen based on the best eventual convergence. The plot shows the typical behavior of AdaGrad with respect to stochastic gradient methods, at least for problems with appropriate geometry: with good initial stepsize choice, AdaGrad often outperforms stochastic gradient descent. (We have been vague about the "right" geometry for problems in which we expect AdaGrad to perform well. Roughly, problems for which the domain C is well-approximated by a box $\{x \in \mathbb{R}^n : \|x\|_\infty \leqslant c\}$ are those for which we expect AdaGrad to succeed, and otherwise, it may exhibit worse performance than standard subgradient methods. As in any problem, some care is needed in the choice of methods.) Figure 4.3.12 shows this somewhat more broadly, plotting the convergence $f(x_k) - f(x^\star)$ versus iteration k for a number of initial stepsize choices for both stochastic gradient descent and AdaGrad on the problem (4.3.11). Roughly, we see that both methods are sensitive to initial stepsize choice, but the best choice for AdaGrad often outperforms the best choice for SGD.

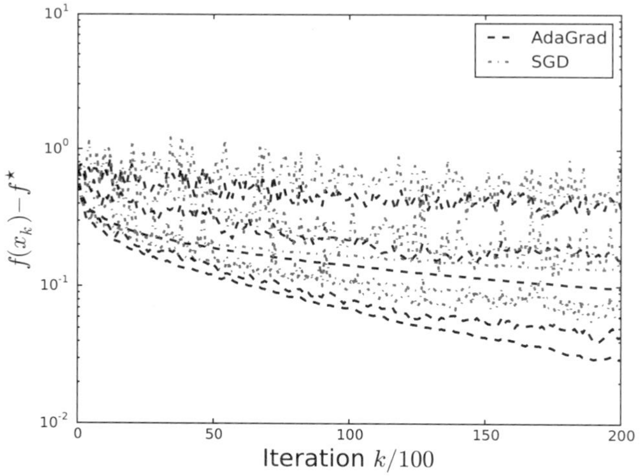

FIGURE 4.3.12. A comparison of the convergence of AdaGrad and SGD on the problem (4.3.11) for various initial stepsize choices $\alpha \in \{10^{-i/2}, i = -2, \ldots, 2\} = \{.1, .316, 1, 3.16, 10\}$. Both methods are sensitive to the initial stepsize choice α, though for each initial stepsize choice, AdaGrad has better convergence than the subgradient method.

Notes and further reading The mirror descent method was originally developed by Nemirovski and Yudin [41] in order to more carefully control the norms of gradients, and associated dual spaces, in first-order optimization methods. Since their original development, a number of researchers have explored variants and extensions of their methods. Beck and Teboulle [5] give an analysis of mirror descent as a non-Euclidean gradient method, which is the approach we take in this lecture. Nemirovski et al. [40] study mirror descent methods in stochastic settings, giving high-probability convergence guarantees similar to those we gave in the previous lecture. Bubeck and Cesa-Bianchi [15] explore the use of mirror descent methods in the context of *bandit* optimization problems, where instead of observing stochastic gradients one observes only random function values $f(x) + \varepsilon$, where ε is mean-zero noise perturbtion.

Variable metric methods have a similarly long history. Our simple results with stepsize selection follow the more advanced techniques of Auer et al. [3] (see especially their Lemma 3.5), and the AdaGrad method (and our development) is due to Duchi, Hazan, and Singer [22] and McMahan and Streeter [38]. More general metric methods include Shor's space dilation methods (of which the ellipsoid method is a celebrated special case), which develop matrices H_k that make new directions of descent somewhat less correlated with previous directions, allowing faster convergence in directions toward $x^\star$; see the books of Shor [55, 56] as well as the thesis of Nedić [39]. Newton methods, which we do not discuss, use scaled multiples of $\nabla^2 f(x_k)$ for H_k, while Quasi-Newton methods approximate $\nabla^2 f(x_k)$ with H_k while using only gradient-based information; for more on these and other more advanced methods for smooth optimization problems, see the books of Nocedal and Wright [46] and Boyd and Vandenberghe [12].

5. Optimality Guarantees

Lecture Summary: In this lecture, we provide a framework for demonstrating the optimality of a number of algorithms for solving stochastic optimization problems. In particular, we introduce minimax lower bounds, showing how techniques for reducing estimation problems to statistical testing problems allow us to prove lower bounds on optimization.

5.1. Introduction The procedures and algorithms we have presented thus far enjoy good performance on a number of statistical, machine learning, and stochastic optimization tasks, and we have provided theoretical guarantees on their performance. It is interesting to ask whether it is possible to improve the algorithms, or in what ways it may be possible to improve them. With that in mind, in this lecture we develop a number of tools for showing optimality—according to certain metrics—of optimization methods for stochastic problems.

Minimax rates We provide optimality guarantees in the *minimax* framework for optimality, which proceeds roughly as follows: we have a collection of possible problems and an error measure for the performance of a procedure, and we measure a procedure's performance by its behavior on the *hardest* (most difficult) member of the problem class. We then ask for the best procedure under this worst-case error measure. Let us describe this more formally in the context of our stochastic optimization problems, where the goal is to understand the difficulty of minimizing a convex function f subject to constraints $x \in C$ while observing only stochastic gradient (or other noisy) information about f. Our bounds build on three objects:

(i) A collection $\mathcal{F}$ of convex functions $f : \mathbb{R}^n \to \mathbb{R}$;
(ii) A closed convex set $C \subset \mathbb{R}^n$ over which we optimize;
(iii) A *stochastic gradient oracle*, which consists of a sample space $\mathcal{S}$, a gradient mapping

$$g : \mathbb{R}^n \times \mathcal{S} \times \mathcal{F} \to \mathbb{R}^n,$$

and (implicitly) a probability distributions P on $\mathcal{S}$. The stochastic gradient oracle may be *queried* at a point x, and when queried, draws $S \sim P$ with the property that

$$\mathbb{E}[g(x, S, f)] \in \partial f(x). \tag{5.1.1}$$

Depending on the scenario of the problem, the optimization procedure may be given access either to S or simply the value of the stochastic gradient $g = g(x, S, f)$, and the goal is to use the sequence of observations $g(x_k, S_k, f)$, for $k = 1, 2, \ldots$, to optimize f.

A simple example of the setting (i)–(iii) is as follows. Let $A \in \mathbb{R}^{n \times n}$ be a fixed positive definite matrix, and let $\mathcal{F}$ be the collection of convex functions of the form $f(x) = \frac{1}{2}x^\top Ax - b^\top x$ for all $b \in \mathbb{R}^n$. Then C may be any convex set, and—for the sake of proving lower bounds, not for real applicability in solving problems—we might take the stochastic gradient

$$g = \nabla f(x) + \xi = Ax - b + \xi \text{ for } \xi \stackrel{\text{iid}}{\sim} \mathsf{N}(0, I_{n \times n}).$$

A somewhat more complex example, but with more fidelity to real problems, comes from the stochastic programming problem (3.4.2) from Lecture 3 on subgradient methods. In this case, there is a known convex function $F : \mathbb{R}^n \times \mathcal{S} \to \mathbb{R}$, which is the instantaneous loss function $F(x; s)$. The problem is then to optimize

$$f_P(x) := \mathbb{E}_P[F(x; S)]$$

where the distribution P on the random variable S is unknown to the method *a priori*; there is then a correspondence between distributions P and functions $f \in \mathcal{F}$. Generally, an optimization is given access to a sample $S_1, \ldots, S_K$ drawn i.i.d. according to the distribution P (in this case, there is no selection of points x_i by the optimization procedure, as the sample $S_1, \ldots, S_K$ contains even more information than the stochastic gradients). A similar variant with a natural stochastic gradient oracle is to set $g(x, s, F) \in \partial F(x; s)$ instead of providing the sample $S = s$.

We focus in this note on the case when the optimization procedure may view only the sequence of subgradients $g_1, g_2, \ldots$ at the points it queries. We note in passing, however, that for many problems we can reconstruct S from a gradient $g \in \partial F(x; S)$. As an example, consider a logistic regression problem with data $s = (a, b) \in \{0, 1\}^n \times \{-1, 1\}$, a typical data case. Then

$$F(x; s) = \log(1 + e^{-b\langle a, x\rangle}), \text{ and } \nabla_x F(x; s) = -\frac{1}{1 + e^{b\langle a, x\rangle}} ba,$$

so that (a, b) is identifiable from any $g \in \partial F(x; s)$. More generally, classical linear models in statistics have gradients that are scaled multiples of the data, so that the sample s is typically identifiable from $g \in \partial F(x; s)$.

Now, given function f and stochastic gradient oracle g, an optimization procedure chooses query points $x_1, x_2, \ldots, x_K$ and observes stochastic subgradients g_k with $\mathbb{E}[g_k] \in \partial f(x_k)$. Based on these stochastic gradients, the optimization procedure outputs $\widehat{x}_K$, and we assess the quality of the procedure in terms of the excess loss

$$\mathbb{E}\left[f\left(\widehat{x}_K(g_1, \ldots, g_K)\right) - \inf_{x^\star \in C} f(x^\star)\right],$$

where the expectation is taken over the subgradients $g(x_i, S_i, f)$ returned by the stochastic oracle and any randomness in the chosen iterates, or query points, $x_1, \ldots, x_K$ of the optimization method. Of course, if we only consider this excess objective value for a fixed function f, then a trivial optimization procedure achieves excess risk 0: simply return some $x \in \operatorname{argmin}_{x \in C} f(x)$. It is thus important to ask for a more uniform notion of risk: we would like the procedure to have good performance *uniformly* across all functions $f \in \mathcal{F}$, leading us to measure the performance of a procedure by its worst-case risk

$$\sup_{f \in \mathcal{F}} \mathbb{E}\left[f(\widehat{x}(g_1, \ldots, g_k)) - \inf_{x \in C} f(x^\star)\right],$$

where the supremum is taken over functions $f \in \mathcal{F}$ (the subgradient oracle g then implicitly depends on f). An optimal estimator for this metric then gives the *minimax risk* for optimizing the family of stochastic optimization problems $\{f\}_{f \in \mathcal{F}}$ over $x \in C \subset \mathbb{R}^n$, which is

$$\mathfrak{M}_K(C, \mathcal{F}) := \inf_{\widehat{x}_K} \sup_{f \in \mathcal{F}} \mathbb{E}\left[f(\widehat{x}_K(g_1, \ldots, g_K)) - \inf_{x^\star \in C} f(x^\star)\right]. \tag{5.1.2}$$

We take the supremum (worst-case) over distributions $f \in \mathcal{F}$ and the infimum over all possible optimization schemes $\widehat{x}_K$ using K stochastic gradient samples.

A criticism of the framework (5.1.2) is that it is too pessimistic: by taking a worst-case over distributions of functions $f \in \mathcal{F}$, one is making the family of problems too challenging. We will not address these challenges except to say that one response is to develop *adaptive* procedures $\widehat{x}$, which are simultaneously optimal for a variety of collections of problems $\mathcal{F}$.

The basic approach There are a variety of techniques giving lower bounds on the minimax risk (5.1.2). Each of them transforms the maximum risk by lower bounding it via a Bayesian problem (e.g. [31,33,34]), then proving a lower bound on the performance of all possible estimators for the Bayesian problem. In particular, let $\{f_v\} \subset \mathcal{F}$ be a collection of functions in $\mathcal{F}$ indexed by some (finite or countable) set $\mathcal{V}$ and π be any probability mass function over $\mathcal{V}$. Let $f^\star = \inf_{x \in C} f(x)$. Then for any procedure $\widehat{x}$, the maximum risk has lower bound

$$\sup_{f \in \mathcal{F}} \mathbb{E}\left[f(\widehat{x}) - f^\star\right] \geqslant \sum_v \pi(v) \mathbb{E}\left[f_v(\widehat{x}) - f_v^\star\right].$$

While trivial, this lower bound serves as the departure point for each of the subsequent techniques for lower bounding the minimax risk. The lower bound also allows us to assume that the procedure $\widehat{x}$ is *deterministic*. Indeed, assume that $\widehat{x}$ is non-deterministic, which we can represent generally as depending on some auxiliary random variable U independent of the observed subgradients. Then we certainly have

$$\mathbb{E}\left[\sum_v \pi(v) \mathbb{E}\left[f_v(\widehat{x}) - f_v^\star \mid U\right]\right] \geqslant \inf_u \sum_v \pi(v) \mathbb{E}\left[f_v(\widehat{x}) - f_v^\star \mid U = u\right],$$

that is, there is some realization of the auxiliary randomness that is at least as good as the average realization. We can simply incorporate this into our minimax optimal procedures $\widehat{x}$, and thus we assume from this point onward that all our optimization procedures are deterministic when proving our lower bounds.

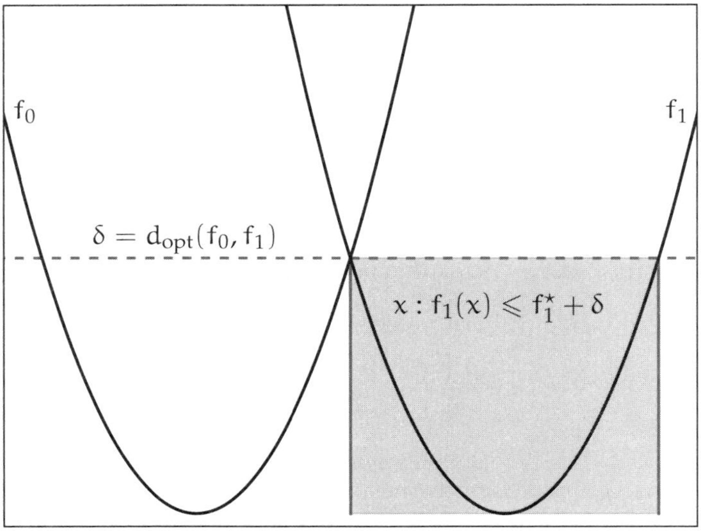

FIGURE 5.1.3. Separation of optimizers of f_0 and f_1. Optimizing one function to accuracy better than $\delta = d_{\mathrm{opt}}(f_0, f_1)$ implies we optimize the other poorly; the gap $f(x) - f^\star$ is at least δ.

The second step in proving minimax bounds is to reduce the optimization problem to a type of statistical test [58, 62, 63]. To perform this reduction, we define a distance-like quantity between functions such that, if we have optimized a function f_v to better than the distance, we cannot have optimized other functions well. In particular, consider two convex functions f_0 and f_1. Let $f_v^\star = \inf_{x \in C} f_v(x)$ for $v \in \{0, 1\}$. We let the *optimization separation* between functions f_0 and f_1 over the set C be

$$d_{\text{opt}}(f_0, f_1; C) :=$$

$$\sup \left\{ \delta \geqslant 0 : \begin{array}{l} f_1(x) \leqslant f_1^\star + \delta \text{ implies } f_0(x) \geqslant f_0^\star + \delta \\ f_0(x) \leqslant f_0^\star + \delta \text{ implies } f_1(x) \geqslant f_1^\star + \delta \end{array} \quad \text{for any } x \in C \right\}. \tag{5.1.4}$$

That is, if we have any point x such that $f_v(x) - f_v^\star \leqslant d_{\text{opt}}(f_0, f_1)$, then x cannot optimize f_{1-v} well, i.e. we can only optimize one of the two functions f_0 and f_1 to accuracy $d_{\text{opt}}(f_0, f_1)$. See Figure 5.1.3 for an illustration of this quantity. For example, if $f_1(x) = (x + c)^2$ and $f_0(x) = (x - c)^2$ for a constant $c \neq 0$, then we have $d_{\text{opt}}(f_1, f_0) = c^2$.

This separation d_{opt} allows us to give a reduction from optimization to testing via the *canonical hypothesis testing problem*, which is as defined as follows:

1. Nature chooses an index $V \in \mathcal{V}$ uniformly at random.
2. Conditional on the choice $V = v$, the procedure observes stochastic subgradients for the function f_v according to the oracle $g(x_k, S_k, f_v)$ for i.i.d. S_k.

Then, given the observed subgradients, the goal is to test which of the random indices v nature chose. Intuitively, if we can optimize f_v well—to better than the separation $d_{\text{opt}}(f_v, f_{v'})$—then we can identify the index v. If we can show this, then we can adapt classical statistical results on optimal hypothesis testing to lower bound the probability of error in testing whether the data was generated conditional on $V = v$.

More formally, we have the following key lower bound. In the lower bound, we say that a collection of functions $\{f_v\}_{v \in \mathcal{V}}$ is *δ-separated*, where $\delta \geqslant 0$, if

$$d_{\text{opt}}(f_v, f_{v'}; C) \geqslant \delta \text{ for each } v, v' \in \mathcal{V} \text{ with } v \neq v'. \tag{5.1.5}$$

Then we have the next proposition.

Proposition 5.1.6. *Let S be drawn uniformly from $\mathcal{V}$, where $|\mathcal{V}| < \infty$, and assume the collection $\{f_v\}_{v \in \mathcal{V}}$ is δ-separated. Then for any optimization procedure $\widehat{x}$ based on the observed subgradients,*

$$\frac{1}{|\mathcal{V}|} \sum_{v \in \mathcal{V}} \mathbb{E}[f_v(\widehat{x}) - f_v^\star] \geqslant \delta \cdot \inf_{\widehat{v}} \mathbb{P}(\widehat{v} \neq V),$$

where the distribution $\mathbb{P}$ is the joint distribution over the random index V and the observed gradients $g_1, \ldots, g_K$ and the infimum is taken over all testing procedures $\widehat{v}$ based on the observed data.

Proof. We let P_v denote the distribution of the subgradients conditional on the choice $V = v$, meaning that $\mathbb{E}[g_k \mid V = v] \in \partial f_v(x_k)$. We observe that for any v, we have

$$\mathbb{E}[f_v(\widehat{x}) - f_v^\star] \geqslant \delta \mathbb{E}[\mathbf{1}\{f_v(\widehat{x}) \geqslant f_v^\star + \delta\}] = \delta P_v(f_v(\widehat{x}) \geqslant f_v^\star + \delta).$$

Now, define the hypothesis test $\widehat{v}$, which is a function of $\widehat{x}$, by

$$\widehat{v} = \begin{cases} v & \text{if } f_v(\widehat{x}) \leqslant f_v^\star + \delta \\ \text{arbitrary in } \mathcal{V} & \text{otherwise.} \end{cases}$$

This is a well-defined mapping, as by the condition that $d_{\text{opt}}(f_v, f_{v'}) \geqslant \delta$, there can be only a *single* index v such that $f_v(x) \leqslant f_v^\star + \delta$. We then note the following implication:

$$\widehat{v} \neq v \text{ implies } f_v(\widehat{x}) \geqslant f_v^\star + \delta.$$

Thus we have

$$P_v(\widehat{v} \neq v) \leqslant P_v(f_v(\widehat{x}) \geqslant f_v^\star + \delta),$$

or, summarizing, we have

$$\frac{1}{|\mathcal{V}|} \sum_{v \in \mathcal{V}} \mathbb{E}[f_v(\widehat{x}) - f_v^\star] \geqslant \delta \cdot \frac{1}{|\mathcal{V}|} \sum_{v \in \mathcal{V}} P_v(\widehat{v} \neq v).$$

But by definition of the distribution $\mathbb{P}$, we have $\frac{1}{|\mathcal{V}|} \sum_{v \in \mathcal{V}} P_v(\widehat{v} \neq v) = \mathbb{P}(\widehat{v} \neq V)$, and taking the best possible test $\widehat{v}$ gives the result of the proposition. □

Proposition 5.1.6 allows us to then bring in the tools of optimal testing in statistics and information theory, which we can use to prove lower bounds. To leverage Proposition 5.1.6, we follow a two phase strategy: we construct a well-separated function collection, and then we show that it is difficult to test which of the functions we observe data from. There is a natural tension in the proposition, as it is easier to distinguish functions that are far apart (i.e. large δ), while hard-to-distinguish functions (i.e. large $\mathbb{P}(\widehat{v} \neq V)$) often have smaller separation. Thus we trade these against one another carefully in constructing our lower bounds on the minimax risk. We also present a variant lower bound in Section 5.3 based on a similar reduction, except that we use multiple binary hypothesis tests.

5.2. Le Cam's Method Our first set of lower bounds is based on Le Cam's method [33], which uses optimality guarantees for simple binary hypothesis tests to provide lower bounds for optimization problems. That is, we let $\mathcal{V} = \{-1, 1\}$ and will construct only pairs of functions and distributions P_1, P_{-1} generating data. In this section, we show how to use these binary hypothesis tests to prove lower bounds on the family of stochastic optimization problems characterized by the following conditions: the domain $C \subset \mathbb{R}^n$ contains an ℓ_2-ball of radius R and the subgradients g_k satisfy the second moment bound

$$\mathbb{E}[\|g_k\|_2^2] \leqslant M^2$$

for all k. We assume that $\mathcal{F}$ consists of M-Lipschitz continuous convex functions.

With the definition (5.1.4) of the separation in terms of optimization value, we can provide a lower bound on optimization in terms of distances between distributions P_1 and P_{-1}. Before we continue, we require a few definitions about distances between distributions.

Definition 5.2.1. Let P and Q be distributions on a space $\mathcal{S}$, and assume that they are both absolutely continuous with respect to a measure μ on $\mathcal{S}$. The *variation distance* between P and Q is

$$\|P - Q\|_{TV} := \sup_{A \subset \mathcal{S}} |P(A) - Q(A)| = \frac{1}{2} \int_{\mathcal{S}} |p(s) - q(s)| d\mu(s).$$

The *Kullback-Leibler divergence* between P and Q is

$$D_{kl}(P|Q) := \int_{\mathcal{S}} p(s) \log \frac{p(s)}{q(s)} d\mu(s).$$

We can connect the variation distance to binary hypothesis tests via the following lemma, due to Le Cam. The lemma states that testing between two distributions is hard precisely when they are close in variation distance.

Lemma 5.2.2. *Let P_1 and P_{-1} be any distributions. Then*

$$\inf_{\widehat{v}} \{P_1(\widehat{v} \neq 1) + P_{-1}(\widehat{v} \neq -1)\} = 1 - \|P_1 - P_{-1}\|_{TV}.$$

Proof. Any testing procedure $\widehat{v} : \mathcal{S} \to \{-1, 1\}$ maps one region of the sample space, call it A, to 1 and the complement A^c to -1. Thus, we have

$$P_1(\widehat{v} \neq 1) + P_{-1}(\widehat{v} \neq -1) = P_1(A^c) + P_{-1}(A) = 1 - P_1(A) + P_{-1}(A).$$

Optimizing over $\widehat{v}$ is then equivalent to optimizing over sets A, yielding

$$\begin{aligned}\inf_{\widehat{v}}\{P_1(\widehat{v} \neq 1) + P_{-1}(\widehat{v} \neq -1)\} &= \inf_A\{1 - P_1(A) + P_{-1}(A)\} \\ &= 1 - \sup_A\{P_1(A) - P_{-1}(A)\} = 1 - \|P_1 - P_{-1}\|_{TV}\end{aligned}$$

as desired. □

As an immediate consequence of Lemma 5.2.2, we obtain the standard minimax lower bound based on binary hypothesis testing. In particular, let f_1 and f_{-1} be δ-separated and belong to $\mathcal{F}$, and assume that the method $\widehat{x}$ receives data (in this case, the data is the K subgradients) from P_v^K when f_v is the true function. Then we immediately have

$$\text{(5.2.3)} \quad \mathfrak{M}_K(C, \mathcal{F}) \geqslant \inf_{\widehat{x}_K} \max_{v \in \{-1,1\}} \{\mathbb{E}_{P_v}[f_v(\widehat{x}_K) - f_v^\star]\} \geqslant \frac{1}{2}\delta \cdot \left[1 - \left\|P_1^K - P_{-1}^K\right\|_{TV}\right].$$

Inequality (5.2.3) gives a quantitative guarantee on an intuitive fact: if we observe data from one of two distributions P_1 and P_{-1} that are close, while the optimizers of the functions f_1 and f_{-1} associated with P_1 and P_{-1} differ, it is difficult to optimize well. Moreover, there is a natural tradeoff—the farther apart the functions f_1 and f_{-1} are (i.e. $\delta = d_{opt}(f_1, f_{-1})$ is large), the bigger the penalty for optimizing one well, but conversely, this usually forces the distributions P_1 and

P_{-1} to be quite different, as they provide subgradient information on f_1 and f_{-1}, respectively.

It is challenging to compute quantities—especially with multiple samples—involving the variation distance, so we now convert our bounds to ones involving the KL-divergence, which is computationally easier when dealing with multiple samples. First, we use Pinsker's inequality (see Appendix A.3, Proposition A.3.2 for a proof): for any distributions P and Q,

$$\|P - Q\|_{TV}^2 \leqslant \frac{1}{2} D_{kl}(P|Q).$$

As we see presently, the KL-divergence *tensorizes* when we have multiple observations from different distributions (see Lemma 5.2.8 to come), allowing substantially easier computation of individual divergence terms. Then we have the following theorem.

Theorem 5.2.4. *Let $\mathcal{F}$ be a collection of convex functions, and let $f_1, f_{-1} \in \mathcal{F}$. Assume that when function f_ν is to be optimized, we observe K subgradients according to P_ν^K. Then*

$$\mathfrak{M}_K(C, \mathcal{P}) \geqslant \frac{d_{opt}(f_{-1}, f_1; C)}{2}\left[1 - \sqrt{\frac{1}{2} D_{kl}\left(P_1^K | P_{-1}^K\right)}\right].$$

What remains to give a concrete lower bound, then, is (1) to construct a family of well-separated functions f_1, f_{-1}, and (2) to construct a stochastic gradient oracle for which we give a small *upper* bound on the KL-divergence between the distributions P_1 and P_{-1} associated with the functions, which means that testing between P_1 and P_{-1} is hard.

Constructing well-separated functions Our first goal is to construct a family of well-separated functions and an associated first-order subgradient oracle that makes the functions hard to distinguish. We parameterize our functions—of which we construct only 2—by a parameter $\delta > 0$ governing their separation. Our construction applies in dimension $n = 1$: let us assume that C contains the interval $[-R, R]$ (this is no loss of generality, as we may simply shift the interval). Then define the M-Lipschitz continuous functions

$$f_1(x) = M\delta|x - R| \text{ and } f_{-1}(x) = M\delta|x + R|. \tag{5.2.5}$$

See Figure 5.2.7 for an example of these functions, which makes clear that their separation (5.1.4) is

$$d_{opt}(f_1, f_{-1}) = \delta M R.$$

We also consider the stochastic oracle for this problem, recalling that we must construct subgradients satisfying $\mathbb{E}[\|g\|_2^2] \leqslant M^2$. We will do slightly more: we will guarantee that $|g| \leqslant M$ always. With this in mind, we assume that $\delta \leqslant 1$, and define the stochastic gradient oracle for the distribution P_ν, $\nu \in \{-1, 1\}$ at the point x to be

$$g_\nu(x) = \begin{cases} M\,\mathrm{sign}(x-\nu R) & \text{with probability } \frac{1+\delta}{2} \\ -M\,\mathrm{sign}(x-\nu R) & \text{with probability } \frac{1-\delta}{2}. \end{cases} \tag{5.2.6}$$

At $x = \nu R$ the oracle simply returns a random sign. Then by inspection, we see that

$$\mathbb{E}[g_\nu(x)] = \frac{M\delta}{2}\,\mathrm{sign}(x-\nu R) - \frac{M\delta}{2}(-\,\mathrm{sign}(x-\nu R)) = M\delta\,\mathrm{sign}(x-\nu R) \in \partial f_\nu(x)$$

for $\nu = -1, 1$. Thus, the combination of the functions (5.2.5) and the stochastic gradient (5.2.6) give us a valid subgradient and well-separated pair of functions.

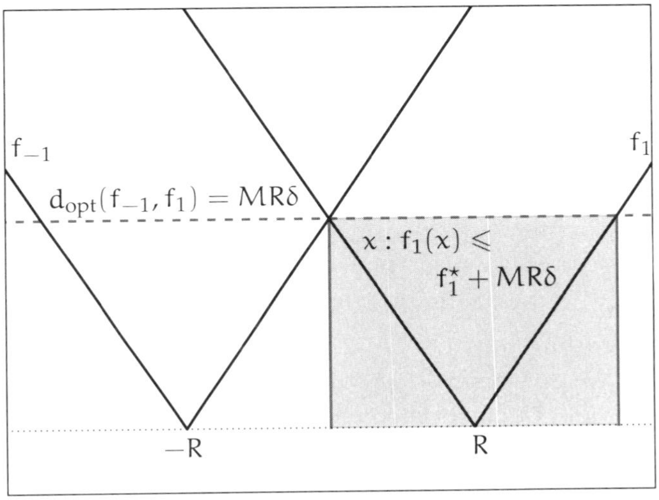

FIGURE 5.2.7. The function construction (5.2.5) with separation $d_{\mathrm{opt}}(f_1, f_{-1}) = MR\delta$.

Bounding the distance between distributions The second step in proving our minimax lower bound is to upper bound the distance between the distributions that generate the subgradients our methods observe. This means that testing which of the functions we are optimizing is challenging, giving us a strong lower bound. At a high level, building off of Theorem 5.2.4, we hope to show an upper bound of the form

$$D_{\mathrm{kl}}\left(P_1^K | P_{-1}^K\right) \leqslant \kappa\delta^2$$

for some κ. This is a local condition, allowing us to scale our problems with δ to achieve minimax bounds. If we have such a quadratic, we may simply choose $\delta^2 = 1/2\kappa$, giving the constant probability of error

$$1 - \left\|P_1^K - P_{-1}^K\right\|_{\mathrm{TV}} \geqslant 1 - \sqrt{\frac{1}{2} D_{\mathrm{kl}}\left(P_1^K | P_{-1}^K\right)/2} \geqslant 1 - \sqrt{\frac{\kappa\delta^2}{2}} \geqslant \frac{1}{2}.$$

To this end, we begin with a standard lemma (the chain rule for KL divergence), which applies when we have K potentially dependent observations from a distribution. The result is an immediate consequence of Bayes' rule.

Lemma 5.2.8. *Let $P(\cdot \mid g_1, \ldots, g_{k-1})$ denote the conditional distribution of g_k given $g_1, \ldots, g_{k-1}$. For each $k \in \mathbb{N}$ let P_1^k and P_{-1}^k be distributions on the K subgradients $g_1, \ldots, g_k$. Then*

$$D_{kl}\left(P_1^K | P_{-1}^K\right) = \sum_{k=1}^{K} \mathbb{E}_{P_1^{k-1}}\left[D_{kl}\left(P_1(\cdot \mid g_1, \ldots, g_{k-1}) | P_{-1}(\cdot \mid g_1, \ldots, g_{k-1})\right)\right].$$

Using Lemma 5.2.8, we have the following upper bound on the KL-divergence between P_1^K and P_{-1}^K for the stochastic gradient (5.2.6).

Lemma 5.2.9. *Let the K observations under distribution P_v come from the stochastic gradient oracle* (5.2.6). *Then for $\delta \leqslant \frac{4}{5}$,*

$$D_{kl}\left(P_1^K | P_{-1}^K\right) \leqslant 3K\delta^2.$$

Proof. We use the chain-rule for KL-divergence, whence we must only provide an upper bound on the individual terms. We first note that x_k is a function of $g_1, \ldots, g_{k-1}$ (because we may assume w.l.o.g. that x_k is deterministic) so that $P_v(\cdot \mid g_1, \ldots, g_{k-1})$ is the distribution of a Bernoulli random variable with distribution (5.2.6), i.e. with probabilities $\frac{1 \pm \delta}{2}$. Thus we have

$$\begin{aligned} D_{kl}\left(P_1(\cdot \mid g_1, \ldots, g_{k-1}) | P_{-1}(\cdot \mid g_1, \ldots, g_{k-1})\right) &\leqslant D_{kl}\left(\frac{1+\delta}{2} \Big| \frac{1-\delta}{2}\right) \\ &= \frac{1+\delta}{2} \log \frac{1+\delta}{1-\delta} + \frac{1-\delta}{2} \log \frac{1-\delta}{1+\delta} \\ &= \delta \log \frac{1+\delta}{1-\delta}. \end{aligned}$$

By a Taylor expansion, we have that

$$\delta \log \frac{1+\delta}{1-\delta} = \delta\left(\delta - \frac{1}{2}\delta^2 + O(\delta^3)\right) - \delta\left(-\delta - \frac{1}{2}\delta^2 + O(\delta^3)\right) = 2\delta^2 + O(\delta^4) \leqslant 3\delta^2$$

for $\delta \leqslant \frac{4}{5}$, or

$$D_{kl}\left(P_1(\cdot \mid g_1, \ldots, g_{k-1}) | P_{-1}(\cdot \mid g_1, \ldots, g_{k-1})\right) \leqslant 3\delta^2$$

for $\delta \leqslant \frac{4}{5}$. Summing over k completes the proof. □

Putting it all together: a minimax lower bound With Lemma 5.2.9 in place along with our construction (5.2.5) of well-separated functions, we can now give the best possible convergence guarantees for a broad family of problems.

Theorem 5.2.10. *Let $C \subset \mathbb{R}^n$ be a convex set containing an ℓ_2 ball of radius R, and let $\mathcal{P}$ denote the collection of distributions generating stochastic subgradients with $\|g\|_2 \leqslant M$ with probability 1. Then, for all $K \in \mathbb{N}$,*

$$\mathfrak{M}_K(C, \mathcal{P}) \geqslant \frac{RM}{4\sqrt{6}\sqrt{K}}$$

Proof. We combine Le Cam's method, Lemma 5.2.2 (and the subsequent Theorem 5.2.4) with our construction (5.2.5) and their stochastic subgradients (5.2.6). Certainly, the class of n-dimensional optimization problems is at least as challenging as a 1-dimensional problem (we may always restrict our functions to depend only on a single coordinate), so that for any $\delta \geqslant 0$ we have

$$\mathfrak{M}_K(C, \mathcal{F}) \geqslant \frac{\delta M R}{2}\left(1 - \sqrt{\frac{1}{2} D_{kl}\left(P_1^K | P_{-1}^K\right)}\right).$$

Next, we use Lemma 5.2.9, which guarantees the further lower bound

$$\mathfrak{M}_K(C, \mathcal{F}) \geqslant \frac{\delta M R}{2}\left(1 - \sqrt{\frac{3K\delta^2}{2}}\right),$$

valid for all $\delta \leqslant \frac{4}{5}$.

Finally, choosing $\delta^2 = \frac{1}{6K} < \frac{4}{5}$, we have that $D_{kl}\left(P_1^K | P_{-1}^K\right) \leqslant \frac{1}{2}$, and

$$\mathfrak{M}_K(C, \mathcal{F}) \geqslant \frac{\delta M R}{4}.$$

Substituting our choice of δ into this expression gives the theorem. □

In short, Theorem 5.2.10 gives a guarantee that matches the upper bounds of the previous lectures to within a numerical constant factor of 10. A more careful inspection of our analysis allows us to prove a lower bound, at least as $K \to \infty$, of $1/8\sqrt{K}$. In particular, by using Theorem 3.4.9 of our lecture on subgradient methods, we find that if the set C contains an ℓ_2-ball of radius R_{inner} and is contained in an ℓ_2-ball of radius R_{outer}, we have

$$\frac{1}{\sqrt{96}} \frac{M R_{\text{inner}}}{\sqrt{K}} \leqslant \mathfrak{M}_K(C, \mathcal{F}) \leqslant \frac{M R_{\text{outer}}}{\sqrt{K}} \tag{5.2.11}$$

for all $K \in \mathbb{N}$, where the upper bound is attained by the stochastic projected subgradient method.

5.3. Multiple dimensions and Assouad's Method The results in Section 5.2 provide guarantees for problems where we can embed much of the difficulty of our family $\mathcal{F}$ in optimizing a pair of only two functions—something reminiscent of problems in classical statistics on the "hardest one-dimensional subproblem" (see, for example, the work of Donoho, Liu, and MacGibbon [19]). In many stochastic optimization problems, the higher-dimension n yields increased difficulty, so that we would like to derive bounds that incorporate dimension more directly. With that in mind, we develop a family of lower bounds, based on Assouad's method [2], that reduce optimization to a collection of binary hypothesis tests, one for each of the n dimensions of the problem.

More precisely, we let $\mathcal{V} = \{-1, 1\}^n$ be the n-dimensional binary hypercube, and for each $v \in \mathcal{V}$, we assume we have a function $f_v \in \mathcal{F}$ where $f_v : \mathbb{R}^n \to \mathbb{R}$. Without loss of generality, we will assume that our constraint set C has the point 0 in its interior. Let $\delta \in \mathbb{R}_+^n$ be an n-dimensional nonnegative vector. Then we say that the functions $\{f_v\}$ induce a *δ-separation in the Hamming metric* if for any

$x \in C \subset \mathbb{R}^n$ we have

$$f_v(x) - f_v^\star \geqslant \sum_{j=1}^{n} \delta_j \mathbf{1}\left\{\mathrm{sign}(x_j) \neq v_j\right\}, \tag{5.3.1}$$

where the subscript j denotes the jth coordinate. For example, if we define the function $f_v(x) = \delta \left\|x - v\right\|_1$ for each $v \in \mathcal{V}$, then certainly $\{f_v\}$ is $\delta\mathbf{1}$-separated in the Hamming metric; more generally, $f_v(x) = \sum_{j=1}^n \delta_j |x_j - v_j|$ is δ-separated. With this definition, we have the following lemma, providing a lower bound for functions $f : \mathbb{R}^n \to \mathbb{R}$.

Lemma 5.3.2 (Generalized Assouad). *Let $\delta \in \mathbb{R}^n_+$ and let $\{f_v\}$ be δ-separated in Hamming metric where $v \in \mathcal{V} = \{-1, 1\}^n$. Let $\widehat{x}$ be any optimization algorithm, and let P_v be the distribution of (all) the subgradients $g_1, \ldots, g_K$ the procedure $\widehat{x}$ observes when optimizing f_v. Define*

$$P_{+j} = \frac{1}{2^{n-1}} \sum_{v : v_j = 1} P_v \text{ and } P_{-j} = \frac{1}{2^{n-1}} \sum_{v : v_j = -1} P_v.$$

Then

$$\frac{1}{2^n} \sum_{v \in \{-1,1\}^n} \mathbb{E}[f_v(\widehat{x}) - f_v^\star] \geqslant \frac{1}{2} \sum_{j=1}^{n} \delta_j \left(1 - \left\|P_{+j} - P_{-j}\right\|_{\mathrm{TV}}\right).$$

Proof. By using the separation condition, we immediately see that

$$\mathbb{E}[f_v(\widehat{x}) - f_v^\star] \geqslant \sum_{j=1}^{d} \delta_j P_v(\mathrm{sign}(\widehat{x}_j) \neq v_j)$$

for any $v \in \mathcal{V}$. Averaging over the vectors $v \in \mathcal{V}$, we obtain

$$\begin{aligned}
&\frac{1}{2^n} \sum_{v \in \mathcal{V}} \mathbb{E}[f_v(\widehat{x}) - f_v^\star] \\
&\geqslant \sum_{j=1}^{d} \frac{1}{|\mathcal{V}|} \sum_{v \in \mathcal{V}} \delta_j P_v(\mathrm{sign}(\widehat{x}_j) \neq v_j) \\
&= \sum_{j=1}^{d} \delta_j \frac{1}{|\mathcal{V}|} \left[\sum_{v : v_j = 1} P_v(\mathrm{sign}(\widehat{x}_j) \neq 1) + \sum_{v : v_j = -1} P_v(\mathrm{sign}(\widehat{x}_j) \neq -1) \right] \\
&= \sum_{j=1}^{d} \frac{\delta_j}{2} \left[P_{+j}(\mathrm{sign}(\widehat{x}_j) \neq 1) + P_{-j}(\mathrm{sign}(\widehat{x}_j) \neq -1) \right].
\end{aligned}$$

Now we use Le Cam's lemma (Lemma 5.2.2) on optimal binary hypothesis tests to see that

$$P_{+j}(\mathrm{sign}(\widehat{x}_j) \neq 1) + P_{-j}(\mathrm{sign}(\widehat{x}_j) \neq -1) \geqslant 1 - \left\|P_{+j} - P_{-j}\right\|_{\mathrm{TV}}$$

which gives the desired result. □

As a nearly immediate consequence of Lemma 5.3.2, we see that if the separation is a constant $\delta > 0$ for each coordinate, we have the following lower bound on the minimax risk.

Proposition 5.3.3. *Let the collection $\{f_v\}_{v \in \mathcal{V}} \subset \mathcal{F}$, where $\mathcal{V} = \{-1, 1\}^n$, be δ-separated in Hamming metric for some $\delta \in \mathbb{R}_+$, and let the conditions of Lemma* 5.3.2 *hold. Then*

$$\mathfrak{M}_K(C, \mathcal{F}) \geqslant \frac{n}{2}\delta\left(1 - \sqrt{\frac{1}{2n}\sum_{j=1}^{n} D_{kl}\left(P_{+j} | P_{-j}\right)}\right).$$

Proof. Lemma 5.3.2 guarantees that

$$\mathfrak{M}_K(C, \mathcal{F}) \geqslant \frac{\delta}{2}\sum_{j=1}^{n}(1 - \|P_{+j} - P_{-j}\|_{TV}).$$

Applying the Cauchy-Schwarz inequality, we have by Pinsker's inequality

$$\sum_{j=1}^{n} \|P_{+j} - P_{-j}\|_{TV} \leqslant \sqrt{n\sum_{j=1}^{n} \|P_{+j} - P_{-j}\|_{TV}^2} \leqslant \sqrt{\frac{n}{2}\sum_{j=1}^{n} D_{kl}\left(P_{+j} | P_{-j}\right)}$$

Substituting this into the previous bound gives the desired result. □

With this proposition, we can give a number of minimax lower bounds. We focus on two concrete cases, which show that the stochastic gradient procedures we have developed are optimal for a variety of problems. We give one result, deferring others to the exercises associated with the lecture notes. For our main result using Assouad's method, we consider optimization problems for which the set $C \subset \mathbb{R}^n$ contains an ℓ_∞ ball of radius R. We also assume that the stochastic gradient oracle satisfies the ℓ_1-bound condition

$$\mathbb{E}[\|g(x, S, f)\|_1^2] \leqslant M^2.$$

This means that all the functions $f \in \mathcal{F}$ are M-Lipschitz continuous with respect to the ℓ_∞-norm, that is, $|f(x) - f(y)| \leqslant M \|x - y\|_\infty$.

Theorem 5.3.4. *Let $\mathcal{F}$ and the stochastic gradient oracle be as above, and assume that $C \supset [-R, R]^n$. Then*

$$\mathfrak{M}_K(C, \mathcal{F}) \geqslant RM \min\left\{\frac{1}{5}, \frac{1}{\sqrt{96}}\frac{\sqrt{n}}{\sqrt{K}}\right\}.$$

Proof. Our proof is similar to our construction of our earlier lower bounds, except that now we must construct functions defined on $\mathbb{R}^n$ so that our minimax lower bound on convergence rate grows with the dimension. Let $\delta > 0$ be fixed for now. For each $v \in \mathcal{V} = \{-1, 1\}^n$, define the function

$$f_v(x) := \frac{M\delta}{n}\|x - Rv\|_1.$$

Then by inspection, the collection $\{f_v\}$ is $\frac{MR\delta}{n}$-separated in Hamming metric, as

$$f_v(x) = \frac{M\delta}{n}\sum_{j=1}^{n}|x_j - Rv_j| \geqslant \frac{M\delta}{n}\sum_{j=1}^{n} R\mathbf{1}\left\{\text{sign}(x_j) \neq v_j\right\}.$$

Now, we must (as before) construct a stochastic subgradient oracle. Let $e_1, \ldots, e_n$ be the n standard basis vectors. For each $v \in \mathcal{V}$, we define the stochastic subgradient as

$$g(x, f_v) = \begin{cases} M e_j \operatorname{sign}(x_j - R v_j) & \text{with probability } \frac{1+\delta}{2n} \\ -M e_j \operatorname{sign}(x_j - R v_j) & \text{with probability } \frac{1-\delta}{2n}. \end{cases} \tag{5.3.5}$$

That is, the oracle randomly chooses a coordinate $j \in \{1, \ldots, n\}$, then conditional on this choice, flips a biased coin and with probability $\frac{1+\delta}{2}$ returns the correctly signed jth coordinate of the subgradient, $M e_j \operatorname{sign}(x_j - R v_j)$, and otherwise returns the negative. Letting $\operatorname{sign}(x)$ denote the vector of signs of x, we then have

$$\mathbb{E}[g(x, f_v)] = M \sum_{j=1}^{n} e_j \left[\frac{1+\delta}{n} - \frac{1-\delta}{n}\right] \operatorname{sign}(x_j - R v_j) = \frac{M\delta}{n} \operatorname{sign}(x - R v).$$

That is, $\mathbb{E}[g(x, f_v)] \in \partial f_v(x)$ as desired.

Now, we apply Proposition 5.3.3, which guarantees that

$$\mathfrak{M}_K(C, \mathcal{F}) \geqslant \frac{MR\delta}{2} \left(1 - \sqrt{\frac{1}{2n} \sum_{j=1}^{n} D_{kl}\left(P_{+j} | P_{-j}\right)}\right). \tag{5.3.6}$$

It remains to upper bound the KL-divergence terms. Let P_v^K denote the distribution of the K subgradients the method observes for the function f_v, and let $v_{(\pm j)}$ denote the vector v except that its jth entry is forced to be ± 1. Then, we may use the convexity of the KL-divergence to obtain that

$$D_{kl}\left(P_{+j} | P_{-j}\right) \leqslant \frac{1}{2^n} \sum_{v \in \mathcal{V}} D_{kl}\left(P^K_{v_{(+j)}} | P^K_{v_{(-j)}}\right).$$

Let us thus bound $D_{kl}\left(P_v^K | P_{v'}^K\right)$ when v and v' differ in only a single coordinate (we let it be the first coordinate with no loss of generality). Let us assume for notational simplicity $M = 1$ for the next calculation, as this only changes the support of the subgradient distribution (5.3.5) but not any divergences. Applying the chain rule (Lemma 5.2.8), we have

$$D_{kl}\left(P_v^K | P_{v'}^K\right) = \sum_{k=1}^{K} \mathbb{E}_{P_v}\left[D_{kl}\left(P_v(\cdot \mid g_{1:k-1}) | P_{v'}(\cdot \mid g_{1:k-1})\right)\right].$$

We consider one of the terms, noting that the kth query x_k is a function of $g_1, \ldots, g_{k-1}$. We have

$$\begin{aligned} & D_{kl}\left(P_v(\cdot \mid x_k) | P_{v'}(\cdot \mid x_k)\right) \\ & = P_v(g = e_1 \mid x_k) \log \frac{P_v(g = e_1 \mid x_k)}{P_{v'}(g = e_1 \mid x_k)} + P_v(g = -e_1 \mid x_k) \log \frac{P_v(g = -e_1 \mid x_k)}{P_{v'}(g = -e_1 \mid x_k)}, \end{aligned}$$

because P_v and $P_{v'}$ assign the same probability to all subgradients except when $g \in \{\pm e_1\}$. Continuing the derivation, we obtain

$$D_{kl}\left(P_v(\cdot \mid x_k) | P_{v'}(\cdot \mid x_k)\right) = \frac{1+\delta}{2n} \log \frac{1+\delta}{1-\delta} + \frac{1-\delta}{2n} \log \frac{1-\delta}{1+\delta} = \frac{\delta}{n} \log \frac{1+\delta}{1-\delta}.$$

Noting that this final quantity is bounded by $\frac{3\delta^2}{n}$ for $\delta \leqslant \frac{4}{5}$ gives that

$$D_{kl}\left(P_\nu^K | P_{\nu'}^K\right) \leqslant \frac{3K\delta^2}{n} \text{ if } \delta \leqslant \frac{4}{5}.$$

Substituting the preceding calculation into the lower bound (5.3.6), we obtain

$$\mathfrak{M}_K(C, \mathcal{F}) \geqslant \frac{MR\delta}{2}\left(1 - \sqrt{\frac{1}{2n}\sum_{j=1}^{n}\frac{3K\delta^2}{n}}\right) = \frac{MR\delta}{2}\left(1 - \sqrt{\frac{3K\delta^2}{2n}}\right).$$

Choosing $\delta^2 = \min\{16/25, \frac{n}{4K}\}$ gives the result of the theorem. □

A few remarks are in order. First, the theorem recovers the 1-dimensional result of Theorem 5.2.10, by simply taking $n = 1$ in its statement. Second, we see that if we wish to optimize over a set larger than the ℓ_2-ball, then there must necessarily be some dimension-dependent penalty, at least in the worst case. Lastly, the result again is sharp. By using Theorem 3.4.9, we obtain the following corollary.

Corollary 5.3.7. *In addition to the conditions of Theorem* 5.3.4, *let* $C \subset \mathbb{R}^n$ *contain an* ℓ_∞ *box of radius* R_{inner} *and be contained in an* ℓ_∞ *box of radius* R_{outer}. *Then*

$$R_{\text{inner}} M \min\left\{\frac{1}{5}, \frac{1}{\sqrt{96}}\frac{\sqrt{n}}{\sqrt{K}}\right\} \leqslant \mathfrak{M}_K(C, \mathcal{F}) \leqslant R_{\text{outer}} M \min\left\{1, \frac{\sqrt{n}}{\sqrt{K}}\right\}.$$

Notes and further reading The minimax criterion for measuring optimality of optimization and estimation procedures has a long history, dating back at least to Wald [59] in 1939. The information-theoretic approach to optimality guarantees was extensively developed by Ibragimov and Has'minskii [31], and this is our approach. Our treatment in this chapter is specifically based off of that by Agarwal et al. [1] for proving lower bounds for stochastic optimization problems, though our results appear to have slightly sharper constants. Notably missing in our treatment is the use of Fano's inequality for lower bounds, which is commonly used to prove converse statements to achievability results in information theory [17,63]. Recent treatments of various techniques for proving lower bounds in statistics can be found in the book of Tsybakov [58] or the lecture notes [21].

Our focus on stochastic optimization problems allows reasonably straightforward reductions from optimization to statistical testing problems, for which information theoretic and statistical tools give elegant solutions. For lower bounds for non-stochastic problems, the classical reference is the book of Nemirovski and Yudin [41] (who also provide optimality guarantees for stochastic problems). The basic idea is to provide lower bounds for the *oracle model* of convex optimization, where we consider optimality in terms of the number of queries to an oracle giving true first- or second-order information (as opposed to the stochastic oracle studied here). More recent work, including the lecture notes [42] and the book [44] provide a somewhat easier guide to such results, while the recent paper of Braun et al. [13] shows how to use information-theoretic tools to guarantee optimality even for non-stochastic optimization problems.

A. Technical Appendices

A.1. Continuity of Convex Functions In this appendix, we provide proofs of the basic continuity results for convex functions. Our arguments are based on those of Hiriart-Urruty and Lemaréchal [27].

Proof of Lemma 2.3.1 We can write $x \in B_1$ as $x = \sum_{i=1}^n x_i e_i$, where e_i are the standard basis vectors and $\sum_{i=1}^n |x_i| \leqslant 1$. Thus, we have

$$\begin{aligned} f(x) = f\left(\sum_{i=1}^n e_i x_i\right) &= f\left(\sum_{i=1}^n |x_i| \operatorname{sign}(x_i) e_i + (1 - \|x\|_1) 0\right) \\ &\leqslant \sum_{i=1}^n |x_i| f(\operatorname{sign}(x_i) e_i) + (1 - \|x\|_1) f(0) \\ &\leqslant \max\{f(e_1), f(-e_1), f(e_2), f(-e_2), \ldots, f(e_n), f(-e_n), f(0)\}. \end{aligned}$$

The first inequality uses the fact that the $|x_i|$ and $(1 - \|x\|_1)$ form a convex combination, since $x \in B_1$, as does the second.

For the lower bound, note by the fact that $x \in \operatorname{int} B_1$ satisfies $x \in \operatorname{int} \operatorname{dom} f$, we have $\partial f(x) \neq \emptyset$ by Theorem 2.4.3. In particular, there is a vector g such that $f(y) \geqslant f(x) + \langle g, y - x\rangle$ for all y, and even more,

$$f(y) \geqslant f(x) + \inf_{y \in B_1} \langle g, y - x\rangle \geqslant f(x) - 2\|g\|_\infty$$

for all $y \in B_1$.

Proof of Theorem 2.3.2 First, let us suppose that for each point $x_0 \in C$, there exists an open ball $B \subset \operatorname{int} \operatorname{dom} f$ such that

$$|f(x) - f(x')| \leqslant L\|x - x'\|_2 \text{ for all } x, x' \in B. \tag{A.1.1}$$

The collection of such balls B covers C, and as C is compact, there exists a finite subcover $B_1, \ldots, B_k$ with associated Lipschitz constants $L_1, \ldots, L_k$. Take $L = \max_i L_i$ to obtain the result.

It thus remains to show that we can construct balls satisfying the Lipschitz condition (A.1.1) at each point $x_0 \in C$.

With that in mind, we use Lemma 2.3.1, which shows that for each point x_0, there is some $\epsilon > 0$ and $-\infty < m \leqslant M < \infty$ such that

$$-\infty < m \leqslant \inf_{v : \|v\|_2 \leqslant 2\epsilon} f(x + v) \leqslant \sup_{v : \|v\|_2 \leqslant 2\epsilon} f(x + v) \leqslant M < \infty.$$

We make the following claim, from which the condition (A.1.1) evidently follows based on the preceding display.

Lemma A.1.2. *Let $\epsilon > 0$, let f be convex, and let $B = \{v : \|v\|_2 \leqslant 1\}$. Suppose that $f(x) \in [m, M]$ for all $x \in x_0 + 2\epsilon B$. Then*

$$|f(x) - f(x')| \leqslant \frac{M - m}{\epsilon}\|x - x'\|_2 \textit{ for all } x, x' \in x_0 + \epsilon B.$$

Proof. Let $x, x' \in x_0 + \epsilon B$. Let

$$x'' = x' + \epsilon \frac{x' - x}{\|x' - x\|_2} \in x_0 + 2\epsilon B,$$

as $(x - x')/\|x - x'\|_2 \in B$. By construction, we have that x' lies in the segment $\{tx + (1-t)x'', t \in [0,1]\}$ between x and x''; explicitly,

$$\left(1 + \frac{\epsilon}{\|x' - x\|_2}\right) x' = x'' + \frac{\epsilon}{\|x' - x\|_2} x \text{ or } x' = \frac{\|x' - x\|_2}{\|x' - x\|_2 + \epsilon} x'' + \frac{\epsilon}{\|x' - x\|_2 + \epsilon} x.$$

Then we find that

$$f(x') \leqslant \frac{\|x - x'\|_2}{\|x - x'\|_2 + \epsilon} f(x'') + \frac{\epsilon}{\|x - x'\|_2 + \epsilon} f(x),$$

or

$$\begin{aligned} f(x') - f(x) &\leqslant \frac{\|x - x'\|_2}{\|x - x'\|_2 + \epsilon} \left[f(x'') - f(x)\right] \leqslant \frac{\|x - x'\|_2}{\|x - x'\|_2 + \epsilon} [M - m] \\ &\leqslant \frac{M - m}{\epsilon} \|x - x'\|_2 . \end{aligned}$$

Swapping the roles of x and x' gives the result. □

A.2. Probability background In this section, we very tersely review a few of the necessary definitions and results that we employ here. We provide a non measure-theoretic treatment, as it is not essential for the basic uses we have.

Definition A.2.1. A sequence $X_1, X_2, \ldots$ of random vectors *converges in probability* to a random vector X_∞ if for all $\epsilon > 0$, we have

$$\limsup_{n \to \infty} \mathbb{P}(\|X_n - X_\infty\| > \epsilon) = 0.$$

Definition A.2.2. A sequence $X_1, X_2, \ldots$ of random vectors is a *martingale* if there is a sequence of random variables $Z_1, Z_2, \ldots$ (which may contain all the information about $X_1, X_2, \ldots$) such that for each n, (i) X_n is a function of Z_n, (ii) Z_{n-1} is a function of Z_n, and (iii) we have the conditional expectation condition

$$\mathbb{E}[X_n \mid Z_{n-1}] = X_{n-1}.$$

When condition (i) is satisfied, we say that X_n is *adapted to* Z. We say that a sequence $X_1, X_2, \ldots$ is a *martingale difference sequence* if $S_n = \sum_{i=1}^n X_i$ is a martingale, or, equivalently, if $\mathbb{E}[X_n \mid Z_{n-1}] = 0$.

We now provide a self-contained proof of the Azuma-Hoeffding inequality. Our first result is an important intermediate result.

Lemma A.2.3 (Hoeffding's Lemma [30]). *Let X be a random variable with $a \leqslant X \leqslant b$. Then*

$$\mathbb{E}\left[\exp(\lambda(X - \mathbb{E}[X]))\right] \leqslant \exp\left(\frac{\lambda^2 (b-a)^2}{8}\right) \text{ for all } \lambda \in \mathbb{R}.$$

Proof. First, we note that if Y is any random variable with $Y \in [c_1, c_2]$, then $\mathrm{Var}(Y) \leqslant \frac{(c_2 - c_1)^2}{4}$. Indeed, we have that $\mathrm{Var}(Y) = \mathbb{E}[(Y - \mathbb{E}[Y])^2]$ and that $\mathbb{E}[Y]$

minimizes $\mathbb{E}[(Y-t)^2]$ over $t \in \mathbb{R}$, so that

$$\text{(A.2.4)} \qquad \operatorname{Var}(Y) \leqslant \mathbb{E}\left[\left(Y - \frac{c_2+c_1}{2}\right)^2\right] \leqslant \left(c_2 - \frac{c_2+c_1}{2}\right)^2 = \frac{(c_2-c_1)^2}{4}.$$

Without loss of generality, we assume that $\mathbb{E}[X] = 0$ and that $0 \in [a,b]$. Let $\psi(\lambda) = \log \mathbb{E}[e^{\lambda X}]$. Then

$$\psi'(\lambda) = \frac{\mathbb{E}[Xe^{\lambda X}]}{\mathbb{E}[e^{\lambda X}]} \text{ and } \psi''(\lambda) = \frac{\mathbb{E}[X^2e^{\lambda X}]}{\mathbb{E}[e^{\lambda X}]} - \frac{\mathbb{E}[Xe^{\lambda X}]^2}{\mathbb{E}[e^{\lambda X}]^2}.$$

Note that $\psi'(0) = \mathbb{E}[X] = 0$. Let P denote the distribution of X, and assume without loss of generality that X has a density p.[5] Define the random variable Y to have the shifted density f defined by

$$f(y) = \frac{e^{\lambda y}}{\mathbb{E}[e^{\lambda X}]} p(y)$$

for $y \in \mathbb{R}$, where $p(y) = 0$ for $y \notin [a,b]$. Then we find $\mathbb{E}[Y] = \psi'(\lambda)$ and $\operatorname{Var}(Y) = \mathbb{E}[Y^2] - \mathbb{E}[Y]^2 = \psi''(\lambda)$. But of course, we know that $Y \in [a,b]$ because the distribution P of X is supported on $[a,b]$, so that

$$\psi''(\lambda) = \operatorname{Var}(Y) \leqslant \frac{(b-a)^2}{4}$$

by inequality (A.2.4). Using Taylor's theorem, we have that

$$\psi(\lambda) = \psi(0) + \underbrace{\psi'(0)}_{=0}\lambda + \frac{\lambda^2}{2}\psi''(\widetilde{\lambda})\psi(\lambda) = \psi(0) + \frac{\lambda^2}{2}\psi''(\widetilde{\lambda})$$

for some $\widetilde{\lambda}$ between 0 and λ. But $\psi''(\widetilde{\lambda}) \leqslant \frac{(b-a)^2}{4}$, so that $\psi(\lambda) \leqslant \frac{\lambda^2}{2}\frac{(b-a)^2}{4}$ as desired. □

Theorem A.2.5 (Azuma-Hoeffding Inequality [4]). *Let $X_1, X_2, \ldots$ be a martingale difference sequence with $|X_i| \leqslant B$ for all $i = 1, 2, \ldots$. Then*

$$\mathbb{P}\left(\sum_{i=1}^n X_i \geqslant t\right) \leqslant \exp\left(-\frac{2t^2}{nB^2}\right)$$

and

$$\mathbb{P}\left(\sum_{i=1}^n X_i \leqslant -t\right) \leqslant \exp\left(-\frac{2t^2}{nB^2}\right)$$

for all $t \geqslant 0$.

Proof. We prove the upper tail, as the lower tail is similar. The proof is a nearly immediate consequence of Hoeffding's lemma (Lemma A.2.3) and the Chernoff bound technique. Indeed, we have

$$\mathbb{P}\left(\sum_{i=1}^n X_i \geqslant t\right) \leqslant \mathbb{E}\left[\exp\left(\lambda \sum_{i=1}^n X_i\right)\right] \exp(-\lambda t)$$

[5]We may assume there is a dominating base measure μ with respect to which P has a density p.

for all $\lambda \geqslant 0$. Now, letting Z_i be the sequence to which the X_i are adapted, we iterate conditional expectations. We have

$$\begin{aligned}\mathbb{E}\left[\exp\left(\lambda\sum_{i=1}^{n}X_i\right)\right] &= \mathbb{E}\left[\mathbb{E}\left[\exp\left(\lambda\sum_{i=1}^{n-1}X_i\right)\exp(\lambda X_n)\mid Z_{n-1}\right]\right]\\ &= \mathbb{E}\left[\exp\left(\lambda\sum_{i=1}^{n-1}X_i\right)\mathbb{E}[\exp(\lambda X_n)\mid Z_{n-1}]\right]\\ &\leqslant \mathbb{E}\left[\exp\left(\lambda\sum_{i=1}^{n-1}X_i\right)e^{\frac{\lambda^2B^2}{8}}\right]\end{aligned}$$

because $X_1,\ldots,X_{n-1}$ are functions of Z_{n-1}. By iteratively applying this calculation, we arrive at

$$\mathbb{E}\left[\exp\left(\lambda\sum_{i=1}^{n}X_i\right)\right] \leqslant \exp\left(\frac{\lambda^2 nB^2}{8}\right). \tag{A.2.6}$$

Now we optimize by choosing $\lambda \geqslant 0$ to minimize the upper bound that inequality (A.2.6) provides, namely

$$\mathbb{P}\left(\sum_{i=1}^{n}X_i \geqslant t\right) \leqslant \inf_{\lambda\geqslant 0}\exp\left(\frac{\lambda^2 nB^2}{8}-\lambda t\right) = \exp\left(-\frac{2t^2}{nB^2}\right)$$

by taking $\lambda = \frac{4t}{Bn}$. □

A.3. Auxiliary results on divergences We present a few standard results on divergences without proof, referring to standard references (e.g. the book of Cover and Thomas [17] or the extensive paper on divergence measures by Liese and Vajda [35]). Nonetheless, we state and prove a few results. The first is known as the *data processing inequality*, and it says that processing a random variable (even adding noise to it) can only make distributions closer together. See Cover and Thomas [17] or Theorem 14 of Liese and Vajda [35] for a proof.

Proposition A.3.1 (Data processing). *Let P_0 and P_1 be distributions on a random variable $S \in \mathcal{S}$, and let $Q(\cdot \mid s)$ denote any conditional probability distribution conditioned on s, and define*

$$Q_v(A) = \int Q(A \mid s)dP_v(s)$$

for $v = 0, 1$ and all sets A. Then

$$\|Q_0 - Q_1\|_{TV} \leqslant \|P_0 - P_1\|_{TV} \text{ and } D_{kl}(Q_0|Q_1) \leqslant D_{kl}(P_0|P_1).$$

This proposition is somewhat intuitive: if we do any processing on a random variable $S \sim P$, then there is less "information" about the initial distribution of P than if we did no further processing. A consequence is Pinsker's inequality.

Proposition A.3.2 (Pinsker's inequality). *Let P and Q be arbitrary distributions. Then*

$$\|P - Q\|_{TV}^2 \leqslant \frac{1}{2}D_{kl}(P|Q).$$

Proof. First, we note that if we show the result assuming that the sample space $\mathcal{S}$ on which P and Q are defined is finite, we have the general result. Indeed, suppose that $A \subset \mathcal{S}$ achieves the supremum

$$\|P - Q\|_{TV} = \sup_{A \subset \mathcal{S}} |P(A) - Q(A)|.$$

(We may assume without loss of generality that such a set exists.) If we define $\widetilde{P}$ and $\widetilde{Q}$ to be the binary distributions with $\widetilde{P}(0) = P(A)$ and $\widetilde{P}(1) = 1 - P(A)$, and similarly for $\widetilde{Q}$, we have $\|P - Q\|_{TV} = \|\widetilde{P} - \widetilde{Q}\|_{TV}$, and Proposition A.3.1 immediately guarantees that

$$D_{kl}(\widetilde{P}|\widetilde{Q}) \leqslant D_{kl}(P|Q).$$

Let us assume then that $|\mathcal{S}| < \infty$.

In this case, Pinsker's inequality is an immediate consequence of the strong convexity of the negative entropy functional $h(p) = \sum_{i=1}^n p_i \log p_i$ with respect to the ℓ_1-norm over the probability simplex. For completeness, let us prove this. Let p and $q \in \mathbb{R}_+^n$ satisfy $\sum_{i=1}^n p_i = \sum_{i=1}^n q_i = 1$. Then Taylor's theorem guarantees that

$$h(q) = h(p) + \langle \nabla h(p), q - p \rangle + \frac{1}{2}(q-p)^\top \nabla^2 h(\widetilde{q})(q-p),$$

where $\widetilde{q} = \lambda p + (1-\lambda) q$ for some $\lambda \in [0,1]$. Now, we note that

$$\nabla^2 h(p) = \operatorname{diag}(1/p_1, \ldots, 1/p_n),$$

and using that $\nabla h(p) = [\log p_i + 1]_{i=1}^n$, we find

$$h(q) = h(p) + \sum_{i=1}^n (q_i - p_i) \log p_i + \frac{1}{2} \sum_{i=1}^n \frac{(q_i - p_i)^2}{\widetilde{q}_i}.$$

Using the Cauchy-Schwarz inequality, we have

$$\left(\sum_{i=1}^n |q_i - p_i|\right)^2 = \left(\sum_{i=1}^n \sqrt{\widetilde{q}_i} \frac{|q_i - p_i|}{\sqrt{\widetilde{q}_i}}\right)^2 \leqslant \left(\sum_{i=1}^n \widetilde{q}_i\right)\left(\sum_{i=1}^n \frac{(q_i - p_i)^2}{\widetilde{q}_i}\right).$$

Of course, this gives

$$h(q) \geqslant h(p) + \sum_{i=1}^n (q_i - p_i) \log p_i + \frac{1}{2} \|p - q\|_1^2.$$

Rearranging this, we have $h(q) - h(p) - \langle \nabla h(p), q - p \rangle = \sum_{i=1}^n q_i \log \frac{q_i}{p_i}$, or that

$$D_{kl}(q|p) \geqslant \frac{1}{2} \|p - q\|_1^2 = 2 \|P - Q\|_{TV}^2.$$

This is the result. □

B. Questions and Exercises

Exercises for Lecture 2

Question B.2.1: Let $\pi_C(x) := \operatorname{argmin}_{y \in C} \|x - y\|_2$ denote the Euclidean projection of x onto the set C, where C is closed convex. Show that the projection is a

Lipschitz mapping, that is, for all vectors x_0, x_1,

$$\|\pi_C(x_0) - \pi_C(x_1)\|_2 \leqslant \|x_0 - x_1\|_2 .$$

Show that, even if C is compact, this inequality cannot (in general) be improved.

Question B.2.2: If $S_n = \{A \in \mathbb{R}^{n \times n} : A = A^T\}$ is the set of symmetric matrices and, for $A \in S_n$, $f(A) = \lambda_{\max}(A)$, show that f is convex and compute $\partial f(A)$.

Question B.2.3: A convex function f is called λ strongly convex with respect to the norm $\|\cdot\|$ on the (convex) domain X if for any $x, y \in X$, we have

$$f(y) \geqslant f(x) + \langle g, y - x \rangle + \frac{\lambda}{2}\|x - y\|^2$$

for all $g \in \partial f(x)$. Recall that a function f is L-Lipschitz continuous with respect to the norm $\|\cdot\|$ on the domain X if

$$|f(x) - f(y)| \leqslant L \|x - y\| \text{ for all } x, y \in X.$$

Let f be λ-strongly convex w.r.t. $\|\cdot\|$ and h_1, h_2 be L-Lipschitz continuous convex functions with respect to the norm $\|\cdot\|$. For $i = 1, 2$ define

$$x_i = \arg\min_{x \in X}\{f(x) + h_i(x)\}.$$

Show that

$$\|x_1 - x_2\| \leqslant \frac{2L}{\lambda}.$$

Hint: You may use the fact, demonstrated in the notes, that if h is L-Lipschitz and convex, then $\|g\|_* \leqslant L$ for all $g \in \partial h(x)$, where $\|\cdot\|_*$ is the dual norm to $\|\cdot\|$.

Question B.2.4 (Hölder's inequality): Let x and y be vectors in $\mathbb{R}^n$ and let $p, q \in (1, \infty)$ be conjugate, that is, satisfy $1/p + 1/q = 1$. In this question, we will show that $\langle x, y \rangle \leqslant \|x\|_p \|y\|_q$, and moreover, that $\|\cdot\|_p$ and $\|\cdot\|_q$ are dual norms. (The result is essentially immediate in the case that $p = 1$ and $q = \infty$.)

(a) Show that for any $a, b \geqslant 0$ and any $\eta \geqslant 0$, we have

$$ab \leqslant \frac{\eta^p}{p} a^p + \frac{1}{\eta^q q} b^q.$$

Hint: use the concavity of the logarithm and that $1/p + 1/q = 1$.

(b) Show that $\langle x, y \rangle \leqslant \frac{\eta^p}{p} \|x\|_p^p + \frac{1}{\eta^q q} \|y\|_q^q$ for all $\eta > 0$.

(c) Using the result of part (b), show that $\langle x, y \rangle \leqslant \|x\|_p \|y\|_q$.

(d) Show that $\|\cdot\|_p$ and $\|\cdot\|_q$ are dual norms.

Exercises for Lecture 3

Question B.3.1: In this question and the next, we perform experiments with (stochastic) subgradient methods to train a handwritten digit recognition classifier (one to recognize the digits $\{0, 1, \ldots, 9\}$). A warning: we use optimization notation here, consistent with Example 3.4.6, which is non-standard for typical machine learning or statistical learning applications.

We represent a multiclass classifier using a matrix

$$X = [x_1 \; x_2 \; \cdots \; x_k] \in \mathbb{R}^{d \times k},$$

where there are k classes, and the predicted class for a data vector $a \in \mathbb{R}^d$ is

$$\underset{l \in [k]}{\text{argmax}} \langle a, x_l \rangle = \underset{l \in [k]}{\text{argmax}} \{[X^T a]_l\}.$$

We represent data as pairs $(a, b) \in \mathbb{R}^d \times \{1, \ldots, k\}$, where a is the data point (features) and b the label of the data point. We use the *multiclass hinge* loss function

$$F(X; (a, b)) = \max_{l \neq b} \left[1 + \langle a, x_l - x_b \rangle\right]_+$$

where $[t]_+ = \max\{t, 0\}$ denotes the positive part. We will use stochastic gradient descent to attempt to minimize

$$f(X) := \mathbb{E}_P[F(X; (A, B))] = \int F(X; (a, b)) dP(a, b),$$

where the expectation is taken over pairs (A, B).

(a) Show that F is convex.

(b) Show that $F(X; (a, b)) = 0$ if and only if the classifer represented by X has a *large margin*, meaning that

$$\langle a, x_b \rangle \geqslant \langle a, x_l \rangle + 1 \text{ for all } l \neq b.$$

(c) For a pair (a, b), give a way to calculate a vector $G \in \partial F(X; (a, b))$ (note that $G \in \mathbb{R}^{d \times k}$).

Question B.3.2: In this problem, you will perform experiments to explore the performance of stochastic subgradient methods for classification problems, specifically, a handwritten digit recognition problem using zip code data from the United States Postal Service (this data is taken from the book [24], originally due to Yann Le Cunn). The data—training data `zip.train`, test data `zip.test`, and information file `zip.inf`—are available for download from the zipped tar file `http://web.stanford.edu/~jduchi/PCMIConvex/ZIPCodes.tgz`. Starter code is available for `julia` and `Matlab` at the following urls.

i. For Julia: `http://web.stanford.edu/~jduchi/PCMIConvex/sgd.jl`

ii. For Matlab: `http://web.stanford.edu/~jduchi/PCMIConvex/matlab.tgz`

There are two methods left un-implemented in the starter code: the `sgd` method and the `MulticlassSVMSubgradient` method. Implement these methods (you may find the code for unit-testing the multiclass SVM subgradient useful to double check your implementation). For the SGD method, your stepsizes should be proportional to $\alpha_i \propto 1/\sqrt{i}$, and you should project X to the Frobenius norm ball

$$B_r := \{X \in \mathbb{R}^{d \times k} : \|X\|_{Fr} \leqslant r\}, \text{ where } \|X\|_{Fr}^2 = \sum_{ij} X_{ij}^2.$$

We have implemented a pre-processing step that also *kernelizes* the data representation. Let the function $K(a, a') = \exp(-\frac{1}{2\tau} \|a - a'\|_2^2)$. Then the kernelized data representation transforms each datapoint $a \in \mathbb{R}^d$ into a vector

$$\phi(a) = \begin{bmatrix} K(a, a_{i_1}) & K(a, a_{i_2}) & \cdots & K(a, a_{i_m}) \end{bmatrix}^\top$$

where $i_1, \ldots, i_m$ is a random subset of $\{1, \ldots, N\}$ (see `GetKernelRepresentation`.)

Once you have implemented the `sgd` and `MulticlassSVMSubgradient` methods, use the method `RunExperiment` (Julia/Matlab). What performance do you get in classification? Which digits is your classifier most likely to confuse?

Question B.3.3: In this problem, we give a simple bound on the rate of convergence for stochastic optimization for minimization of strongly convex functions. Let C denote a compact convex set and f denote a λ-strongly convex function with respect to the ℓ_2-norm on C, meaning that

$$f(y) \geqslant f(x) + \langle g, y - x \rangle + \frac{\lambda}{2} \|x - y\|_2^2 \text{ for all } g \in \partial f(x),\ x, y \in C.$$

Consider the following stochastic gradient method: at iteration k, we

i. receive a noisy subgradient g_k with $\mathbb{E}[g_k \mid x_k] \in \partial f(x_k)$;
ii. perform the projected subgradient step

$$x_{k+1} = \pi_C(x_k - \alpha_k g_k).$$

Show that if $\mathbb{E}[\|g_k\|_2^2] \leqslant M^2$ for all k, then with the stepsize choice $\alpha_k = \frac{1}{\lambda k}$, we have the convergence guarantee

$$\mathbb{E}\left[\sum_{k=1}^{K} (f(x_k) - f(x^*))\right] \leqslant \frac{M^2}{2\lambda}(\log K + 1).$$

Exercises for Lecture 4

Question B.4.1: We saw in the lecture that if we use mirror descent,

$$x_{k+1} = \operatorname*{argmin}_{x \in C} \left\{ \langle g_k, x \rangle + \frac{1}{\alpha_k} D_h(x, x_k) \right\},$$

in the stochastic setting with $\mathbb{E}[g_k \mid x_k] \in \partial f(x_k)$ then we have the *regret* bound

$$\mathbb{E}\left[\sum_{k=1}^{K} (f(x_k) - f(x^\star))\right] \leqslant \mathbb{E}\left[\frac{1}{\alpha_K} R^2 + \frac{1}{2} \sum_{k=1}^{K} \alpha_k \|g_k\|_*^2\right].$$

Here we have assumed that $D_h(x^\star, x_k) \leqslant R^2$ for all k. We now use this inequality to prove Corollary 4.3.3. Choose the stepsize α_k adaptively at the kth step by optimizing the convergence bound up to the current iterate, that is, set

$$\alpha_k = R\Big(\sum_{i=1}^{k} \|g_i\|_*^2\Big)^{-\frac{1}{2}},$$

based on the previous subgradients. Prove that in this case one has

$$\mathbb{E}\left[\sum_{k=1}^{K} (f(x_k) - f(x^*))\right] \leqslant 3R\mathbb{E}\left[\Big(\sum_{k=1}^{K} \|g_k\|_*^2\Big)^{\frac{1}{2}}\right]$$

Conclude Corollary 4.3.3.

Hint: An intermediate step, which may be useful, is to prove the following inequality: for any non-negative sequence $a_1, a_2, \ldots, a_k$, one has

$$\sum_{i=1}^{k} \frac{a_i}{\sqrt{\sum_{j=1}^{i} a_j}} \leqslant 2\sqrt{\sum_{i=1}^{k} a_i}.$$

Induction is one natural strategy.

Question B.4.2 (Strong convexity of ℓ_p-norms): Prove the claim of Example 4.2.7. That is, for some fixed $p \in (1, 2]$, if $h(x) = \frac{1}{2(p-1)} \|x\|_p^2$, show that h is strongly convex with respect to the ℓ_p-norm.

Hint: Let $\Psi(t) = \frac{1}{2(p-1)} t^{2/p}$ and $\phi(t) = |t|^p$, noting that $h(x) = \Psi(\sum_{j=1}^{n} \phi(x_j))$. Then by a Taylor expansion, this question is equivalent to showing that for any $w, x \in \mathbb{R}^n$, we have

$$x^\top \nabla^2 h(w) x \geqslant \|x\|_p^2$$

where, defining the shorthand vector $\nabla \phi(w) = [\phi'(w_1) \ \cdots \ \phi'(w_n)]^\top$, we have

$$\begin{aligned}\nabla^2 h(w) &= \Psi''\Big(\sum_{j=1}^{n} \phi(w_j)\Big) \nabla \phi(w) \nabla \phi(w)^\top \\ &\quad + \Psi'\Big(\sum_{j=1}^{n} \phi(w_j)\Big) \operatorname{diag}\left(\phi''(w_1), \ldots, \phi''(w_n)\right).\end{aligned}$$

Now apply an argument similar to that used in Example 4.2.5 to show the strong convexity of $h(x) = \sum_j x_j \log x_j$, but applying Hölder's inequality instead of Cauchy-Schwarz.

Question B.4.3 (Variable metric methods and AdaGrad): Consider the following variable-metric method for minimizing a convex function f on a convex subset $C \subset \mathbb{R}^n$:

$$x_{k+1} = \operatorname*{argmin}_{x \in C} \left\{ \langle g_k, x \rangle + \frac{1}{2}(x - x_k)^\top H_k (x - x_k) \right\},$$

where $\mathbb{E}[g_k] \in \partial f(x_k)$. In the lecture, we showed that

$$\mathbb{E}\left[\sum_{k=1}^{K} (f(x_k) - f(x^\star))\right] \leqslant$$

$$\frac{1}{2}\mathbb{E}\left[\sum_{k=2}^{K} \left(\|x_k - x^\star\|_{H_k}^2 - \|x_k - x^\star\|_{H_{k-1}}^2 \right) + \|x_1 - x^\star\|_{H_1}^2 \right] + \frac{1}{2}\mathbb{E}\left[\sum_{k=1}^{K} \|g_k\|_{H_k^{-1}}^2\right].$$

(a) Let

$$H_k = \operatorname{diag}\left(\sum_{i=1}^{k} g_i g_i^\top\right)^{\frac{1}{2}}$$

be the diagonal matrix whose entries are the square roots of the sum of the squares of the gradient coordinates. (This is the AdaGrad method.) Show that

$$\|x_k - x^\star\|_{H_k}^2 - \|x_k - x^\star\|_{H_{k-1}}^2 \leqslant \|x_k - x^\star\|_\infty \operatorname{tr}(H_k - H_{k-1}),$$

where $\operatorname{tr}(A) = \sum_{i=1}^n A_{ii}$ is the trace of the matrix

(b) Assume that $R_\infty = \sup_{x \in C} \|x - x^\star\|_\infty$ is finite. Show that with any choice of diagonal matrix H_k, we obtain

$$\mathbb{E}\left[\sum_{k=1}^K (f(x_k) - f(x^\star))\right] \leqslant \frac{1}{2} R_\infty \mathbb{E}[\operatorname{tr}(H_K)] + \frac{1}{2}\mathbb{E}\left[\sum_{k=1}^K \|g_k\|_{H_k^{-1}}^2\right].$$

(c) Let $g_{k,j}$ denote the jth coordinate of the kth subgradient. Let H_k be chosen as above. Show that

$$\mathbb{E}\left[\sum_{k=1}^K (f(x_k) - f(x^\star))\right] \leqslant \frac{3}{2} R_\infty \sum_{j=1}^n \mathbb{E}\left[\left(\sum_{k=1}^K g_{k,j}^2\right)^{\frac{1}{2}}\right].$$

(d) Suppose that the domain $C = \{x : \|x\|_\infty \leqslant 1\}$. What is the expected regret of AdaGrad? Show that (to a numerical constant factor we ignore) this expected regret is *always* smaller than the expected regret bound for standard projected gradient descent, which is

$$\mathbb{E}\left[\sum_{k=1}^K (f(x_k) - f(x^\star))\right] \leqslant O(1) \sup_{x \in C} \|x - x^\star\|_2 \, \mathbb{E}\left[\sum_{k=1}^K \|g_k\|_2^2\right]^{\frac{1}{2}}.$$

Hint: Use Cauchy-Schwarz.

(e) As in the previous sub-question, assume that $C = \{x : \|x\|_\infty \leqslant 1\}$. Suppose that the subgradients are such that $g_k \in \{-1, 0, 1\}^n$ for all k, and that for each coordinate j we have $\mathbb{P}(g_{k,j} \neq 0) = p_j$. Show that AdaGrad has convergence guarantee

$$\mathbb{E}\left[\sum_{k=1}^K (f(x_k) - f(x^\star))\right] \leqslant \frac{3\sqrt{K}}{2} \sum_{j=1}^n \sqrt{p_j}.$$

What is the corresponding bound for standard projected gradient descent? How much better can AdaGrad be?

Exercises for Lecture 5

Question B.5.1: In this problem, we prove a lower bound for strongly convex optimization problems. Suppose at each iteration of the optimization procedure, we receive a noisy subgradient g_k satisfying

$$g_k = \nabla f(x_k) + \xi_k, \ \xi_k \stackrel{\text{iid}}{\sim} N(0, \sigma^2).$$

To prove a lower bound for optimization procedures, we use the functions

$$f_\nu(x) = \frac{\lambda}{2}(x - \nu\delta)^2, \; \nu \in \{\pm 1\}.$$

Let $f_\nu^\star = 0$ denote the minimum function values for f_ν on $\mathbb{R}$ for $\nu = \pm 1$.

(a) Recall the separation between two functions f_1 and f_{-1} as defined previously (5.1.4),

$$d_{\text{opt}}(f_{-1}, f_1; C) :=$$

$$\sup\left\{\delta \geqslant 0 : \begin{array}{l} f_1(x) \leqslant f_1^\star + \delta \text{ implies } f_{-1}(x) \geqslant f_{-1}^\star + \delta \\ f_{-1}(x) \leqslant f_{-1}^\star + \delta \text{ implies } f_1(x) \geqslant f_1^\star + \delta \end{array} \text{ for any } x \in C. \right\}.$$

When $C = \mathbb{R}$ (or, more generally, as long as $C \supset [-\delta, \delta]$), show that

$$d_{\text{opt}}(f_{-1}, f_1; C) \geqslant \frac{\lambda}{2}\delta^2.$$

(b) Show that the Kullback-Leibler divergence between two normal distributions $P_1 = \mathsf{N}(\mu_1, \sigma^2)$ and $P_2 = \mathsf{N}(\mu_2, \sigma^2)$ is

$$D_{\text{kl}}(P_1 | P_{-1}) = \frac{(\mu_1 - \mu_2)^2}{2\sigma^2}.$$

(c) Use Le Cam's method to show the following lower bound for stochastic optimization: for any optimization procedure $\widehat{x}_K$ using K noisy gradient evaluations,

$$\max_{\nu \in \{-1,1\}} \mathbb{E}_{P_\nu}[f_\nu(\widehat{x}_K) - f_\nu^\star] \geqslant \frac{\sigma^2}{32\lambda K}.$$

Compare the result with the regret upper bound in problem B.3.3. *Hint:* If P_ν^K denotes the distribution of the K noisy gradients for function f_ν, show that

$$D_{\text{kl}}\left(P_1^K | P_{-1}^K\right) \leqslant \frac{2K\lambda^2\delta^2}{\sigma^2}.$$

Question B.5.2: Let $C = \{x \in \mathbb{R}^n : \|x\|_\infty \leqslant 1\}$, and consider the collection of functions $\mathcal{F}$ where the stochastic gradient oracle $g : \mathbb{R}^n \times \mathcal{S} \times \mathcal{F} \to \{-1, 0, 1\}^n$ satisfies

$$\mathbb{P}(g_j(x, S, f) \neq 0) \leqslant p_j$$

for each coordinate $j = 1, 2, \ldots, n$. Show that, for large enough $K \in \mathbb{N}$, a minimax lower bound for this class of functions and the given stochastic oracle is

$$\mathfrak{M}_K(C, \mathcal{F}) \geqslant c\frac{1}{\sqrt{K}}\sum_{j=1}^n \sqrt{p_j},$$

where $c > 0$ is a numerical constant. How does this compare to the convergence guarantee that AdaGrad gives?

References

[1] A. Agarwal, P. L. Bartlett, P. Ravikumar, and M. J. Wainwright, *Information-theoretic lower bounds on the oracle complexity of stochastic convex optimization*, IEEE Trans. Inform. Theory **58** (2012), no. 5, 3235–3249, DOI 10.1109/TIT.2011.2182178. MR2952543 ←171

[2] P. Assouad, *Deux remarques sur l'estimation* (French, with English summary), C. R. Acad. Sci. Paris Sér. I Math. **296** (1983), no. 23, 1021–1024. MR777600 ←167

[3] P. Auer, N. Cesa-Bianchi, and C. Gentile, *Adaptive and self-confident on-line learning algorithms*, J. Comput. System Sci. **64** (2002), no. 1, 48–75, DOI 10.1006/jcss.2001.1795. Special issue on COLT 2000 (Palo Alto, CA). MR1896142 ←157

[4] K. Azuma, *Weighted sums of certain dependent random variables*, Tôhoku Math. J. (2) **19** (1967), 357–367, DOI 10.2748/tmj/1178243286. MR0221571 ←174

[5] A. Beck and M. Teboulle, *Mirror descent and nonlinear projected subgradient methods for convex optimization*, Oper. Res. Lett. **31** (2003), no. 3, 167–175, DOI 10.1016/S0167-6377(02)00231-6. MR1967286 ←157

[6] A. Ben-Tal, L. El Ghaoui, and A. Nemirovski, *Robust optimization*, Princeton Series in Applied Mathematics, Princeton University Press, Princeton, NJ, 2009. MR2546839 ←102

[7] D. P. Bertsekas, *Stochastic optimization problems with nondifferentiable cost functionals*, J. Optimization Theory Appl. **12** (1973), 218–231, DOI 10.1007/BF00934819. MR0329725 ←120

[8] D. P. Bertsekas, *Convex optimization theory*, Athena Scientific, Nashua, NH, 2009. MR2830150 ←101, 122

[9] D. P. Bertsekas, *Nonlinear programming*, 2nd ed., Athena Scientific Optimization and Computation Series, Athena Scientific, Belmont, MA, 1999. MR3444832 ←101

[10] S. Boyd, J. Duchi, and L. Vandenberghe, *Subgradients*, 2015. Course notes for Stanford Course EE364b. ←140

[11] S. Boyd and A. Mutapcic, *Stochastic subgradient methods*, 2007. Course notes for EE364b at Stanford, available at http://www.stanford.edu/class/ee364b/notes/stoch_subgrad_notes.pdf. ←140

[12] S. Boyd and L. Vandenberghe, *Convex optimization*, Cambridge University Press, Cambridge, 2004. MR2061575 ←101, 119, 122, 123, 157

[13] G. Braun, C. Guzmán, and S. Pokutta, *Lower bounds in the oracle complexity of nonsmooth convex optimization via information theory*, IEEE Trans. Inform. Theory **63** (2017), no. 7, 4709–4724, DOI 10.1109/TIT.2017.2701343. MR3666985 ←171

[14] P. Brucker, *An* $O(n)$ *algorithm for quadratic knapsack problems*, Oper. Res. Lett. **3** (1984), no. 3, 163–166, DOI 10.1016/0167-6377(84)90010-5. MR761510 ←130, 143, 151

[15] S. Bubeck and N. Cesa-Bianchi, *Regret analysis of stochastic and nonstochastic multi-armed bandit problems*, Foundations and Trends in Machine Learning **5** (2012), no. 1, 1–122. ←157

[16] N. Cesa-Bianchi, A. Conconi, and C. Gentile, *On the generalization ability of on-line learning algorithms*, IEEE Trans. Inform. Theory **50** (2004), no. 9, 2050–2057, DOI 10.1109/TIT.2004.833339. MR2097190 ←140

[17] T. M. Cover and J. A. Thomas, *Elements of information theory*, 2nd ed., Wiley-Interscience [John Wiley & Sons], Hoboken, NJ, 2006. MR2239987 ←171, 175

[18] A. Defazio, F. Bach, and S. Lacoste-Julien, *SAGA: A fast incremental gradient method with support for non-strongly convex composite objectives*, Advances in neural information processing systems 27, 2014. ←140

[19] D. L. Donoho, R. C. Liu, and B. MacGibbon, *Minimax risk over hyperrectangles, and implications*, Ann. Statist. **18** (1990), no. 3, 1416–1437, DOI 10.1214/aos/1176347758. MR1062717 ←167

[20] D. L. Donoho, *Compressed sensing*, IEEE Trans. Inform. Theory **52** (2006), no. 4, 1289–1306, DOI 10.1109/TIT.2006.871582. MR2241189 ←129

[21] J. C. Duchi, *Stats311/EE377: Information theory and statistics*, 2015. ←171

[22] J. Duchi, E. Hazan, and Y. Singer, *Adaptive subgradient methods for online learning and stochastic optimization*, J. Mach. Learn. Res. **12** (2011), 2121–2159. MR2825422 ←153, 157

[23] J. C. Duchi, S. Shalev-Shwartz, Y. Singer, and T. Chandra, *Efficient projections onto the* ℓ_1*-ball for learning in high dimensions*, Proceedings of the 25th international conference on machine learning, 2008. ←130, 143, 151

[24] T. Hastie, R. Tibshirani, and J. Friedman, *The elements of statistical learning*, 2nd ed., Springer Series in Statistics, Springer, New York, 2009. Data mining, inference, and prediction. MR2722294 ←178

[25] E. Hazan, *The convex optimization approach to regret minimization*, Optimization for machine learning, 2012. ←140
[26] E. Hazan, *Introduction to online convex optimization*, Foundations and Trends in Optimization **2** (2016), no. 3–4, 157–325. ←102
[27] J. Hiriart-Urruty and C. Lemaréchal, *Convex analysis and minimization algorithms I*, Springer, New York, 1993. ←101, 119, 122, 172
[28] J. Hiriart-Urruty and C. Lemaréchal, *Convex Analysis and Minimization Algorithms II*, Springer, New York, 1993. ←101, 122
[29] J.-B. Hiriart-Urruty and C. Lemaréchal, *Fundamentals of convex analysis*, Springer, 2001. ←122
[30] W. Hoeffding, *Probability inequalities for sums of bounded random variables*, J. Amer. Statist. Assoc. **58** (1963), 13–30. MR0144363 ←173
[31] I. A. Ibragimov and R. Z. Has′minskiĭ, *Statistical estimation*, Applications of Mathematics, vol. 16, Springer-Verlag, New York-Berlin, 1981. Asymptotic theory; Translated from the Russian by Samuel Kotz. MR620321 ←102, 160, 171
[32] R. Johnson and T. Zhang, *Accelerating stochastic gradient descent using predictive variance reduction*, Advances in neural information processing systems 26, 2013. ←140
[33] L. Le Cam, *Asymptotic methods in statistical decision theory*, Springer Series in Statistics, Springer-Verlag, New York, 1986. MR856411 ←102, 160, 162
[34] E. L. Lehmann and G. Casella, *Theory of point estimation*, 2nd ed., Springer Texts in Statistics, Springer-Verlag, New York, 1998. MR1639875 ←160
[35] F. Liese and I. Vajda, *On divergences and informations in statistics and information theory*, IEEE Trans. Inform. Theory **52** (2006), no. 10, 4394–4412, DOI 10.1109/TIT.2006.881731. MR2300826 ←175
[36] D. G. Luenberger, *Optimization by vector space methods*, John Wiley & Sons, Inc., New York-London-Sydney, 1969. MR0238472 ←122
[37] J. E. Marsden, *Elementary classical analysis*, W. H. Freeman and Co., San Francisco, 1974. With the assistance of Michael Buchner, Amy Erickson, Adam Hausknecht, Dennis Heifetz, Janet Macrae and William Wilson, and with contributions by Paul Chernoff, István Fáry and Robert Gulliver. MR0357693 ←101
[38] B. McMahan and M. Streeter, *Adaptive bound optimization for online convex optimization*, Proceedings of the twenty third annual conference on computational learning theory, 2010. ←157
[39] A. Nedić, *Subgradient methods for convex minimization*, Ph.D. Thesis, 2002. ←157
[40] A. Nemirovski, A. Juditsky, G. Lan, and A. Shapiro, *Robust stochastic approximation approach to stochastic programming*, SIAM J. Optim. **19** (2008), no. 4, 1574–1609, DOI 10.1137/070704277. MR2486041 ←140, 157
[41] A. S. Nemirovsky and D. B. Yudin, *Problem complexity and method efficiency in optimization*, A Wiley-Interscience Publication, John Wiley & Sons, Inc., New York, 1983. Translated from the Russian and with a preface by E. R. Dawson; Wiley-Interscience Series in Discrete Mathematics. MR702836 ←102, 123, 157, 171
[42] A. Nemirovski, *Efficient methods in convex programming*, 1994. Technion: The Israel Institute of Technology. ←171
[43] A. Nemirovski, *Lectures on modern convex optimization*, 2005. Georgia Institute of Technology. ←140
[44] Y. Nesterov, *Introductory lectures on convex optimization*, Applied Optimization, vol. 87, Kluwer Academic Publishers, Boston, MA, 2004. A basic course. MR2142598 ←124, 140, 171
[45] Y. Nesterov and A. Nemirovskii, *Interior-point polynomial algorithms in convex programming*, SIAM Studies in Applied Mathematics, vol. 13, Society for Industrial and Applied Mathematics (SIAM), Philadelphia, PA, 1994. MR1258086 ←123
[46] J. Nocedal and S. J. Wright, *Numerical optimization*, 2nd ed., Springer Series in Operations Research and Financial Engineering, Springer, New York, 2006. MR2244940 ←101, 157
[47] B. T. Polyak, *Introduction to optimization*, Translations Series in Mathematics and Engineering, Optimization Software, Inc., Publications Division, New York, 1987. Translated from the Russian; With a foreword by Dimitri P. Bertsekas. MR1099605 ←101, 140
[48] B. T. Polyak and A. B. Juditsky, *Acceleration of stochastic approximation by averaging*, SIAM J. Control Optim. **30** (1992), no. 4, 838–855, DOI 10.1137/0330046. MR1167814 ←140
[49] R. T. Rockafellar, *Convex analysis*, Princeton University Press, 1970. ←101, 104, 121
[50] W. Rudin, *Principles of mathematical analysis*, 3rd ed., McGraw-Hill Book Co., New York-Auckland-Düsseldorf, 1976. International Series in Pure and Applied Mathematics. MR0385023 ←101

[51] S. Shalev-Shwartz, *Online learning: Theory, algorithms, and applications*, Ph.D. Thesis, 2007. ←144

[52] O. Shamir and S. Shalev-Shwartz, *Matrix completion with the trace norm: learning, bounding, and transducing*, J. Mach. Learn. Res. **15** (2014), 3401–3423. MR3277164 ←102

[53] S. Shalev-Shwartz and T. Zhang, *Stochastic dual coordinate ascent methods for regularized loss minimization*, J. Mach. Learn. Res. **14** (2013), 567–599. MR3033340 ←140

[54] A. Shapiro, D. Dentcheva, and A. Ruszczyński, *Lectures on stochastic programming*, MPS/SIAM Series on Optimization, vol. 9, Society for Industrial and Applied Mathematics (SIAM), Philadelphia, PA; Mathematical Programming Society (MPS), Philadelphia, PA, 2009. Modeling and theory. MR2562798 ←102

[55] N. Z. Shor, *Minimization methods for nondifferentiable functions*, Springer Series in Computational Mathematics, vol. 3, Springer-Verlag, Berlin, 1985. Translated from the Russian by K. C. Kiwiel and A. Ruszczyński. MR775136 ←157

[56] N. Z. Shor, *Nondifferentiable optimization and polynomial problems*, Nonconvex Optimization and its Applications, vol. 24, Kluwer Academic Publishers, Dordrecht, 1998. MR1620179 ←153, 157

[57] R. Tibshirani, *Regression shrinkage and selection via the lasso*, J. Roy. Statist. Soc. Ser. B **58** (1996), no. 1, 267–288. MR1379242 ←129

[58] A. B. Tsybakov, *Introduction to nonparametric estimation*, Springer, 2009. ←161, 171

[59] A. Wald, *Contributions to the theory of statistical estimation and testing hypotheses*, Ann. Math. Statistics **10** (1939), 299–326. MR0000932 ←102, 171

[60] A. Wald, *Statistical decision functions which minimize the maximum risk*, Ann. of Math. (2) **46** (1945), 265–280, DOI 10.2307/1969022. MR0012402 ←102

[61] S. Wright, *Optimization Algorithms for Data Analysis*, 2018. ←100

[62] Y. Yang and A. Barron, *Information-theoretic determination of minimax rates of convergence*, Ann. Statist. **27** (1999), no. 5, 1564–1599, DOI 10.1214/aos/1017939142. MR1742500 ←102, 161

[63] B. Yu, *Assouad, Fano, and Le Cam*, Festschrift for Lucien Le Cam, Springer, New York, 1997, pp. 423–435. MR1462963 ←102, 161, 171

[64] M. Zinkevich, *Online convex programming and generalized infinitesimal gradient ascent*, Proceedings of the twentieth international conference on machine learning, 2003. ←140

Stanford University, Stanford CA 94305
Email address: jduchi@stanford.edu

IAS/Park City Mathematics Series
Volume 25, Pages 187–229
https://doi.org/10.1090/pcms/025/00832

Randomized Methods for Matrix Computations

Per-Gunnar Martinsson

Contents

1. Introduction

1.1. Scope and objectives The objective of this chapter is to describe a set of randomized methods for efficiently computing a low rank approximation to a given matrix. In other words, given an $m \times n$ matrix A, we seek to compute factors E and F such that

$$\underset{m \times n}{\mathsf{A}} \approx \underset{m \times k}{\mathsf{E}} \; \underset{k \times n}{\mathsf{F}}, \tag{1.1.1}$$

where the rank k of the approximation is a number we assume to be much smaller than either m or n. In some situations, the rank k is given to us in advance, while in others, it is part of the problem to determine a rank such that the approximation satisfies a bound of the type

$$\|\mathsf{A} - \mathsf{EF}\| \leqslant \varepsilon$$

where ε is a given tolerance, and $\|\cdot\|$ is some specified matrix norm (in this chapter, we will discuss only the spectral and the Frobenius norms).

An approximation of the form (1.1.1) is useful for storing the matrix A more frugally (we can store E and F using $k(m+n)$ numbers, as opposed to mn numbers for storing A), for efficiently computing a matrix vector product $\mathsf{z} = \mathsf{Ax}$ (via $\mathsf{y} = \mathsf{Fx}$ and $\mathsf{z} = \mathsf{Ey}$), for data interpretation, and much more. Low-rank approximation problems of this type form a cornerstone of data analysis and scientific computing, and arise in a broad range of applications, including principal component analysis (PCA) in computational statistics, spectral methods for clustering high-dimensional data and finding structure in graphs, image and video compression, model reduction in physical modeling, and many more.

In performing low-rank approximation, one is typically interested in specific factorizations where the factors E and F satisfy additional constraints. When A is a symmetric $n \times n$ matrix, one is commonly interested in finding an approximate rank-k eigenvalue decomposition (EVD), which takes the form

$$\underset{n \times n}{A} \approx \underset{n \times k}{U} \; \underset{k \times k}{D} \; \underset{k \times n}{U^*}, \tag{1.1.2}$$

where the columns of U form an orthonormal set, and where D is diagonal. For a general $m \times n$ matrix A, we would typically be interested in an approximate rank-k singular value decomposition (SVD), which takes the form

$$\underset{m \times n}{A} \approx \underset{m \times k}{U} \; \underset{k \times k}{D} \; \underset{k \times n}{V^*}, \tag{1.1.3}$$

where U and V have orthonormal columns, and D is diagonal. In this chapter, we will discuss both the EVD and the SVD in depth. We will also describe factorizations such as the *interpolative decomposition (ID)* and the *CUR decomposition* which are highly useful for data interpretation, and for certain applications in scientific computing. In these, we seek to determine a subset of the columns (rows) of A itself that form a good approximate basis for the column (row) space.

While most of the chapter is aimed at computing low rank factorizations where the target rank k is much smaller than the dimensions of the matrix m and n, we will in the last couple of sections of the chapter also discuss how randomization can be used to speed up factorization of *full* matrices, such as a full column pivoted QR factorization, or various relaxations of the SVD that are useful for solving least-squares problems, etc.

1.2. The key ideas of randomized low-rank approximation To quickly introduce the central ideas of the current chapter, let us describe a simple prototypical randomized algorithm: Let A be a matrix of size $m \times n$ that is approximately of low rank. In other words, we assume that for some integer $k < \min(m, n)$, there exists an approximate low rank factorization of the form (1.1.1). Then a natural question is how do you in a computationally efficient manner construct the factors E and F? In [39], it was observed that random matrix theory provides a simple solution: Draw a *Gaussian random matrix* G of size $n \times k$, form the *sampling matrix*

$$E = AG,$$

and then compute the factor F via

$$F = E^{\dagger}A,$$

where $E^{\dagger}$ is the Moore-Penrose pseudo-inverse of A, as in Subsection 2.4. (Then $EF = EE^{\dagger}A$, where $EE^{\dagger}$ is the orthogonal projection onto the linear space spanned by the k columns in E.) Then in many important situations, the approximation

$$\underset{m \times n}{A} \approx \underset{m \times k}{E} \; \underset{k \times n}{(E^{\dagger}A)}, \tag{1.2.1}$$

is close to optimal. With this observation as a starting point, we will construct highly efficient algorithms for computing approximate spectral decompositions of A, for solving certain least-squares problems, for doing principal component analysis of large data sets, etc.

1.3. Advantages of randomized methods The algorithms that result from using randomized sampling techniques are computationally efficient, and are simple to implement as they rely on standard building blocks (matrix-matrix multiplication, unpivoted QR factorization, etc.) that are readily available for most computing environments (multicore CPU, GPU, distributed memory machines, etc). As an illustration, we invite the reader to peek ahead at Algorithm 4.0.1, which provides a complete Matlab code for a randomized algorithm that computes an approximate singular value decomposition of a matrix. Examples of improvements enabled by these randomized algorithms include:

- Given an $m \times n$ matrix A, the cost of computing a rank-k approximant by classical methods is $O(mnk)$. Randomized algorithms can attain complexity $O(mn \log k + k^2(m+n))$, cf. [26, Sec. 6.1], and Section 6.
- Algorithms for performing principal component analysis (PCA) of large data sets have been greatly accelerated, in particular when the data is stored out-of-core, cf. [25].
- Randomized methods tend to require less communication than traditional methods, and can be efficiently implemented on severely communication constrained environments such as GPUs [38] and distributed computing platforms, cf. [24, Ch. 4] and [15, 18].
- Randomized algorithms have enabled the development of *single-pass* matrix factorization algorithms in which the matrix is "streamed" and never stored, cf. [26, Sec. 6.3] and Section 5.

1.4. Relation to other chapters and the broader literature Our focus in this chapter is to describe randomized methods that attain high practical computational efficiency. In particular, we use randomization mostly as a tool for minimizing *communication*, rather than minimizing the flop count (although we do sometimes improve asymptotic flop counts as well). The methods described were first published in [39] (which was inspired by [17], and later led to [31, 40]; see also [45]). Our presentation largely follows that in the 2011 survey [26], but with a focus more on practical usage, rather than theoretical analysis. We have also included material from more recent work, including [49] on factorizations that allow for better data interpretation, [38] on blocking and adaptive error estimation, and [35, 36] on full factorizations.

The idea of using randomization to improve algorithms for low-rank approximation of matrices has been extensively investigated within the theoretical computer science community, with early work including [7,17,45]. The focus of these texts has been to develop algorithms with optimal or close to optimal theoretical performance guarantees in terms of asymptotic flop counts and error bounds. This is the scope of the lectures of Drineas and Mahoney [11] in this volume to which we refer the reader for further details. The surveys [32] and [52] also provide excellent introductions to this literature.

2. Notation

2.1. Notation Throughout the chapter, we measure vectors in $\mathbb{R}^n$ using their Euclidean norm, $\|v\| = \sqrt{\sum_{j=1}^n |v(i)|^2}$. We measure matrices using the spectral and the Frobenius norms, defined by

$$\|A\| = \sup_{\|x\|=1} \|Ax\|, \qquad \text{and} \qquad \|A\|_{\text{Fro}} = \left(\sum_{i,j} |A(i,j)|^2\right)^{1/2},$$

respectively. We use the notation of Golub and Van Loan [20] to specify submatrices. In other words, if B is an $m \times n$ matrix with entries $B(i,j)$, and if $I = [i_1, i_2, \ldots, i_k]$ and $J = [j_1, j_2, \ldots, j_\ell]$ are two index vectors, then $B(I,J)$ denotes the $k \times \ell$ matrix

$$B(I,J) = \begin{bmatrix} B(i_1,j_1) & B(i_1,j_2) & \cdots & B(i_1,j_\ell) \\ B(i_2,j_1) & B(i_2,j_2) & \cdots & B(i_2,j_\ell) \\ \vdots & \vdots & & \vdots \\ B(i_k,j_1) & B(i_k,j_2) & \cdots & B(i_k,j_\ell) \end{bmatrix}.$$

We let $B(I,:)$ denote the matrix $B(I,[1, 2, \ldots, n])$, and define $B(:,J)$ analogously.

The transpose of B is denoted B^*, and we say that a matrix U is *orthonormal (ON)* if its columns form an orthonormal set, so that $U^*U = I$.

2.2. The singular value decomposition (SVD) The SVD was introduced briefly in the introduction. Here we define it again, with some more detail added. Let A denote an $m \times n$ matrix, and set $r = \min(m,n)$. Then A admits a factorization

$$\underset{m\times n}{A} = \underset{m\times r}{U}\;\underset{r\times r}{D}\;\underset{r\times n}{V^*}, \tag{2.2.1}$$

where the matrices U and V are orthonormal, and D is diagonal. We let $\{u_i\}_{i=1}^r$ and $\{v_i\}_{i=1}^r$ denote the columns of U and V, respectively. These vectors are the left and right singular vectors of A. The diagonal elements $\{\sigma_j\}_{j=1}^r$ of D are the singular values of A. We order these so that $\sigma_1 \geqslant \sigma_2 \geqslant \cdots \geqslant \sigma_r \geqslant 0$.

We let A_k denote the truncation of the SVD to its first k terms,

$$A_k = U(:,1:k)D(1:k,1:k)\big(V(:,1:k)\big)^* = \sum_{j=1}^{k} \sigma_j u_j v_j^*.$$

It is easily verified that

$$\|A - A_k\| = \sigma_{k+1},$$

and that

$$\|A - A_k\|_{\text{Fro}} = \left(\sum_{j=k+1}^{\min(m,n)} \sigma_j^2\right)^{1/2},$$

where $\|A\|$ denotes the operator norm of A and $\|A\|_{\text{Fro}}$ denotes the Frobenius norm of A. Moreover, the Eckart-Young theorem [14] states that these errors are the smallest possible errors that can be incurred when approximating A by a matrix of rank k.

2.3. Orthonormalization Given an $m \times \ell$ matrix X, with $m \geqslant \ell$, we introduce the function

$$Q = \texttt{orth}(X)$$

to denote orthonormalization of the columns of X. In other words, the matrix Q will be an $m \times \ell$ orthonormal matrix whose columns form a basis for the column space of X.

In practice, this step is typically achieved most efficiently by a call to a packaged QR factorization; e.g., in Matlab, we would write $[Q, \sim] = \texttt{qr}(X, 0)$. However, all calls to `orth` in this manuscript can be implemented *without pivoting*, which makes efficient implementation much easier.

2.4. The Moore-Penrose pseudoinverse The Moore-Penrose pseudoinverse is a generalization of the concept of an inverse for a non-singular square matrix. To define it, let A be a given $m \times n$ matrix. Let k denote its actual rank, so that its singular value decomposition (SVD) takes the form

$$A = \sum_{j=1}^{k} \sigma_j u_j v_j^* = U_k D_k V_k^*,$$

where $\sigma_1 \geqslant \sigma_2 \geqslant \sigma_k > 0$. Then the pseudoinverse of A is the $n \times m$ matrix defined via

$$A^\dagger = \sum_{j=1}^{k} \frac{1}{\sigma_j} v_j u_j^* = V_k D_k^{-1} U_k^*.$$

For any matrix A, the matrices

$$A^\dagger A = V_k V_k^*, \qquad \text{and} \qquad A A^\dagger = U_k U_k^*,$$

are the orthogonal projections onto the row and column spaces of A, respectively. If A is square and non-singular, then $A^\dagger = A^{-1}$.

3. A two-stage approach

The problem of computing an approximate low-rank factorization to a given matrix can conveniently be split into two distinct "stages." For concreteness, we describe the split for the specific task of computing an approximate singular value decomposition. To be precise, given an $m \times n$ matrix A and a target rank k, we seek to compute factors U, D, and V such that

$$\underset{m\times n}{A} \approx \underset{m\times k}{U}\ \underset{k\times k}{D}\ \underset{k\times n}{V^*}.$$

The factors U and V should be orthonormal, and D should be diagonal. (For now, we assume that the rank k is known in advance, techniques for relaxing this assumption are described in Section 12.) Following [26], we split this task into two computational stages:

Stage A — find an approximate range: Construct an $m \times k$ matrix Q with orthonormal columns such that $A \approx QQ^*A$. (In other words, the columns of Q form an approximate basis for the column space of A.) This step will be executed via a randomized process described in Section 4.

Stage B — form a specific factorization: Given the matrix Q computed in Stage A, form the factors U, D, and V using classical deterministic techniques. For instance, this stage can be executed via the following steps:

(1) Form the $k \times n$ matrix $B = Q^*A$.
(2) Compute the SVD of the (small) matrix B so that $B = \hat{U}DV^*$.
(3) Form $U = Q\hat{U}$.

The point here is that in a situation where $k \ll \min(m, n)$, the difficult part of the computation is all in Stage A. Once that is finished, the post-processing in Stage B is easy, as all matrices involved have at most k rows or columns.

Remark 3.0.1. Stage B is exact up to floating point arithmetic so all errors in the factorization process are incurred at Stage A. To be precise, we have

$$Q\underbrace{Q^*A}_{=B} = Q\underbrace{B}_{=\hat{U}DV^*} = \underbrace{Q\hat{U}}_{=U}DV^* = UDV^*.$$

In other words, if the factor Q satisfies $\|A - QQ^*A\| \leqslant \varepsilon$, then automatically

$$\|A - UDV^*\| = \|A - QQ^*A\| \leqslant \varepsilon \tag{3.0.2}$$

unless ε is close to the machine precision.

Remark 3.0.3. A bound of the form (3.0.2) implies that the diagonal elements $\{D(i,i)\}_{i=1}^k$ of D are accurate approximations to the singular values of A in the sense that $|\sigma_i - D(i,i)| \leqslant \varepsilon$ for $i = 1, 2, \ldots, k$. However, a bound like (3.0.2) does not provide assurances on the *relative errors* in the singular values; nor does it, in the general case, provide strong assurances that the columns of U and V are good approximations to the singular vectors of A.

4. A randomized algorithm for "Stage A" — the range finding problem

This section describes a randomized technique for solving the range finding problem introduced as "Stage A" in Section 3. As a preparation for this discussion, let us recall that an "ideal" basis matrix Q for the range of a given matrix A is the matrix U_k formed by the k leading left singular vectors of A. Letting $\sigma_j(A)$ denote the jth singular value of A, the Eckart-Young theorem [47] states that

$$\inf\{\|A - C\| : \text{ C has rank } k\} = \|A - U_k U_k^* A\| = \sigma_{k+1}(A).$$

Now consider a simplistic randomized method for constructing a spanning set with k vectors for the range of a matrix A: Draw k random vectors $\{g_j\}_{j=1}^{k}$ from a Gaussian distribution, map these to vectors $y_j = Ag_j$ in the range of A, and then use the resulting set $\{y_j\}_{j=1}^{k}$ as a basis. Upon orthonormalization via, e.g., Gram-Schmidt, an orthonormal basis $\{q_j\}_{j=1}^{k}$ would be obtained. For the special case where the matrix A has *exact* rank k, one can prove that the vectors $\{Ag_j\}_{j=1}^{k}$ would with probability 1 be linearly independent, and the resulting orthonormal (ON) basis $\{q_j\}_{j=1}^{k}$ would therefore exactly span the range of A. This would in a sense be an ideal algorithm. The problem is that in practice, there are almost always many non-zero singular values beyond the first k ones. The left singular vectors associated with these modes all "pollute" the sample vectors $y_j = Ag_j$ and will therefore shift the space spanned by $\{y_j\}_{j=1}^{k}$ so that it is no longer aligned with the ideal space spanned by the k leading singular vectors of A. In consequence, the process described can (and frequently does) produce a poor basis. Luckily, there is a fix: Simply take a few extra samples. It turns out that if we take, say, $k + 10$ samples instead of k, then the process will with probability almost 1 produce a basis that is comparable to the best possible basis.

To summarize this discussion, the randomized sampling algorithm for constructing an approximate basis for the range of a given $m \times n$ matrix A proceeds as follows: First pick a small integer p representing how much "over-sampling" we do. (The choice $p = 10$ is often good.) Then execute the following steps:

(1) Form a set of $k + p$ random Gaussian vectors $\{g_j\}_{j=1}^{k+p}$.
(2) Form a set $\{y_j\}_{j=1}^{k+p}$ of samples from the range where $y_j = Ag_j$.
(3) Perform Gram-Schmidt on the set $\{y_j\}_{j=1}^{k+p}$ to form the ON-set $\{q_j\}_{j=1}^{k+p}$.

Now observe that the $k + p$ matrix-vector products are independent and can advantageously be executed in parallel. A full algorithm for computing an approximate SVD using this simplistic sampling technique for executing "Stage A" is summarized in Algorithm 4.0.1.

The error incurred by the randomized range finding method described in this section is a random variable. There exist rigorous bounds for both the expectation of this error, and for the likelihood of a large deviation from this expectation. These bounds demonstrate that when the singular values of A decay "reasonably fast," the error incurred is close to the theoretically optimal one. We provide more details in Section 7.

ALGORITHM: RSVD — BASIC RANDOMIZED SVD

Inputs: An $m \times n$ matrix A, a target rank k, and an over-sampling parameter p (say $p = 10$).

Outputs: Matrices U, D, and V in an approximate rank-$(k+p)$ SVD of A (so that U and V are orthonormal, D is diagonal, and $A \approx UDV^*$.)

Stage A:

(1) Form an $n \times (k+p)$ Gaussian random matrix G.

```
G = randn(n,k+p)
```

(2) Form the sample matrix $Y = A\,G$.

```
Y = A*G
```

(3) Orthonormalize the columns of the sample matrix $Q = \mathtt{orth}(Y)$.

```
[Q,~] = qr(Y,0)
```

Stage B:

(4) Form the $(k+p) \times n$ matrix $B = Q^* A$.

```
B = Q'*A
```

(5) Form the SVD of the small matrix B: $B = \hat{U}DV^*$.

```
[Uhat,D,V] = svd(B,'econ')
```

(6) Form $U = Q\hat{U}$.

```
U = Q*Uhat
```

Algorithm 4.0.1. A basic randomized algorithm. If a factorization of precisely rank k is desired, the factorization in Step 5 can be truncated to the k leading terms. The sans-serif text below each line is Matlab code for executing it.

Remark 4.0.2 (How many basis vectors?). The reader may have observed that while our stated goal was to find a matrix Q that holds k orthonormal columns, the randomized process discussed in this section and summarized in Algorithm 4.0.1 results in a matrix with $k+p$ columns instead. The p extra vectors are needed to ensure that the basis produced in "Stage A" accurately captures the k dominant left singular vectors of A. In a situation where an approximate SVD with precisely k modes is sought, one can drop the last p components when executing Stage B. Using Matlab notation, we would after Step (5) run the commands

```
Uhat = Uhat(:,1:k); D = D(1:k,1:k); V = V(:,1:k);.
```

From a practical point of view, the cost of carrying around a few extra samples in the intermediate steps is often entirely negligible.

5. Single pass algorithms

The randomized algorithm described in Algorithm 4.0.1 accesses the matrix A twice, first in "Stage A" where we build an orthonormal basis for the column space, and then in "Stage B" where we project A on to the space spanned by the computed basis vectors. It turns out to be possible to modify the algorithm in such a way that each entry of A is accessed only *once*. This is important because it allows us to compute the factorization of a matrix that is too large to be stored.

For *Hermitian* matrices, the modification to Algorithm 4.0.1 is very minor and we describe it in Subsection 5.1. Subsection 5.2 then handles the case of a general matrix.

Remark (Loss of accuracy). The single-pass algorithms described in this section tend to produce a factorization of lower accuracy than what Algorithm 4.0.1 would yield. In situations where one has a choice between using either a one-pass or a two-pass algorithm, the latter is generally preferable since it yields higher accuracy, at only moderately higher cost.

Remark (Streaming Algorithms). We say that an algorithm for processing a matrix is a *streaming algorithm* if each entry of the matrix is accessed only once, and if, in addition, entries can be fed in any order. (In other words, the algorithm is not allowed to dictate the order in which elements are viewed.) The algorithms described in this section satisfy both of these conditions.

5.1. Hermitian matrices Suppose that $A = A^*$, and that our objective is to compute an approximate eigenvalue decomposition

$$\begin{array}{ccccc} A & \approx & U & D & U^* \\ n\times n & & n\times k & k\times k & k\times n \end{array} \tag{5.1.1}$$

with U an orthonormal matrix and D diagonal. (Note that for a Hermitian matrix, the EVD and the SVD are essentially equivalent, and that the EVD is the more natural factorization.) Then execute Stage A with an over-sampling parameter p to compute an orthonormal matrix Q whose columns form an approximate basis for the column space of A:

(1) Draw a Gaussian random matrix G of size $n \times (k+p)$.
(2) Form the sampling matrix $Y = AG$.
(3) Orthonormalize the columns of Y to form Q, in other words $Q = \mathtt{orth}(Y)$.

Then

$$A \approx QQ^*A. \tag{5.1.2}$$

Since A is Hermitian, its row and column spaces are identical, so we also have

$$A \approx AQQ^*. \tag{5.1.3}$$

Inserting (5.1.2) into (5.1.3), we (informally!) find that

$$A \approx QQ^*AQQ^*. \tag{5.1.4}$$

We define

$$C = Q^*AQ. \tag{5.1.5}$$

If C is known, then the post-processing is straight-forward: Simply compute the EVD of C to obtain $\mathsf{C} = \hat{\mathsf{U}}\mathsf{D}\hat{\mathsf{U}}^*$, then define $\mathsf{U} = \mathsf{Q}\hat{\mathsf{U}}$, to find that

$$\mathsf{A} \approx \mathsf{QCQ}^* = \mathsf{Q}\hat{\mathsf{U}}\mathsf{D}\hat{\mathsf{U}}^*\mathsf{Q}^* = \mathsf{UDU}^*.$$

The problem now is that since we are seeking a single-pass algorithm, we are not in position to evaluate C directly from formula (5.1.5). Instead, we will derive a formula for C that can be evaluated without revisiting A. To this end, multiply (5.1.5) by $\mathsf{Q}^*\mathsf{G}$ to obtain

$$\mathsf{C}(\mathsf{Q}^*\mathsf{G}) = \mathsf{Q}^*\mathsf{AQQ}^*\mathsf{G}. \tag{5.1.6}$$

We use that $\mathsf{AQQ}^* \approx \mathsf{A}$ (cf. (5.1.3)), to approximate the right hand side in (5.1.6):

$$\mathsf{Q}^*\mathsf{AQQ}^*\mathsf{G} \approx \mathsf{Q}^*\mathsf{AG} = \mathsf{Q}^*\mathsf{Y}. \tag{5.1.7}$$

Combining, (5.1.6) and (5.1.7), and ignoring the approximation error, we define C as the solution of the linear system (recall that $\ell = k + p$)

$$\underset{\ell \times \ell}{\mathsf{C}} \quad \underset{\ell \times \ell}{(\mathsf{Q}^*\mathsf{G})} \quad = \quad \underset{\ell \times \ell}{(\mathsf{Q}^*\mathsf{Y})}. \tag{5.1.8}$$

At first, it may appear that (5.1.8) is perfectly balanced in that there are ℓ^2 equations for ℓ^2 unknowns. However, we need to enforce that C is Hermitian, so the system is actually over-determined by roughly a factor of two. Putting everything together, we obtain the method summarized in Algorithm 5.1.9.

ALGORITHM: SINGLE-PASS RANDOMIZED EVD FOR A HERMITIAN MATRIX

Inputs: An $n \times n$ Hermitian matrix A, a target rank k, and an over-sampling parameter p (say $p = 10$).

Outputs: Matrices U and D in an approximate rank-k EVD of A (so that U is an orthonormal $n \times k$ matrix, D is a diagonal $k \times k$ matrix, and $\mathsf{A} \approx \mathsf{UDU}^*$).

Stage A:

(1) Form an $n \times (k + p)$ Gaussian random matrix G.
(2) Form the sample matrix $\mathsf{Y} = \mathsf{A}\,\mathsf{G}$.
(3) Let Q denote the orthonormal matrix formed by the k dominant left singular vectors of Y.

Stage B:

(4) Let C denote the $k \times k$ least squares solution of $\mathsf{C}\,(\mathsf{Q}^*\mathsf{G}) = (\mathsf{Q}^*\mathsf{Y})$ obtained by enforcing that C should be Hermitian.
(5) Compute that eigenvalue decomposition of C so that $\mathsf{C} = \hat{\mathsf{U}}\mathsf{D}\hat{\mathsf{U}}^*$.
(6) Form $\mathsf{U} = \mathsf{Q}\hat{\mathsf{U}}$.

Algorithm 5.1.9. A basic randomized algorithm single-pass algorithm suitable for a Hermitian matrix.

The procedure described in this section is less accurate than the procedure described in Algorithm 4.0.1 for two reasons: (1) The approximation error in formula (5.1.4) tends to be larger than the error in (5.1.2). (2) While the matrix Q^*G is invertible, it tends to be very ill-conditioned.

Remark 5.1.10 (Extra over-sampling)**.** To combat the problem that Q^*G tends to be ill-conditioned, it is helpful to over-sample more aggressively when using a single pass algorithm, even to the point of setting $p = k$ if memory allows. Once the sampling stage is completed, we form Q as the leading k left singular vectors of Y (compute these by forming the full SVD of Y, and then discard the last p components). Then C will be of size $k \times k$, and the equation that specifies C reads

$$\underset{k\times k}{C}\ \underset{k\times \ell}{(QG)} = \underset{k\times \ell}{Q^*Y.} \tag{5.1.11}$$

Since (5.1.11) is over-determined, we solve it using a least-squares technique. Observe that we are now looking for less information (a $k \times k$ matrix rather than an $\ell \times \ell$ matrix), and have more information in order to determine it.

5.2. General matrices We next consider a general $m \times n$ matrix A. In this case, we need to apply randomized sampling to both its row space and its column space simultaneously. We proceed as follows:

(1) Draw two Gaussian random matrices G_c of size $n \times (k+p)$ and G_r of size $m \times (k+p)$.
(2) Form two sampling matrices $Y_c = AG_c$ and $Y_r = A^*G_r$.
(3) Compute two basis matrices $Q_c = \mathtt{orth}(Y_c)$ and $Q_r = \mathtt{orth}(Y_r)$.

Now define the small projected matrix via

$$C = Q_c^* A Q_r. \tag{5.2.1}$$

We will derive two relationships that together will determine C in a manner that is analogous to (5.1.6). First left multiply (5.2.1) by $G_r^*Q_c$ to obtain

$$G_r^*Q_cC = G_r^*Q_cQ_c^*AQ_r \approx G_r^*AQ_r = Y_r^*Q_r. \tag{5.2.2}$$

Next we right multiply (5.2.1) by $Q_r^*G_c$ to obtain

$$CQ_r^*G_c = Q_c^*AQ_rQ_r^*G_c \approx Q_c^*AG_c = Q_c^*Y_c. \tag{5.2.3}$$

We now define C as the least-square solution of the two equations

$$\left(G_r^*Q_c\right)C = Y_r^*Q_r \qquad \text{and} \qquad C\left(Q_r^*G_c\right) = Q_c^*Y_c.$$

Again, the system is over-determined by about a factor of 2, and it is advantageous to make it further over-determined by more aggressive over-sampling, cf. Remark 5.1.10. Algorithm 5.2.4 summarizes the single-pass method for a general matrix.

ALGORITHM: SINGLE-PASS RANDOMIZED SVD FOR A GENERAL MATRIX

Inputs: An $m \times n$ matrix A, a target rank k, and an over-sampling parameter p (say $p = 10$).

Outputs: Matrices U, V, and D in an approximate rank-k SVD of A (so that U and V are orthonormal with k columns each, D is diagonal, and $A \approx UDV^*$.)

Stage A:

(1) Form two Gaussian random matrices G_c and G_r of sizes $n \times (k+p)$ and $m \times (k+p)$, respectively.
(2) Form the sample matrices $Y_c = A\,G_c$ and $Y_r = A^*\,G_r$.
(3) Form orthonormal matrices Q_c and Q_r consisting of the k dominant left singular vectors of Y_c and Y_r.

Stage B:

(4) Let C denote the $k \times k$ least squares solution of the joint system of equations formed by $\left(G_r^* Q_c\right) C = Y_r^* Q_r$ and $C\left(Q_r^* G_c\right) = Q_c^* Y_c$.
(5) Compute the SVD of C so that $C = \hat{U}D\hat{V}^*$.
(6) Form $U = Q_c\hat{U}$ and $V = Q_r\hat{V}$.

Algorithm 5.2.4. A basic randomized algorithm single-pass algorithm suitable for a general matrix.

6. A method with complexity $O(mn \log k)$ for general dense matrices

The Randomized SVD (RSVD) algorithm given in Algorithm 4.0.1 is highly efficient when we have access to fast methods for evaluating matrix-vector products $x \mapsto Ax$. For the case where A is a general $m \times n$ matrix given simply as an array or real numbers, the cost of evaluating the sample matrix $Y = AG$ (in Step (2) of Algorithm 4.0.1) is $O(mnk)$. RSVD is still often faster than classical methods since the matrix-matrix multiply can be highly optimized, but it does not have an edge in terms of asymptotic complexity. However, it turns out to be possible to modify the algorithm by replacing the Gaussian random matrix G with a different random matrix Ω that has two seemingly contradictory properties:

(1) Ω is sufficiently *structured* that $A\Omega$ can be evaluated in $O(mn\log(k))$ flops;

(2) Ω is sufficiently *random* that the columns of $A\Omega$ accurately span the range of A.

For instance, a good choice of random matrix Ω is

$$\underset{n \times \ell}{\Omega} = \underset{n \times n}{D}\;\underset{n \times n}{F}\;\underset{n \times \ell}{S}, \tag{6.0.1}$$

where D is a diagonal matrix whose diagonal entries are complex numbers of modulus one drawn from a uniform distribution on the unit circle in the complex plane, where F is the discrete Fourier transform,

$$F(p,q) = n^{-1/2}\, e^{-2\pi i (p-1)(q-1)/n}, \qquad p,q \in \{1, 2, 3, \ldots, n\},$$

and where S is a matrix consisting of a random subset of ℓ columns from the $n \times n$ unit matrix (drawn without replacement). In other words, given an arbitrary matrix X of size $m \times n$, the matrix XS consists of a randomly drawn subset of ℓ columns of X. For the matrix Ω specified by (6.0.1), the product $X\Omega$ can be evaluated via a subsampled FFT in $O(mn\log(\ell))$ operations. The parameter ℓ should be chosen slightly larger than the target rank k; the choice $\ell = 2k$ is often good. (A transform of this type was introduced in [1] under the name "Fast Johnson-Lindenstrauss Transform" and was applied to the problem of low-rank approximation in [45, 53]. See also [2, 29, 30].)

By using the structured random matrix described in this section, we can reduce the complexity of "Stage A" in the RSVD from $O(mnk)$ to $O(mn\log(k))$. In order to attain overall cost $O(mn\log(k))$, we must also modify "Stage B" to eliminate the need to compute Q^*A (since direct evaluation of Q^*A has cost $O(mnk)$). One option is to use the single pass algorithm described in 5.2.4, using the structured random matrix to approximate both the row and the column spaces of A. A second, and typically better, option is to use a so called *row-extraction* technique for Stage B; we describe the details in Section 10.

Our theoretical understanding of the errors incurred by the accelerated range finder is not as satisfactory as what we have for Gaussian random matrices, cf. [26, Sec. 11]. In the general case, only quite weak results have been proven. In practice, the accelerated scheme is often as accurate as the Gaussian one, but we do not currently have good theory to predict precisely when this happens.

7. Theoretical performance bounds

In this section, we will briefly summarize some proven results concerning the error in the output of the basic RSVD algorithm in Algorithm 4.0.1. Observe that the factors U, D, V depend not only on A, but also on the draw of the random matrix G. This means that the error that we try to bound is a *random variable.* It is therefore natural to seek bounds on first the expected value of the error, and then on the likelihood of large deviations from the expectation.

Before we start, let us recall from Remark 3.0.1 that all the error incurred by the RSVD algorithm in Algorithm 4.0.1 is incurred in Stage A. The reason is that the "post-processing" in Stage B is exact (up to floating point arithmetic). Consequently, we can (and will) restrict ourselves to giving bounds on $\|A - QQ^*A\|$.

Remark 7.0.1. The theoretical investigation of errors resulting from randomized methods in linear algebra is an active area of research that draws heavily on random matrix theory, theoretical computer science, classical numerical linear algebra, and many other fields. Our objective here is merely to state a couple of representative results, without providing any proofs or details about their derivation. Both results are taken from [26], where the interested reader can find an in-depth treatment of the subject. More recent results pertaining to the RSVD can be found in, e.g., [22, 51], while a detailed discussion of a related method for low-rank approximation can be found in [11]

7.1. Bounds on the expectation of the error A basic result on the *typical* error observed is Theorem 10.6 of [26], which states:

Theorem 7.1.1. *Let* A *be an* $m \times n$ *matrix with singular values* $\{\sigma_j\}_{j=1}^{\min(m,n)}$. *Let* k *be a target rank, and let* p *be an over-sampling parameter such that* $p \geqslant 2$ *and such that* $k + p \leqslant \min(m, n)$. *Let* G *be a Gaussian random matrix of size* $n \times (k+p)$ *and set* $\mathsf{Q} = \mathtt{orth}(\mathsf{AG})$. *Then the average error, as measured in the Frobenius norm, satisfies*

$$(7.1.2)\qquad \mathbb{E}\big[\|\mathsf{A} - \mathsf{QQ}^*\mathsf{A}\|_{\mathrm{Fro}}\big] \leqslant \left(1 + \frac{k}{p-1}\right)^{1/2} \left(\sum_{j=k+1}^{\min(m,n)} \sigma_j^2\right)^{1/2},$$

where $\mathbb{E}$ *refers to expectation with respect to the draw of* G. *The corresponding result for the spectral norm reads*

$$(7.1.3)\qquad \mathbb{E}\big[\|\mathsf{A} - \mathsf{QQ}^*\mathsf{A}\|\big] \leqslant \left(1 + \sqrt{\frac{k}{p-1}}\right)\sigma_{k+1} + \frac{e\sqrt{k+p}}{p}\left(\sum_{j=k+1}^{\min(m,n)} \sigma_j^2\right)^{1/2}.$$

When errors are measured in the *Frobenius norm,* Theorem 7.1.1 is very gratifying. For our standard recommendation of $p = 10$, we are basically within a factor of $\sqrt{1 + k/9}$ of the theoretically minimal error. (Recall that the Eckart-Young theorem states that $\left(\sum_{j=k+1}^{\min(m,n)} \sigma_j^2\right)^{1/2}$ is a lower bound on the residual for any rank-k approximant.) If you over-sample more aggressively and set $p = k+1$, then we are within a distance of $\sqrt{2}$ of the theoretically minimal error.

When errors are measured in the *spectral norm,* the situation is much less rosy. The first term in the bound in (7.1.3) is perfectly acceptable, but the second term is unfortunate in that it involves the minimal error in the Frobenius norm, which can be much larger, especially when m or n are large. The theorem is quite sharp, as it turns out, so the sub-optimality expressed in (7.1.3) reflects a true limitation on the accuracy to be expected from the basic randomized scheme.

The extent to which the error in (7.1.3) is problematic depends on how rapidly the "tail" singular values $\{\sigma_j\}_{j=1}^{\min(m,n)}$ decay. If they decay fast, then the spectral norm error and the Frobenius norm error are similar, and the RSVD works well. If they decay slowly, then the RSVD performs fine when errors are measured in the Frobenius norm, but not very well when the spectral norm is the one of interest. To illustrate the difference, let us consider two situations:

Case 1 — fast decay: Suppose that the tail singular values decay exponentially fast, so that for some $\beta \in (0, 1)$, we have $\sigma_j \approx \sigma_{k+1}\,\beta^{j-k-1}$ for $j > k$. Then we get an estimate for the tail singular values of

$$\left(\sum_{j=k+1}^{\min(m,n)} \sigma_j^2\right)^{1/2} \approx \sigma_{k+1}\left(\sum_{j=k+1}^{\min(m,n)} \beta^{2(j-k-1)}\right)^{1/2} \leqslant \sigma_{k+1}(1-\beta^2)^{-1/2}.$$

As long as β is not very close to 1, we see that the contribution from the tail singular values is modest in this case.

Case 2 — no decay: Suppose that the tail singular values exhibit *no* decay, so that $\sigma_j = \sigma_{k+1}$ for $j > k$. Now

$$\left(\sum_{j=k+1}^{\min(m,n)} \sigma_j^2\right)^{1/2} = \sigma_{k+1}\sqrt{\min(m,n) - k}.$$

This represents the worst case scenario and, since we want to allow for n and m to be very large, represents devastating suboptimality.

Fortunately, it is possible to modify the RSVD in such a way that the errors produced are close to optimal in both the spectral and the Frobenius norms. The price to pay is a modest increase in the computational cost. See Section 8 and [26, Sec. 4.5].

7.2. Bounds on the likelihood of large deviations One can prove that (perhaps surprisingly) the likelihood of a large deviation from the mean depends only on the over-sampling parameter p, and decays extraordinarily fast. For instance, one can prove that if $p \geqslant 4$, then

$$(7.2.1) \qquad \|A - QQ^*A\| \leqslant \left(1 + 17\sqrt{1 + k/p}\right)\sigma_{k+1} + \frac{8\sqrt{k+p}}{p+1}\left(\sum\nolimits_{j>k} \sigma_j^2\right)^{1/2},$$

with failure probability at most $3\,e^{-p}$, see [26, Cor. 10.9].

8. An accuracy enhanced randomized scheme

8.1. The key idea — power iteration We saw in Section 7 that the basic randomized scheme (see, e.g., Algorithm 4.0.1) gives accurate results for matrices whose singular values decay rapidly, but tends to produce suboptimal results when they do not. To recap, suppose that we compute a rank-k approximation to an $m \times n$ matrix A with singular values $\{\sigma_j\}_{j=1}^{\min(m,n)}$. The theory shows that the error measured in the spectral norm is bounded only by a factor that scales with $\left(\sum_{j>k} \sigma_j^2\right)^{1/2}$. When the singular values decay slowly, this quantity can be much larger than the theoretically minimal approximation error (which is σ_{k+1}).

Recall that the objective of the randomized sampling is to construct a set of orthonormal vectors $\{q_j\}_{j=1}^{\ell}$ that capture to high accuracy the space spanned by the k dominant left singular vectors $\{u_j\}_{j=1}^{k}$ of A. The idea is now to sample not A, but the matrix $A^{(q)}$ defined by

$$A^{(q)} = (AA^*)^q A,$$

where q is a small positive integer (say, $q = 1$ or $q = 2$). A simple calculation shows that if A has the SVD $A = UDV^*$, then the SVD of $A^{(q)}$ is

$$A^{(q)} = U\,D^{2q+1}\,V^*.$$

In other words, $A^{(q)}$ has the same left singular vectors as A, while its singular values are $\{\sigma_j^{2q+1}\}_j$. Even when the singular values of A decay slowly, the singular values of $A^{(q)}$ tend to decay fast enough for our purposes.

The accuracy enhanced scheme now consists of drawing a Gaussian matrix G and then forming a sample matrix

$$Y = (AA^*)^q AG.$$

Then orthonormalize the columns of Y to obtain $Q = \texttt{orth}(Y)$, and proceed as before. The resulting scheme is shown in Algorithm 8.1.1.

ALGORITHM: ACCURACY ENHANCED RANDOMIZED SVD

Inputs: An $m \times n$ matrix A, a target rank k, an over-sampling parameter p (say $p = 10$), and a small integer q denoting the number of steps in the power iteration.

Outputs: Matrices U, D, and V in an approximate rank-$(k+p)$ SVD of A. (I.e. U and V are orthonormal and D is diagonal.)

(1) $G = \texttt{randn}(n, k+p)$;
(2) $Y = AG$;
(3) **for** $j = 1 : q$
(4) $\quad Z = A^*Y$;
(5) $\quad Y = AZ$;
(6) **end for**
(7) $Q = \texttt{orth}(Y)$;
(8) $B = Q^*A$;
(9) $[\hat{U}, D, V] = \texttt{svd}(B, \texttt{'econ'})$;
(10) $U = Q\hat{U}$;

Algorithm 8.1.1. The accuracy enhanced randomized SVD. If a factorization of precisely rank k is desired, the factorization in Step 9 can be truncated to the k leading terms.

Remark 8.1.2. The scheme described in Algorithm 8.1.1 can lose accuracy due to round-off errors. The problem is that as q increases, all columns in the sample matrix $Y = (AA^*)^q AG$ tend to align closer and closer to the dominant left singular vector. This means that essentially all information about the singular values and singular vectors associated with smaller singular values get lots to round-off errors. Roughly speaking, if

$$\frac{\sigma_j}{\sigma_1} \leqslant \epsilon_{\text{mach}}^{1/(2q+1)},$$

where ϵ_{mach} is machine precision, then all information associated with the jth singular value and beyond is lost (see Section 3.2 of [34]). This problem can be fixed by orthonormalizing the columns between each iteration, as shown in

Algorithm 8.1.3. The modified scheme is more costly due to the extra calls to orth. (However, note that orth can be executed using *unpivoted* Gram-Schmidt, which is quite fast.)

ALGORITHM: ACCURACY ENHANCED RANDOMIZED SVD
(WITH ORTHONORMALIZATION)

(1) $\mathsf{G} = \mathtt{randn}(n, k+p);$
(2) $\mathsf{Q} = \mathtt{orth}(\mathsf{AG});$
(3) **for** $j = 1 : q$
(4) $\quad \mathsf{W} = \mathtt{orth}(\mathsf{A}^*\mathsf{Q});$
(5) $\quad \mathsf{Q} = \mathtt{orth}(\mathsf{AW});$
(6) **end for**
(7) $\mathsf{B} = \mathsf{Q}^*\mathsf{A};$
(8) $[\hat{\mathsf{U}}, \mathsf{D}, \mathsf{V}] = \mathtt{svd}(\mathsf{B}, \mathtt{'econ'});$
(9) $\mathsf{U} = \mathsf{Q}\hat{\mathsf{U}};$

Algorithm 8.1.3. This algorithm takes the same inputs and outputs as the method in Algorithm 8.1.1. The only difference is that orthonormalization is carried out between each step of the power iteration, to avoid loss of accuracy due to rounding errors.

8.2. Theoretical results A detailed error analysis of the scheme described in Algorithm 8.1.1 is provided in [26, Sec. 10.4]. In particular, the key theorem states:

Theorem 8.2.1. *Let* A *denote an* $m \times n$ *matrix, let* $p \geqslant 2$ *be an over-sampling parameter, and let* q *denote a small integer. Draw a Gaussian matrix* G *of size* $n \times (k+p)$*, set* $\mathsf{Y} = (\mathsf{AA}^*)^q\mathsf{AG}$*, and let* Q *denote an* $m \times (k+p)$ *orthonormal matrix resulting from orthonormalizing the columns of* Y*. Then*

$$\mathbb{E}\big[\|\mathsf{A} - \mathsf{QQ}^*\mathsf{A}\|\big]$$

$$\leqslant \left[\left(1 + \sqrt{\frac{k}{p-1}}\right)\sigma_{k+1}^{2q+1} + \frac{e\sqrt{k+p}}{p}\left(\sum_{j=k+1}^{\min(m,n)} \sigma_j^{2(2q+1)}\right)^{1/2}\right]^{1/(2q+1)}. \tag{8.2.2}$$

The bound in (8.2.2) is slightly opaque. To simplify it, let us consider a worst case scenario in which there is no decay in the singular values beyond the truncation point, so that we have $\sigma_{k+1} = \sigma_{k+2} = \cdots = \sigma_{\min(m,n)}$. Then (8.2.2) simplifies to

$$\mathbb{E}\big[\|\mathsf{A} - \mathsf{QQ}^*\mathsf{A}\|\big] \leqslant \left[1 + \sqrt{\frac{k}{p-1}} + \frac{e\sqrt{k+p}}{p}\cdot\sqrt{\min\{m,n\} - k}\right]^{1/(2q+1)} \sigma_{k+1}.$$

In other words, as we increase the exponent q, the power scheme drives the factor that multiplies σ_{k+1} to one exponentially fast. This factor represents the degree of "sub-optimality" you can expect to see.

8.3. Extended sampling matrix The scheme given in Subsection 8.1 is slightly wasteful in that it does not directly use all the sampling vectors computed. To further improve accuracy, one could, for a symmetric matrix A and a small positive integer q, form an "extended" sampling matrix

$$Y = \left[AG,\, A^2G,\, \dots,\, A^qG\right].$$

Observe that this new sampling matrix Y has $q\ell$ columns. Then proceed as before:

$$Q = \mathtt{qr}(Y), \qquad B = Q^*A, \qquad [\hat{U}, D, V] = \mathtt{svd}(B, \mathtt{'econ'}), \qquad U = Q\hat{U}. \tag{8.3.1}$$

The computations in (8.3.1) can be quite expensive since the "tall thin" matrices being operated on now have $q\ell$ columns, rather than the tall thin matrices in, e.g., Algorithm 8.1.1, which have only ℓ columns. This results in an increase in cost for all operations (QR factorization, matrix-matrix multiply, SVD) by a factor of $O(q^2)$.

Consequently, the scheme described here is primarily useful in situations in which the computational cost is dominated by applications of A and A^*, and we want to maximally leverage all interactions with A. An early discussion of this idea can be found in [44, Sec. 4.4], with a more detailed discussion in [42].

9. The Nyström method for positive symmetric definite matrices

When the input matrix A is positive semidefinite (psd), the *Nyström method* can be used to improve the quality of standard factorizations at almost no additional cost; see [9] and its bibliography. To describe the idea, we first recall from Subsection 5.1 that when A is Hermitian (which of course every psd matrix is), then it is natural to use the approximation

$$A \approx Q\,(Q^*AQ)\,Q^*. \tag{9.0.1}$$

In contrast, the so called "Nyström scheme" relies on the rank-k approximation

$$A \approx (AQ)\,\left(Q^*AQ\right)^{-1}(AQ)^*. \tag{9.0.2}$$

For both stability and computational efficiency, we typically rewrite (9.0.2) as

$$A \approx FF^*,$$

where F is an approximate *Cholesky* factor of A of size $n \times k$, defined by

$$F = (AQ)\,\left(Q^*AQ\right)^{-1/2}.$$

To compute the factor F numerically, we may first form the matrices $B_1 = AQ$ and $B_2 = Q^*B_1$. Observe that B_2 is necessarily psd, so that we can compute its Cholesky factorization $B_2 = C^*C$. Finally compute the factor $F = B_1C^{-1}$ by performing a triangular solve. The low-rank factorization (9.0.2) can be converted to a standard decomposition using the techniques from Section 3.

The Nyström technique for computing an approximate eigenvalue decomposition is given in Algorithm 9.0.3. Let us compare the cost of this method to the more straight-forward method resulting from using the formula (9.0.1). In both cases, we need to twice apply A to a set of $k + p$ vectors (first in computing AG, then in computing AQ). But the Nyström method tends to result in substantially more accurate results. Informally speaking, the reason is that by exploiting the psd property of A, we can take one step of power iteration "for free." For a more formal analysis of the cost and accuracy of the Nyström method, we refer the reader to [19, 43].

ALGORITHM: EIGENVALUE DECOMPOSITION VIA THE NYSTRÖM METHOD

Given an $n \times n$ non-negative matrix A, a target rank k and an over-sampling parameter p, this procedure computes an approximate eigenvalue decomposition $A \approx U \Lambda U^*$, where U is orthonormal, and Λ is nonnegative and diagonal.

(1) Draw a Gaussian random matrix $G = \texttt{randn}(n, k + p)$.

(2) Form the sample matrix $Y = AG$.

(3) Orthonormalize the columns of the sample matrix to obtain the basis matrix $Q = \texttt{orth}(Y)$.

(4) Form the matrices $B_1 = AQ$ and $B_2 = Q^* B_1$.

(5) Perform a Cholesky factorization $B_2 = C^* C$.

(6) Form $F = B_1 C^{-1}$ using a triangular solve.

(7) Compute a Singular Value Decomposition of the Cholesky factor $[U, \Sigma, \sim] = \texttt{svd}(F, \texttt{'econ'})$.

(8) Set $\Lambda = \Sigma^2$.

Algorithm 9.0.3. The Nyström method for low-rank approximation of self-adjoint matrices with non-negative eigenvalues. It involves two applications of A to matrices with $k + p$ columns, and has comparable cost to the basic RSVD in Algorithm 4.0.1. However, it exploits the symmetry of A to boost the accuracy.

10. Randomized algorithms for computing Interpolatory Decompositions

10.1. Structure preserving factorizations Any matrix A of size $m \times n$ and rank $k < \min(m, n)$, admits a so called "interpolative decomposition (ID)" which takes the form

$$\underset{m \times n}{A} = \underset{m \times k}{C} \; \underset{k \times n}{Z}, \tag{10.1.1}$$

where the matrix C is given by a subset of the columns of A and where Z is well-conditioned in a sense that we will make precise shortly. The ID has several advantages, as compared to, e.g., the QR or SVD factorizations:

- If A is sparse or non-negative, then C shares these properties.
- The ID requires less memory to store than either the QR or the singular value decomposition.
- Finding the indices associated with the spanning columns is often helpful in *data interpretation.*
- In the context of numerical algorithms for discretizing PDEs and integral equations, the ID often preserves "the physics" of a problem in a way that the QR or SVD do not.

One shortcoming of the ID is that when A is not of precisely rank k, then the approximation error by the best possible rank-k ID can be substantially larger than the theoretically minimal error. (In fact, the ID and the column pivoted QR factorizations are closely related, and they attain *exactly* the same minimal error.)

For future reference, let J_s be an index vector in $\{1, 2, \ldots, n\}$ that identifies the k columns in C so that

$$C = A(:, J_s).$$

One can easily show (see, e.g., [34, Thm. 9]) that any matrix of rank k admits a factorization (10.1.1) that is well-conditioned in the sense that each entry of Z is bounded in modulus by one. However, any algorithm that is guaranteed to find such an optimally conditioned factorization must have combinatorial complexity. Polynomial time algorithms with high practical efficiency are discussed in [6, 23]. Randomized algorithms are described in [26, 49].

Remark 10.1.2. The interpolative decomposition is closely related to the so called CUR decomposition which has been studied extensively in the context of randomized algorithms [5, 8, 10, 50]. We will return to this point in Section 11.

10.2. Three flavors of ID: row, column, and double-sided ID Subsection 10.1 describes a factorization where we use a subset of the *columns* of A to span its *column space*. Naturally, this factorization has a sibling which uses the *rows* of A to span its *row space*. In other words A also admits the factorization

$$\underset{m \times n}{A} = \underset{m \times k}{X} \; \underset{k \times n}{R}, \tag{10.2.1}$$

where R is a matrix consisting of k rows of A, and where X is a matrix that contains the $k \times k$ identity matrix. We let I_s denote the index vector of length k that marks the "skeleton" rows so that $R = A(I_s, :)$.

Finally, there exists a so called *double-sided ID* which takes the form

$$\underset{m \times n}{A} = \underset{m \times k}{X} \; \underset{k \times k}{A_s} \; \underset{k \times n}{Z}, \tag{10.2.2}$$

where X and Z are the same matrices as those that appear in (10.1.1) and (10.2.1), and where A_s is the $k \times k$ submatrix of A given by

$$A_s = A(I_s, J_s).$$

10.3. Deterministic techniques for computing the ID In this section we demonstrate that there is a close connection between the column ID and the classical column pivoted QR factorization (CPQR). The end result is that standard software used to compute the CPQR can with some light post-processing be used to compute the column ID.

As a starting point, recall that for a given $m \times n$ matrix A, with $m \geqslant n$, the QR factorization can be written as

$$\underset{m\times n}{A}\ \underset{n\times n}{P} = \underset{m\times n}{Q}\ \underset{n\times n}{S}, \tag{10.3.1}$$

where P is a permutation matrix, where Q has orthonormal columns and where S is upper triangular. [1] Since our objective here is to construct a rank-k approximation to A, we split off the leading k columns from Q and S to obtain partitions

$$Q = m \overset{\quad k \qquad n-k}{\begin{bmatrix} Q_1 & Q_2 \end{bmatrix}}, \quad \text{and} \quad S = \begin{matrix} k \\ m-k \end{matrix} \overset{\quad k \qquad n-k}{\begin{bmatrix} S_{11} & S_{12} \\ 0 & S_{22} \end{bmatrix}}. \tag{10.3.2}$$

Combining (10.3.1) and (10.3.2), we then find that

$$AP = [Q_1 \mid Q_2] \begin{bmatrix} S_{11} & S_{12} \\ 0 & S_{22} \end{bmatrix} = [Q_1 S_{11} \mid Q_1 S_{12} + Q_2 S_{22}]. \tag{10.3.3}$$

Equation (10.3.3) tells us that the $m \times k$ matrix $Q_1 S_{11}$ consists precisely of first k columns of AP. These columns were the first k columns that were chosen as "pivots" in the QR-factorization procedure. They typically form a good (approximate) basis for the column space of A. We consequently define our $m \times k$ matrix C as this matrix holding the first k pivot columns. Letting J denote the permutation vector associated with the permutation matrix P, so that

$$AP = A(:, J),$$

we define $J_s = J(1 : k)$ as the index vector identifying the first k pivots, and set

$$C = A(:, J_s) = Q_1 S_{11}. \tag{10.3.4}$$

Now let us rewrite (10.3.3) by extracting the product $Q_1 S_{11}$

$$AP = Q_1[S_{11} \mid S_{12}] + Q_2[0 \mid S_{22}] = Q_1 S_{11}[I_k \mid S_{11}^{-1} S_{12}] + Q_2[0 \mid S_{22}]. \tag{10.3.5}$$

(Remark 10.3.8 discusses why S_{11} must be invertible.) Now define

$$T = S_{11}^{-1} S_{12}, \qquad \text{and} \qquad Z = [I_k \mid T]P^* \tag{10.3.6}$$

---

[1] We use the letter S instead of the traditional R to avoid confusion with the "R"-factor in the row ID, (10.2.1).

so that (10.3.5) can be rewritten (upon right-multiplication by P^*, which equals P^{-1} since P is unitary) as

$$A = C[I_k \mid T]P^* + Q_2[0 \mid S_{22}]P^* = CZ + Q_2[0 \mid S_{22}]P^*. \tag{10.3.7}$$

Equation (10.3.7) is precisely the column ID we sought, with the additional bonus that the remainder term is explicitly identified. Observe that when the spectral or Frobenius norms are used, the error term is of *exactly* the same size as the error term obtained from a truncated QR factorization:

$$\|A - CZ\| = \|Q_2[0 \mid S_{22}]P^*\| = \|S_{22}\| = \|A - Q_1[S_{11} \mid S_{12}]P^*\|.$$

Remark 10.3.8 (Conditioning). Equation (10.3.5) involves the quantity S_{11}^{-1} which prompts the question of whether S_{11} is necessarily invertible, and what its condition number might be. It is easy to show that whenever the rank of A is at least k, the CPQR algorithm is guaranteed to result in a matrix S_{11} that is non-singular. (If the rank of A is j, where $j < k$, then the QR factorization process can detect this and halt the factorization after j steps.)

Unfortunately, S_{11} is typically quite ill-conditioned. The saving grace is that even though one should expect S_{11} to be poorly conditioned, it is often the case that the linear system

$$S_{11}T = S_{12} \tag{10.3.9}$$

still has a solution T whose entries are of moderate size. Informally, one could say that the directions where S_{11} and S_{12} are small "line up." For standard column pivoted QR, the system (10.3.9) will in practice be observed to almost always have a solution T of small size [6], but counter-examples can be constructed. More sophisticated pivot selection procedures have been proposed that are *guaranteed* to result in matrices S_{11} and S_{12} such that (10.3.9) has a good solution; but these are harder to code and take longer to execute [23].

Of course, the row ID can be computed via an entirely analogous process that starts with a CPQR of the *transpose* of A. In other words, we execute a pivoted Gram-Schmidt orthonormalization process on the rows of A.

Finally, to obtain the double-sided ID, we start with using the CPQR-based process to build the column ID (10.1.1). Then compute the row ID by performing Gram-Schmidt on the rows of the tall thin matrix C.

The three deterministic algorithms described for computing the three flavors of ID are summarized in Algorithm 10.3.11.

Remark 10.3.10 (Partial factorization). The algorithms for computing interpolatory decompositions in Algorithm 10.3.11 are wasteful when $k \ll \min(m, n)$ since they involve a *full* QR factorization, which has complexity $O(mn\min(m, n))$. This problem is very easily remedied by replacing the full QR factorization by a *partial* QR factorization, which has cost $O(mnk)$. Such a partial factorization could take as input either a preset rank k, or a tolerance ε. In the latter case, the factorization would stop once the residual error $\|A - Q(:,1:k)S(1:k,:)\| = \|S_{22}\| \leqslant \varepsilon$. When the QR factorization is interrupted after k steps, the output would still be a factorization of the form (10.3.1), but in this case, S_{22} would not be upper triangular. This is immaterial since S_{22} is never used. To further accelerate the computation, one can advantageously use a *randomized CPQR* algorithm, cf. Sections 14 and 15 or [36,38].

Compute a column ID so that $A \approx A(:, J_s)\, Z$.
function $[J_s,\ Z] = \mathtt{ID\_col}(A, k)$
  $[Q, S, J] = \mathtt{qr}(A, 0);$
  $T = (S(1:k, 1:k))^{-1} S(1:k, (k+1):n);$
  $Z = \mathtt{zeros}(k, n)$
  $Z(:, J) = [I_k\ T];$
  $J_s = J(1:k);$

Compute a row ID so that $A \approx X\, A(I_s, :)$.
function $[I_s,\ X] = \mathtt{ID\_row}(A, k)$
  $[Q, S,\ J] = \mathtt{qr}(A^*, 0);$
  $T = (S(1:k, 1:k))^{-1} S(1:k, (k+1):m);$
  $X = \mathtt{zeros}(m, k)$
  $X(J, :) = [I_k\ T]^*;$
  $I_s = J(1:k);$

Compute a double-sided ID so that $A \approx X\, A(I_s, J_s)\, Z$.
function $[I_s, J_s, X, Z] = \mathtt{ID\_double}(A, k)$
  $[J_s,\ Z] = \mathtt{ID\_col}(A, k);$
  $[I_s,\ X] = \mathtt{ID\_row}(A(:, J_s), k);$

Algorithm 10.3.11. Deterministic algorithms for computing the column, row, and double-sided ID via the column pivoted QR factorization. The input is in every case an $m \times n$ matrix A and a target rank k. Since the algorithms are based on the CPQR, it is elementary to modify them to the situation where a tolerance rather than a rank is given. (Recall that the errors resulting from these ID algorithms are *identical* to the error in the first CPQR factorization executed.)

10.4. Randomized techniques for computing the ID The ID is particularly well suited to being computed via randomized algorithms. To describe the ideas, suppose temporarily that A is an $m \times n$ matrix of *exact* rank k, and that we have by some means computed an approximate rank-k factorization

$$\underset{m\times n}{\mathsf{A}} = \underset{m\times k}{\mathsf{Y}}\ \underset{k\times n}{\mathsf{F}}. \tag{10.4.1}$$

Once the factorization (10.4.1) is available, let us use the algorithm ID_row described in Algorithm 10.3.11 to compute a row ID $[I_s, \mathsf{X}] = \texttt{ID\_row}(\mathsf{Y}, k)$ of Y so that

$$\underset{m\times k}{\mathsf{Y}} = \underset{m\times k}{\mathsf{X}}\ \underset{k\times k}{\mathsf{Y}(I_s,:)}. \tag{10.4.2}$$

It then turns out that $\{I_s, \mathsf{X}\}$ is automatically (!) a row ID of A as well. To see this, simply note that

$$\begin{aligned} & \mathsf{XA}(I_s,:) \\ = \ & \mathsf{XY}(I_s,:)\mathsf{F} && \{\text{Use (10.4.1) restricted to the rows in } I_s.\} \\ = \ & \mathsf{YF} && \{\text{Use (10.4.2).}\} \\ = \ & \mathsf{A} && \{\text{Use (10.4.1).}\} \end{aligned}$$

The key insight here is very simple, but powerful, so let us spell it out explicitly:

> *Observation:* In order to compute a row ID of a matrix A, the only information needed is a matrix Y whose columns span the column space of A.

ALGORITHM: RANDOMIZED ID

Inputs: An $m \times n$ matrix A, a target rank k, an over-sampling parameter p (say $p = 10$), and a small integer q denoting the number of power iterations taken.

Outputs: An $m \times k$ interpolation matrix X and an index vector $I_s \in \mathbb{N}^k$ such that $\mathsf{A} \approx \mathsf{XA}(I_s, :)$.

(1) $\mathsf{G} = \mathtt{randn}(n, k+p)$;
(2) $\mathsf{Y} = \mathsf{AG}$;
(3) **for** $j = 1 : q$
(4) $\quad \mathsf{Y}' = \mathsf{A}^*\mathsf{Y}$;
(5) $\quad \mathsf{Y} = \mathsf{AY}'$;
(6) **end for**
(7) Form an ID of the $n \times (k+p)$ sample matrix: $[I_s, \mathsf{X}] = \texttt{ID\_row}(\mathsf{Y}, k)$.

Algorithm 10.4.3. An $O(mnk)$ algorithm for computing an interpolative decomposition of A via randomized sampling.

As we have seen, the task of finding a matrix Y whose columns form a good basis for the column space of a matrix is ideally suited to randomized sampling. To be precise, we showed in Section 4 that given a matrix A, we can find a matrix Y whose columns approximately span the column space of A via the formula $\mathsf{Y} = \mathsf{AG}$, where G is a tall thin Gaussian random matrix. The scheme that results from combining these two insights is summarized in Algorithm 10.4.3.

The randomized algorithm for computing a row ID shown in Algorithm 10.4.3 has complexity $O(mnk)$. We can reduce this complexity to $O(mn \log k)$ by using a structured random matrix instead of a Gaussian, cf. Section 6. The result is summarized in Algorithm 10.4.4.

ALGORITHM: FAST RANDOMIZED ID

Inputs: An $m \times n$ matrix A, a target rank k, and an over-sampling parameter p (say $p = k$).

Outputs: An $m \times k$ interpolation matrix X and an index vector $I_s \in \mathbb{N}^k$ such that $A \approx XA(I_s, :)$.

(1) Form an $n \times (k+p)$ SRFT Ω.
(2) Form the sample matrix $Y = A\,\Omega$.
(3) Form an ID of the $n \times (k+p)$ sample matrix: $[I_s, X] = \texttt{ID\_row}(Y, k)$.

Algorithm 10.4.4. An $O(mn \log k)$ algorithm for computing an interpolative decomposition of A.

11. Randomized algorithms for computing the CUR decomposition

11.1. The CUR decomposition The so called *CUR-factorization* [10] is a "structure preserving" factorization that is similar to the Interpolative Decomposition described in Section 10. The CUR factorization approximates an $m \times n$ matrix A as a product

$$\underset{m\times n}{A} \approx \underset{m\times k}{C}\;\underset{k\times k}{U}\;\underset{k\times n}{R}, \tag{11.1.1}$$

where C contains a subset of the columns of A and R contains a subset of the rows of A. Like the ID, the CUR decomposition offers the ability to preserve properties like sparsity or non-negativity in the factors of the decomposition, the prospect to reduce memory requirements, and excellent tools for data interpretation.

The CUR decomposition is often obtained in three steps [10, 41]: (1) Some scheme is used to assign a weight or the so called leverage score (of importance) [27] to each column and row in the matrix. This is typically done either using the ℓ_2 norms of the columns and rows or by using the leading singular vectors of A [12, 50]. (2) The matrices C and R are constructed via a randomized sampling procedure, using the leverage scores to assign a sampling probability to each column and row. (3) The U matrix is computed via:

$$U \approx C^{\dagger} A R^{\dagger}, \tag{11.1.2}$$

with $C^{\dagger}$ and $R^{\dagger}$ being the pseudoinverses of C and R. Non-randomized approaches to computing the CUR decomposition are discussed in [46,49].

Remark 11.1.3 (Conditioning of CUR)**.** For matrices whose singular values experience substantial decay, the accuracy of the CUR factorization can deteriorate due to effects of ill-conditioning. To simplify slightly, one would normally expect the leading k singular values of C and R to be rough approximations to the leading k singular values of A, so that the condition numbers of C and R would be roughly $\sigma_1(A)/\sigma_k(A)$. Since low-rank factorizations are most useful when applied to matrices whose singular values decay reasonably rapidly, we would *typically* expect the ratio $\sigma_1(A)/\sigma_k(A)$ to be large, which is to say that C and R would be ill-conditioned. Hence, in the typical case, evaluation of the formula (11.1.2) can be expected to result in substantial loss of accuracy due to accumulation of round-off errors. Observe that the ID does not suffer from this problem; in (10.2.2), the matrix A_{skel} tends to be ill-conditioned, but it does not need to be inverted. (The matrices X and Z are well-conditioned.)

11.2. Converting a double-sided ID to a CUR decomposition The next algorithm converts a double-sided ID to a CUR decomposition.

To explain this algorithm, we begin by assuming that the factorization (10.2.2) has been computed using the procedures described in Section 10 (either the deterministic or the randomized ones). In other words, we assume that the index vectors I_s and J_s, and the basis matrices X and Z, are all available. We then define C and R in the natural way as

$$C = A(:, J_s) \quad \text{and} \quad R = A(I_s, :). \tag{11.2.1}$$

Consequently, C and R are respectively subsets of columns and of rows of A. The index vectors I_s and J_s are determined by the column pivoted QR factorizations, possibly combined with a randomized projection step for computational efficiency. It remains to construct a $k \times k$ matrix U such that

$$A \approx C\,U\,R. \tag{11.2.2}$$

Now recall that, cf. (10.1.1),

$$A \approx C\,Z. \tag{11.2.3}$$

By inspecting (11.2.3) and (11.2.2), we find that we achieve our objective if we determine a matrix U such that

$$\underset{k \times k}{U} \quad \underset{k \times n}{R} \quad = \quad \underset{k \times n}{Z}. \tag{11.2.4}$$

Unfortunately, (11.2.4) is an over-determined system, but at least intuitively, it seems plausible that it should have a fairly accurate solution, given that the rows of R and the rows of Z should, by construction, span roughly the same space (namely, the space spanned by the k leading right singular vectors of A). Solving (11.2.4) in the least-square sense, we arrive at our definition of U:

$$\mathsf{U} := \mathsf{Z}\mathsf{R}^{\dagger}. \tag{11.2.5}$$

These steps are summarized in Algorithm 11.2.6.

ALGORITHM: RANDOMIZED CUR

Inputs: An $m \times n$ matrix A, a target rank k, an over-sampling parameter p (say $p = 10$), and a small integer q denoting the number of power iterations taken.

Outputs: A $k \times k$ matrix U and index vectors I_s and J_s of length k such that $\mathsf{A} \approx \mathsf{A}(:, J_s)\, \mathsf{U}\, \mathsf{A}(I_s, :)$.

(1) $\mathsf{G} = \mathtt{randn}(k + p, m)$;
(2) $\mathsf{Y} = \mathsf{GA}$;
(3) **for** $j = 1 : q$
(4) $\quad \mathsf{Z} = \mathsf{YA}^*$;
(5) $\quad \mathsf{Y} = \mathsf{ZA}$;
(6) **end for**
(7) Form an ID of the $(k + p) \times n$ sample matrix: $[I_s, \mathsf{Z}] = \mathtt{ID\_col}(\mathsf{Y}, k)$.
(8) Form an ID of the $m \times k$ matrix of chosen columns: $[I_s, \sim] = \mathtt{ID\_row}(\mathsf{A}(:, J_s), k)$.
(9) Find the matrix U by solving the least squares equation $\mathsf{U}\mathsf{A}(I_s, :) = \mathsf{Z}$.

Algorithm 11.2.6. A randomized algorithm for computing a CUR decomposition of A via randomized sampling. For $q = 0$, the scheme is fast and accurate for matrices whose singular values decay rapidly. For matrices whose singular values decay slowly, one should pick a larger q (say $q = 1$ or 2) to improve accuracy at the cost of longer execution time. If accuracy better than $\epsilon_{\text{mach}}^{1/(2q+1)}$ is desired, then the scheme should be modified to incorporate orthonormalization as described in Remark 8.1.2.

12. Adaptive rank determination with updating of the matrix

12.1. Problem formulation Up to this point, we have assumed that the rank k is given as an input variable to the factorization algorithm. In practical usage, it is common that we are given instead a matrix A and a computational tolerance ε, and our task is then to determine a matrix A_k of rank k such that $\|\mathsf{A} - \mathsf{A}_k\| \leqslant \varepsilon$.

The techniques described in this section are designed for dense matrices stored in RAM. They directly update the matrix, and come with a firm guarantee that the computed low rank approximation is within distance ε of the original matrix. There are many situations where direct updating is not feasible and we can in practice only interact with the matrix via the matrix-vector multiplication (e.g., very large matrices stored out-of-core, sparse matrices, matrices that are defined implicitly). Section 13 describes algorithms designed for this environment that use randomized sampling techniques to *estimate* the approximation error.

Recall that for the case where a computational tolerance is given (rather than a rank), the optimal solution is given by the SVD. Specifically, let $\{\sigma_j\}_{j=1}^{\min(m,n)}$ be the singular values of A, and let ε be a given tolerance. Then the minimal rank k for which there exists a matrix B of rank k that is within distance ε of A, is the is the smallest integer k such that $\sigma_{k+1} \leqslant \varepsilon$. The algorithms described here will determine a k that is not necessarily optimal, but is typically fairly close.

12.2. A greedy updating algorithm Let us start by describing a general algorithmic template for how to compute an approximate rank-k approximate factorization of a matrix. To be precise, suppose that we are given an $m \times n$ matrix A, and a computational tolerance ε. Our objective is then to determine an integer $k \in \{1, 2, \dots, \min(m, n)\}$, an $m \times k$ orthonormal matrix Q_k, and a $k \times n$ matrix B_k such that $\|A - Q_k B_k\| \leqslant \varepsilon$. Algorithm 12.2.1 outlines how one might in a greedy fashion build the matrices Q_k and B_k, adding one column to Q_k and one row to B_k at each step.

(1) $Q_0 = [\,];\ B_0 = [\,];\ A_0 = A;\ k = 0;$
(2) **while** $\|A_k\| > \varepsilon$
(3) $\quad k = k + 1$
(4) $\quad$ Pick a vector $y \in \text{Ran}(A_{k-1})$
(5) $\quad q = y/\|y\|;$
(6) $\quad b = q^* A_{k-1};$
(7) $\quad Q_k = [Q_{k-1}\ q];$
(8) $\quad B_k = \begin{bmatrix} B_{k-1} \\ b \end{bmatrix};$
(9) $\quad A_k = A_{k-1} - qb;$
(10) **end for**

Algorithm 12.2.1. A greedy algorithm for building a low-rank approximation to a given $m \times n$ matrix A that is accurate to within a given precision ε. To be precise, the algorithm determines an integer k, an $m \times k$ orthonormal matrix Q_k and a $k \times n$ matrix $B_k = Q_k^* A$ such that $\|A - Q_k B_k\| \leqslant \varepsilon$. One can easily verify that after step j, we have $A = Q_j B_j + A_j$.

Algorithm 12.2.1 is a generalization of the classical Gram-Schmidt procedure. The key to understanding how the algorithm works is provided by the identity

$$A = Q_j B_j + A_j, \qquad j = 0, 1, 2, \ldots, k.$$

The computational efficiency and accuracy of the algorithm depend crucially on the strategy for picking the vector y on line (4). Let us consider three possibilities:

Pick the largest remaining column Suppose we instantiate line (4) by letting y be simply the largest column of the remainder matrix A_{k-1}.

(4) Set $j_k = \text{argmax}\{\|A_{k-1}(:, j)\| : j = 1, 2, \ldots, n\}$ and then $y = A_{k-1}(:, j_k)$.

With this choice, Algorithm 12.2.1 is *precisely* column pivoted Gram-Schmidt (CPQR). This algorithm is reasonably efficient, and often leads to fairly close to optimal low-rank approximation. For instance, when the singular values of A decay rapidly, CPQR determines a numerical rank k that is typically reasonably close to the theoretically exact ε-rank. However, this is not always the case even when the singular values decay rapidly, and the results can be quite poor when the singular values decay slowly. (A striking illustration of how suboptimal CPQR can be for purposes of low-rank approximation is provided by the famous "Kahan counter-example," see [28, Sec. 5].)

Pick the locally optimal vector A choice that is natural and conceptually simple is to pick the vector y by solving the obvious minimization problem:

(4) $$y = \text{argmin}\{\|A_{k-1} - yy^* A_{k-1}\| : \|y\| = 1\}.$$

With this choice, the algorithm will produce matrices that attain the theoretically optimal precision

$$\|A - Q_j B_j\| = \sigma_{j+1}.$$

This tells us that the greediness of the algorithm is not a problem. However, this strategy is impractical since solving the local minimization problem is computationally hard.

A randomized selection strategy Suppose now that we pick y by forming a linear combination of the columns of A_{k-1} with the expansion weights drawn from a normalized Gaussian distribution:

(4) Draw a Gaussian random vector $g \in \mathbb{R}^n$ and set $y = A_{k-1} g$.

With this choice, the algorithm becomes logically equivalent to the basic randomized SVD given in Algorithm 4.0.1. This means that this choice often leads to a factorization that is close to optimally accurate, and is also computationally efficient.

One can attain higher accuracy by trading away some computational efficiency to incorporate a couple of steps of power iteration, and, for some small integer q—say $q = 1$ or $q = 2$—choosing $y = (A_{k-1} A_{k-1}^*)^q A_{k-1} g$.

12.3. A blocked updating algorithm A key benefit of the randomized greedy algorithm described in Subsection 12.2 is that it can easily be *blocked*. In other words, given a block size b, we can at each step of the iteration draw a set of b Gaussian random vectors, compute the corresponding sample vectors, and then extend the factors Q and B by adding b columns and b rows at a time, respectively. The result is shown in Algorithm 12.3.1.

(1) $Q = [\,];\ B = [\,];$
(2) **while** $\|A\| > \varepsilon$
(3)   Draw an $n \times b$ Gaussian random matrix G.
(4)   Compute the $m \times b$ matrix $Q_{\text{new}} = \text{qr}(AG, 0)$.
(5) $B_{\text{new}} = Q_{\text{new}}^{*} A$
(6) $Q = [Q\ Q_{\text{new}}]$
(7) $B = \begin{bmatrix} B \\ B_{\text{new}} \end{bmatrix}$
(8) $A = A - Q_{\text{new}} B_{\text{new}}$
(9) **end while**

Algorithm 12.3.1. A greedy algorithm for building a low-rank approximation to a given $m \times n$ matrix A that is accurate to within a given precision ε. This algorithm is a blocked analogue of the method described in Algorithm 12.2.1 and takes as input a block size b. Its output is an orthonormal matrix Q of size $m \times k$ (where k is a multiple of b) and a $k \times n$ matrix B such that $\|A - QB\| \leqslant \varepsilon$. For higher accuracy, one can incorporate a couple of steps of power iteration and set $Q_{\text{new}} = \text{qr}((AA^{*})^{q}AR, 0)$ on Line (4).

12.4. Evaluating the norm of the residual The algorithms described in this section contain one step that could be computationally expensive unless some care is exercised. The potential problem concerns the evaluation of the norm of the remainder matrix $\|A_k\|$ (cf. Line (2) in Algorithm 12.2.1) at each step of the iteration. When the Frobenius norm is used, this evaluation can be done very efficiently, as follows: When the computation starts, evaluate the norm of the input matrix

$$a = \|A\|_{\text{Fro}}.$$

Then observe that after step j completes, we have

$$A = \underbrace{Q_j B_j}_{=Q_j Q_j^{*} A} + \underbrace{A_j}_{=(I - Q_j Q_j^{*})A}.$$

Since the columns in the first term all lie in $\text{Col}(Q_j)$, and the columns of the second term all lie in $\text{Col}(Q_j)^{\perp}$, we now find that

$$\|A\|_{\text{Fro}}^2 = \|Q_j B_j\|_{\text{Fro}}^2 + \|A_j\|_{\text{Fro}}^2 = \|B_j\|_{\text{Fro}}^2 + \|A_j\|_{\text{Fro}}^2,$$

where in the last step we used that $\|Q_j B_j\|_{\text{Fro}} = \|B_j\|_{\text{Fro}}$ since Q_j is orthonormal. In other words, we can easily compute $\|A_j\|_{\text{Fro}}$ via the identity

$$\|A_j\|_{\text{Fro}} = \sqrt{a^2 - \|B_j\|_{\text{Fro}}^2}.$$

The idea here is related to "down-dating" schemes for computing column norms when executing a column pivoted QR factorization, as described in, e.g., [48, Chapter 5, Section 2.1].

When the spectral norm is used, one could use a power iteration to compute an estimate of the norm of the matrix. Alternatively, one can use the randomized procedure described in Section 13 which is faster, but less accurate.

13. Adaptive rank determination without updating the matrix

The techniques described in Section 12 for computing a low rank approximation to a matrix A that is valid to a *given tolerance* (as opposed to a *given rank*) are highly computationally efficient whenever the matrix A itself can be easily updated (e.g. a dense matrix stored in RAM). In this section, we describe algorithms for solving the "given tolerance" problem that do not need to explicitly update the matrix; this comes in handy for sparse matrices, for matrices stored out-of-core, for matrices defined implicitly, etc. In such a situation, it often works well to use randomized methods to estimate the norm of the residual matrix $A - QB$. The framing for such a randomized estimator is that given a tolerated risk probability p, we can cheaply compute a bound for $\|A - QB\|$ that is valid with probability at least $1 - p$.

As a preliminary step in deriving the update-free scheme, let us reformulate the basic RSVD in Algorithm 4.0.1 as the sequential algorithm shown in Algorithm 13.0.1 that builds the matrices Q and B one vector at a time. We observe that this method is similar to the greedy template shown in Algorithm 12.2.1, except that there is not an immediately obvious way to tell when $\|A - Q_j B_j\|$ becomes small enough.

However, it is possible to *estimate* this quantity quite easily. The idea is that once Q_j becomes large enough to capture "most" of the range of A, then the sample vectors y_j drawn will all approximately lie in the span of Q_j, which is to say that the projected vectors z_j will become very small. In other words, once we start to see a sequence of vectors z_j that are all very small, we can reasonably deduce that the basis we have on hand very likely covers most of the range of A.

(1)	$Q_0 = [\,]; B_0 = [\,];$	
(2)	**for** $j = 1, 2, 3, \ldots$	
(3)		Draw a Gaussian random vector $g_j \in \mathbb{R}^n$ and set $y_j = Ag_j$
(4)		Set $z_j = y_j - Q_{j-1}Q_{j-1}^* y_j$ and then $q_j = z_j/\|z_j\|$.
(5)		$Q_j = [Q_{j-1}\ q_j]$
(6)		$B_j = \begin{bmatrix} B_{j-1} \\ q_j^* A \end{bmatrix}$
(7)	**end for**	

Algorithm 13.0.1. A randomized range finder that builds an orthonormal basis $\{q_1, q_2, q_3, \ldots\}$ for the range of A one vector at a time. This algorithm is mathematically equivalent to the basic RSVD in Algorithm 4.0.1 in the sense that if $G = [g_1\ g_2\ g_3\ \ldots]$, then the vectors $\{q_j\}_{j=1}^p$ form an orthonormal basis for $AG(:, 1 : p)$ for both methods. Observe that $z_j = (A - Q_{j-1}Q_{j-1}^* A)g_j$, cf. (13.0.2).

To make the discussion in the previous paragraph more mathematically rigorous, let us first observe that each projected vector z_j satisfies the relation

$$z_j = y_j - Q_{j-1}Q_{j-1}^* y_j = Ag_j - Q_{j-1}Q_{j-1}^* Ag_j = (A - Q_{j-1}Q_{j-1}^* A)g_j. \tag{13.0.2}$$

Next, we use the fact that if T is any matrix, then by looking at the magnitude of $\|Tg\|$, where g is a Gaussian random vector, we can deduce information about the spectral norm of T. The precise result that we need is the following, cf. [53, Sec. 3.4] and [26, Lemma 4.1].

Lemma 13.0.3. *Let T be a real $m \times n$ matrix. Fix a positive integer r and a real number $\alpha \in (0, 1)$. Draw an independent family $\{g_i : i = 1, 2, \ldots, r\}$ of standard Gaussian vectors. Then*

$$\|T\| \leqslant \frac{1}{\alpha}\sqrt{\frac{2}{\pi}} \max_{i=1,\ldots,r} \|Tg_i\|$$

with probability at least $1 - \alpha^r$.

In applying this result, we set $\alpha = 1/10$, whence it follows that if $\|z_j\|$ is smaller than the resulting threshold for r vectors in a row, then $\|A - Q_jB_j\| \leqslant \varepsilon$ with probability at least $1 - 10^{-r}$. The result is shown in Algorithm 13.0.4. Observe that choosing $\alpha = 1/10$ will work well only if the singular values decay reasonably fast.

(1) Draw standard Gaussian vectors $g_1, \ldots, g_r$ of length n.
(2) For $i = 1, 2, \ldots, r$, compute $y_i = A g_i$.
(3) $j = 0$.
(4) $Q_0 = [\,]$, the $m \times 0$ empty matrix.
(5) **while** $\max\left\{\|y_{j+1}\|, \|y_{j+2}\|, \ldots, \|y_{j+r}\|\right\} > \varepsilon/(10\sqrt{2/\pi})$,
(6) $j = j + 1$.
(7)   Overwrite y_j by $y_j - Q_{j-1}(Q_{j-1})^* y_j$.
(8) $q_j = y_j / |y_j|$.
(9) $Q_j = [Q_{j-1}\ q_j]$.
(10)   Draw a standard Gaussian vector g_{j+r} of length n.
(11) $y_{j+r} = \left(I - Q_j (Q_j)^*\right) A g_{j+r}$.
(12)   **for** $i = (j+1), (j+2), \ldots, (j+r-1)$,
(13)   Overwrite y_i by $y_i - q_j \langle q_j, y_i \rangle$.
(14)   **end for**
(15) **end while**
(16) $Q = Q_j$.

Algorithm 13.0.4. A randomized range finder. Given an $m \times n$ matrix A, a tolerance ε, and an integer r, the algorithm computes an orthonormal matrix Q such that $\|A - QQ^*A\| \leqslant \varepsilon$ holds with probability at least $1 - \min\{m, n\} 10^{-r}$. Line (7) is mathematically redundant but improves orthonormality of Q in the presence of round-off errors. (Adapted from Algorithm 4.2 of [26].)

Remark 13.0.5. While the proof of Lemma 13.0.3 is outside the scope of these lectures, it is perhaps instructive to prove a much simpler related result that says that if T is any $m \times n$ matrix, and $g \in \mathbb{R}^n$ is a standard Gaussian vector, then

$$\mathbb{E}[\|Tg\|^2] = \|T\|_{\text{Fro}}^2. \tag{13.0.6}$$

To prove (13.0.6), let T have the singular value decomposition $T = UDV^*$, and set $\tilde{g} = V^* g$. Then

$$\|Tg\|^2 = \|UDV^*g\|^2 = \|UD\tilde{g}\|^2 = \{U \text{ is unitary}\} = \|D\tilde{g}\|^2 = \sum_{j=1}^{n} \sigma_j^2 \tilde{g}_j^2.$$

Then observe that since the distribution of Gaussian vectors is rotationally invariant, the vector $\tilde{g} = V^* g$ is also a standardized Gaussian vector, and so $\mathbb{E}[\tilde{g}_j^2] = 1$. Since the variables $\{\tilde{g}_j\}_{j=1}^n$ are independent, it follows that

$$\mathbb{E}[\|Tg\|^2] = \mathbb{E}\left[\sum_{j=1}^{n} \sigma_j^2 \tilde{g}_j^2\right] = \sum_{j=1}^{n} \sigma_j^2 \mathbb{E}[\tilde{g}_j^2] = \sum_{j=1}^{n} \sigma_j^2 = \|T\|_{\text{Fro}}^2.$$

14. Randomized algorithms for computing a rank-revealing QR decomposition

Up until now, all methods discussed have concerned the problems of computing a low-rank approximation to a given matrix. These methods were designed explicitly for the case where the rank k is substantially smaller than the matrix dimensions m and n. In the last two sections, we will describe some recent developments that illustrate how randomized projections can be used to accelerate matrix factorization algorithms for any rank k, including *full* factorizations where $k = \min(m, n)$. These new algorithms offer "one stop shopping" in that they are faster than traditional algorithms in essentially every computational regime. We start in this section with a randomized algorithm for computing full and partial column pivoted QR (CPQR) factorizations.

The material in this section assumes that the reader is familiar with the classical Householder QR factorization procedure, and with the concept of *blocking* to accelerate matrix factorization algorithms. It is intended as a high-level introduction to randomized algorithms for computing full factorizations of matrices. For details, we refer the reader to [33, 36].

14.1. Column pivoted QR decomposition Given an $m \times n$ matrix A, with $m \geqslant n$, we recall that the column pivoted QR factorization (CPQR) takes the form, cf. (10.3.1),

$$\underset{m\times n}{A}\;\underset{n\times n}{P} = \underset{m\times n}{Q}\;\underset{n\times n}{R}, \tag{14.1.1}$$

where P is a permutation matrix, where Q is an orthogonal matrix, and where R is upper triangular. A standard technique for computing the CPQR of a matrix is the Householder QR process, which we illustrate in Figure 14.1.2. The process requires $n-1$ steps to drive A to upper triangular form from a starting point of $A_0 = A$. At the ith step, we form

$$A_i = Q_i^* A_{i-1} P_i$$

where P_i is a permutation matrix that flips the ith column of A_{i-1} with the column of largest magnitude in $A_{i-1}(:, i : n)$. The column moved into the ith place is called the *pivot column*. The matrix Q_i is a so called *Householder reflector* that zeros out all elements beneath the diagonal in the pivot column. In other words, if we let c_i denote the pivot column, then [2]

$$Q_i^* c_i = \begin{bmatrix} c_i(1 : (i-1)) \\ \pm\|c_i(i : m)\| \\ 0 \end{bmatrix}.$$

[2]The matrix Q_i is in fact symmetric, so $Q_i = Q_i^*$, but we keep the transpose symbol in the formula for consistency with the remainder of the section.

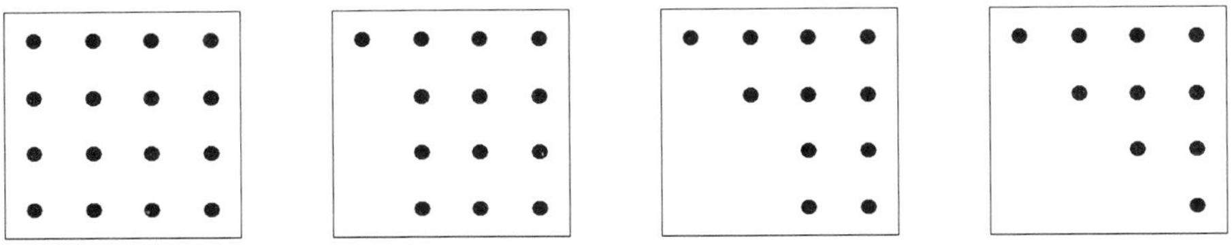

Figure 14.1.2. Basic QR factorization process. The $n \times n$ matrix A is driven to upper triangular form in $n-1$ steps. (Shown for $n = 4$.) At step i, we form $A_i = Q_i^* A_{i-1} P_i$ where P_i is a permutation matrix that moves the largest column in $A_{i-1}(:, 1 : n)$ to the ith position, and Q_i is a Householder reflector that zeros out all elements under the diagonal in the pivot column.

Once the process has completed, the matrices Q and P in (14.1.1) are given by

$$Q = Q_{n-1} Q_{n-2} \cdots Q_1, \qquad \text{and} \qquad P = P_{n-1} P_{n-2} \cdots P_1.$$

For details, see, e.g., [20, Sec. 5.2].

The Householder QR factorization process is a celebrated algorithm that is exceptionally stable and accurate. However, it has a serious weakness in that it executes rather slowly on modern hardware, in particular on systems involving many cores, when the matrix is stored in distributed memory or on a hard drive, etc. The problem is that it inherently consists of a sequence of $n-1$ rank-1 updates (so called BLAS2 operations), which makes the process very communication intensive. In principle, the resolution to this problem is to *block* the process, as shown in Figure 14.1.3.

Let b denote a block size; then in a blocked Householder QR algorithm, we would find groups of b pivot vectors that are moved into the active b slots at once, then b Householder reflectors would be determined by processing the b pivot columns, and then the remainder of the matrix would be updated jointly. Such a blocked algorithm would expend most of its flops on matrix-matrix multiplications (so called BLAS3 operations), which execute very rapidly on a broad range of computing hardware. Many techniques for blocking Householder QR have been proposed over the years, including, e.g., [3,4].

It was recently observed [36] that randomized sampling is ideally suited for resolving the long-standing problem of how to find groups of pivot vectors. The key observation is that a measure of quality for a group of b pivot vectors is its spanning volume in $\mathbb{R}^m$. This turns out to be closely related to how good of a basis these vectors form for the column space of the matrix [21,23,37]. As we saw in Section 10, this task is particularly well suited to randomized sampling. Precisely, consider the task of identifying a group of b good pivot vectors in the first step of the blocked QR process shown in Figure 14.1.3. Using the procedures described in Subsection 10.4, we proceed as follows: Fix an over-sampling parameter p, say $p = 10$. Then draw a Gaussian random matrix G of size $(b + p) \times m$, and form

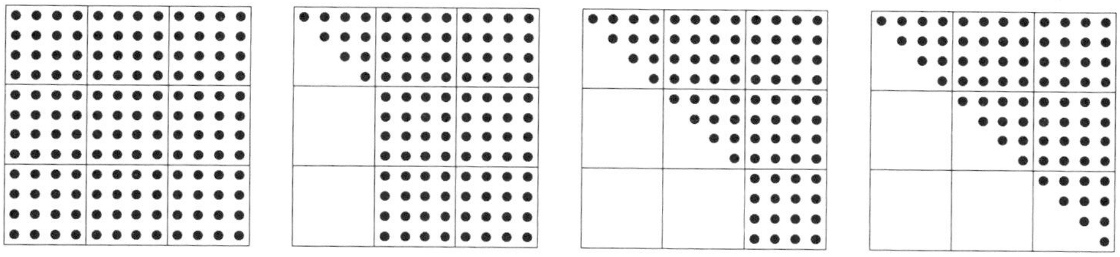

$A_0 = A$ $A_1 = Q_1^* A_0 P_1$ $A_2 = Q_2^* A_1 P_2$ $A_3 = Q_3^* A_2 P_3$

Figure 14.1.3. Blocked QR factorization process. The matrix A consists of $p \times p$ blocks of size $b \times b$ (shown for $p = 3$ and $b = 4$). The matrix is driven to upper triangular form in p steps. At step i, we form $A_i = Q_i^* A_{i-1} P_i$ where P_i is a permutation matrix, and Q_i is a product of b Householder reflectors.

the sampling matrix $Y = GA$. Then simply perform column pivoted QR on the columns of Y. To summarize, we determine P_1 as follows:

$$\begin{aligned} G &= \mathsf{randn}(b + p, m), \\ Y &= GA, \\ [\sim, \sim, P_1] &= \mathsf{qr}(Y, 0). \end{aligned}$$

Observe that the QR factorization of Y is affordable since Y is small, and fits in fast memory close to the processor. For the remaining steps, we simply apply the same idea to find the best spanning columns for the lower right block in A_i that has not yet been driven to upper triangular form. The resulting algorithm is called *Householder QR with Randomization for Pivoting (HQRRP)*; it is described in detail in [36], and is available at `https://github.com/flame/hqrrp/`. (The method described in this section was first published in [33], but is closely related to the independently discovered results in [13].)

To maximize performance, it turns out to be possible to "downdate" the sampling matrix from one step of the factorization to the next, in a manner similar to how downdating of the pivot weights are done in classical Householder QR[48, Ch.5, Sec. 2.1]. This obviates the need to draw a new random matrix at each step [13, 36], and reduces the leading term in the asymptotic flop count of HQRRP to $2mn^2 - (4/3)n^3$, which is identical to classical Householder QR.

15. A strongly rank-revealing UTV decomposition

This section describes a randomized algorithm `randUTV` that is very similar to the randomized QR factorization process described in Section 14 but that results in a so called "UTV factorization." The new algorithm has several advantages:

- `randUTV` provides close to optimal low-rank approximation, and highly accurate estimates for the singular values of a matrix.

- The algorithm randUTV builds the factorization (15.1.1) *incrementally,*. This means that when it is applied to a matrix of numerical rank k, the algorithm can be stopped early and incur an overall cost of $O(mnk)$.
- Like HQRRP, the algorithm randUTV is *blocked*, which enables it to execute fast on modern hardware.
- The algorithm randUTV is not an iterative algorithm. In this regard, it is closer to the CPQR than standard SVD algorithms, which substantially simplifies software optimization.

15.1. The UTV decomposition Given an $m \times n$ matrix A, with $m \geqslant n$, a "UTV decomposition" of A is a factorization the form

$$\underset{m\times n}{A} = \underset{m\times m}{U} \; \underset{m\times n}{T} \; \underset{n\times n}{V^*}, \tag{15.1.1}$$

where U and V are unitary matrices, and T is a triangular matrix (either lower or upper triangular). The UTV decomposition can be viewed as a generalization of other standard factorizations such as, e.g., the *Singular Value Decomposition (SVD)* or the *Column Pivoted QR decomposition (CPQR)*. (To be precise, the SVD is the special case where T is diagonal, and the CPQR is the special case where V is a permutation matrix.) The additional flexibility inherent in the UTV decomposition enables the design of efficient updating procedures, see [48, Ch. 5, Sec. 4] and [16].

15.2. An overview of randUTV The algorithm randUTV follows the same general pattern as HQRRP, as illustrated in Figure 15.2.1. Like HQRRP, it drives a given matrix $A = A_0$ to upper triangular form via a sequence of steps

$$A_i = U_i^* A_{i-1} V_i,$$

where each U_i and V_i is a unitary matrix. As in Section 14, we let b denote a block size, and let $p = \lceil n/b \rceil$ denote the number of steps taken. The key difference from

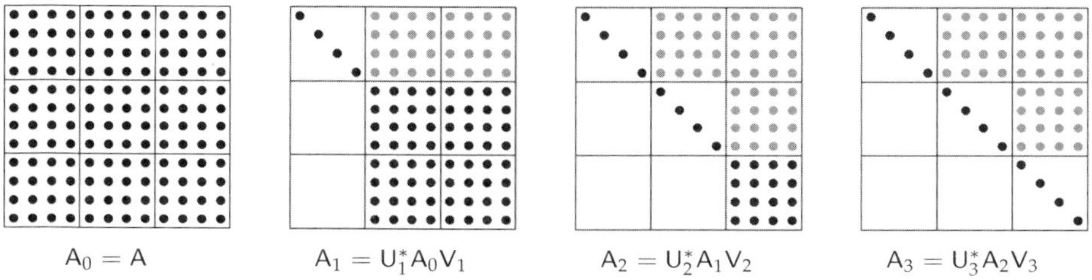

Figure 15.2.1. Blocked UTV factorization process. The matrix A consists of $p \times p$ blocks of size $b \times b$ (shown for $p = 3$ and $b = 4$). The matrix is driven to upper triangular form in p steps. At step i, we form $A_i = U_i^* A_{i-1} V_i$ where U_i and V_i consist (mostly) of a product of b Householder reflectors. The elements shown in grey are not zero, but are very small in magnitude.

HQRRP is that we in randUTV allow the matrices V_i to consist (mostly) of a product of b Householder reflectors. This added flexibility over CPQR allows us to drive more mass onto the diagonal entries, and thereby render the off-diagonal entries in the final matrix T very small in magnitude.

15.3. A single step block factorization The algorithm randUTV consists of repeated application of a randomized technique for building approximations to the spaces spanned by the dominant b left and right singular vectors, where b is a given block size. To be precise, given an $m \times n$ matrix A, we seek to build unitary matrices U_1 and V_1 such that

$$A = U_1 A_1 V_1^*$$

where A_1 has the block structure

$$A_1 = \begin{bmatrix} A_{1,11} & A_{1,12} \\ 0 & A_{1,22} \end{bmatrix} =$$

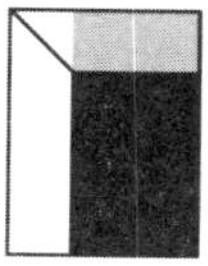

so that $A_{1,11}$ is diagonal, and $A_{1,12}$ has entries of small magnitude.

We first build V_1. To this end, we use the randomized power iteration described in Section 8 to build a sample matrix Y of size $b \times n$ whose columns approximately span the same subspace as the b dominant right singular vectors of A. To be precise, we draw a $b \times m$ Gaussian random matrix G and form the sample matrix

$$Y = GA(A^*A)^q,$$

where q is a parameter indicating the number of steps of power iteration taken (in randUTV, the gain from over-sampling is minimal and is generally not worth the bother). Then we form a unitary matrix $\tilde{V}$ whose first b columns form an orthonormal basis for the column space of Y. (The matrix V consists of a product of b Householder reflectors, which are determined by executing the standard Householder QR procedure on the columns of Y^*.) We then execute b steps of Householder QR on the matrix $A\tilde{V}$ to form a matrix $\tilde{U}$ consisting of a product of b Householder reflectors. This leaves us with a new matrix

$$\tilde{A} = (\tilde{U})^* A \tilde{V}$$

that has the block structure

$$\tilde{A} = \begin{bmatrix} \tilde{A}_{11} & \tilde{A}_{12} \\ 0 & \tilde{A}_{22} \end{bmatrix} =$$

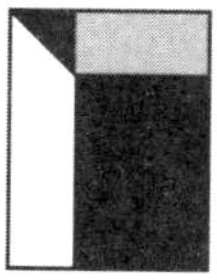

In other words, the top left $b \times b$ block is upper triangular, and the bottom left block is zero. One can also show that all entries of $\tilde{A}_{12}$ are typically small in magnitude.

```
function [U,T,V] = randUTV(A,b,q)
    T = A;
    U = eye(size(A,1));
    V = eye(size(A,2));
    for i = 1 : ceil(size(A,2)/b)
        I1 = 1 : (b(i-1));
        I2 = (b(i-1)+1) : size(A,1);
        J2 = (b(i-1)+1) : size(A,2);
        if (length(J2) > b)
            [Û,T̂,V̂] = stepUTV(T(I2,J2),b,q);
        else
            [Û,T̂,V̂] = svd(T(I2,J2));
        end if
        U(:,I2) = U(:,I2) * Û;
        V(:,J2) = V(:,J2) * V̂;
        T(I2,J2) = T̂;
        T(I1,J2) = T̂(I1,J2) * V̂;
    end for
return
function [U,T,V] = stepUTV(A,b,q)
    G = randn(size(A,1),b);
    Y = A*G;
    for i = 1 : q
        Y = A*(AY);
    end for
    [V,~] = qr(Y);
    [U,D,W] = svd(AV(:,1 : b));
    T = [D, U*AV(:,(b+1) : end)];
    V(:,1 : b) = V(:,1 : b) * W;
return
```

Algorithm 15.3.1. The algorithm randUTV (described in Section 15) that given an $m \times n$ matrix A computes its UTV factorization $A = UTV^*$, cf. (15.1.1). The input parameters b and q reflect the block size and the number of steps of power iteration, respectively. The single step function stepUTV is described in Subsection 15.3. (Observe that most of the unitary matrices that arise consist of products of Householder reflectors; this property must be exploited to attain computational efficiency.)

Next, we compute a full SVD of the block $\tilde{A}_{11} = \hat{U} D_{11} \hat{V}^*$. This step is affordable since $\tilde{A}_{11}$ is of size $b \times b$, where b is small. Then we form the transformation matrices

$$U_1 = \tilde{U} \begin{bmatrix} \hat{U} & 0 \\ 0 & I_{m-b} \end{bmatrix}, \qquad \text{and} \qquad V_1 = \tilde{V} \begin{bmatrix} \hat{V} & 0 \\ 0 & I_{n-b} \end{bmatrix},$$

and set

$$A_1 = U_1^* A V_1.$$

One can demonstrate that the diagonal entries of D_{11} typically form accurate approximations to first b singular values of A, and that

$$\|A_{1,22}\| \approx \inf\{\|A - B\| : B \text{ has rank } b\}.$$

Once the first b columns and rows of A have been processed as described in this Section, randUTV then applies the same procedure to the remaining block $A_{1,22}$, of size $(m-b) \times (n-b)$, and then continues in the same fashion to process all remaining blocks, as outlined in Subsection 15.2. The full algorithm is summarized in Algorithm 15.3.1.

For more information about the UTV factorization, including careful numerical experiments that illustrate how it compares in terms of speed and accuracy to competitors such as column pivoted QR and the traditional SVD, see [35]. Codes are available for download from https://github.com/flame/randutv.

Remark 15.3.2. As discussed in Section 8, the rows of Y approximately span the linear space spanned by the b dominant right singular vectors of A. The first b columns of V_1 simply form an orthonormal basis for $\text{Row}(Y^*)$. Analogously, the first b columns of U_1 approximately form an orthonormal basis for the b dominant left singular vectors of A. However, the additional application of A that is implicit in forming the matrix $A\tilde{V}$ provides a boost in accuracy, and the rank-b approximation resulting from one step of randUTV is similar to the accuracy obtained from the randomized algorithm in Section 8, but with "$q + 1/2$ steps" of power iteration, rather than q steps. (We observe that if the first b columns of $\tilde{U}$ and $\tilde{V}$ spanned *exactly* the spaces spanned by the dominant b left and right singular vectors of A, then we would have $\tilde{A}_{12} = 0$, and the singular values of $\tilde{A}_{11}$ would be identical to the top b singular values of A.)

References

[1] N. Ailon and B. Chazelle, *Approximate nearest neighbors and the fast johnson-lindenstrauss transform*, Proceedings of the thirty-eighth annual acm symposium on theory of computing, 2006, pp. 557–563. ←200

[2] N. Ailon and E. Liberty, *An almost optimal unrestricted fast Johnson-Lindenstrauss transform*, ACM Transactions on Algorithms (TALG) **9** (2013), no. 3, 21. ←200

[3] C. H. Bischof and G. Quintana-Ortí, *Algorithm 782: Codes for rank-revealing QR factorizations of dense matrices*, ACM Transactions on Mathematical Software (TOMS) **24** (1998), no. 2, 254–257. ←222

[4] C. H. Bischof and G. Quintana-Ortí, *Computing rank-revealing QR factorizations of dense matrices*, ACM Transactions on Mathematical Software (TOMS) **24** (1998), no. 2, 226–253. ←222

[5] C. Boutsidis, M. W Mahoney, and P. Drineas, *An improved approximation algorithm for the column subset selection problem*, Proceedings of the twentieth annual acm-siam symposium on discrete algorithms, 2009, pp. 968–977. ←207

[6] H. Cheng, Z. Gimbutas, P.-G. Martinsson, and V. Rokhlin, *On the compression of low rank matrices*, SIAM Journal of Scientific Computing **26** (2005), no. 4, 1389–1404. ←207, 209

[7] P. Drineas, R. Kannan, and M. W. Mahoney, *Fast Monte Carlo algorithms for matrices. II. Computing a low-rank approximation to a matrix*, SIAM J. Comput. **36** (2006), no. 1, 158–183 (electronic). MR2231644 (2008a:68243) ←191

[8] P. Drineas, M. Magdon-Ismail, M. W Mahoney, and D. P Woodruff, *Fast approximation of matrix coherence and statistical leverage*, Journal of Machine Learning Research **13** (2012), no. Dec, 3475–3506. ←207

[9] P. Drineas and M. W. Mahoney, *On the Nyström method for approximating a Gram matrix for improved kernel-based learning*, J. Mach. Learn. Res. **6** (2005), 2153–2175. ←205

[10] P. Drineas and M. W. Mahoney, *CUR matrix decompositions for improved data analysis*, Proceedings of the National Academy of Sciences **106** (2009), no. 3, 697–702. ←207, 212

[11] P. Drineas and M. W. Mahoney, *Lectures on randomized numerical linear algebra*, The mathematics of data, 2018. ←191, 201

[12] P. Drineas, M. W. Mahoney, and S. Muthukrishnan, *Relative-error* CUR *matrix decompositions*, SIAM J. Matrix Anal. Appl. **30** (2008), no. 2, 844–881. MR2443975 (2009k:68269) ←212

[13] J. Duersch and M. Gu, *True blas-3 performance qrcp using random sampling*, 2015. ←223

[14] C. Eckart and G. Young, *The approximation of one matrix by another of lower rank*, Psychometrika **1** (1936), no. 3, 211–218. ←192

[15] Inc. Facebook, *Fast randomized svd*, 2016. ←190

[16] R. D Fierro, P. C. Hansen, and P. S. K. Hansen, *Utv tools: Matlab templates for rank-revealing utv decompositions*, Numerical Algorithms **20** (1999), no. 2-3, 165–194. ←224

[17] A. Frieze, R. Kannan, and S. Vempala, *Fast Monte Carlo algorithms for finding low-rank approximations*, J. ACM **51** (2004), no. 6, 1025–1041. (electronic). MR2145262 (2005m:65006) ←190, 191

[18] A. Gittens, A. Devarakonda, E. Racah, M. Ringenburg, L. Gerhardt, J. Kottalam, J. Liu, K. Maschhoff, S. Canon, J. Chhugani, P. Sharma, J. Yang, J. Demmel, J. Harrell, V. Krishnamurthy, M. W. Mahoney, and Prabhat, *Matrix factorizations at scale: A comparison of scientific data analytics in spark and C+MPI using three case studies*, 2016 IEEE International Conference on Big Data, 2016, pp. 204–213. ←190

[19] A. Gittens and M. W. Mahoney, *Revisiting the nyström method for improved large-scale machine learning*, J. Mach. Learn. Res. **17** (January 2016), no. 1, 3977–4041. ←206

[20] G. H. Golub and C. F. Van Loan, *Matrix computations*, Third, Johns Hopkins Studies in the Mathematical Sciences, Johns Hopkins University Press, Baltimore, MD, 1996. ←191, 222

[21] S. A. Goreinov, N. L. Zamarashkin, and E. E. Tyrtyshnikov, *Pseudo-skeleton approximations by matrices of maximal volume*, Mathematical Notes **62** (1997). ←222

[22] M. Gu, *Subspace iteration randomization and singular value problems*, SIAM Journal on Scientific Computing **37** (2015), no. 3, A1139–A1173. ←201

[23] M. Gu and S. C. Eisenstat, *Efficient algorithms for computing a strong rank-revealing QR factorization*, SIAM J. Sci. Comput. **17** (1996), no. 4, 848–869. MR97h:65053 ←207, 209, 222

[24] N. Halko, *Randomized methods for computing low-rank approximations of matrices*, Ph.D. Thesis, 2012. ←190

[25] N. Halko, P.-G. Martinsson, Y. Shkolnisky, and M. Tygert, *An algorithm for the principal component analysis of large data sets*, SIAM Journal on Scientific Computing **33** (2011), no. 5, 2580–2594. ←190

[26] N. Halko, P.-G. Martinsson, and J. A. Tropp, *Finding structure with randomness: Probabilistic algorithms for constructing approximate matrix decompositions*, SIAM Review **53** (2011), no. 2, 217–288. ←190, 193, 200, 201, 202, 204, 207, 219, 220

[27] D. C. Hoaglin and R. E. Welsch, *The Hat matrix in regression and ANOVA*, The American Statistician **32** (1978), no. 1, 17–22. ←212

[28] W. Kahan, *Numerical linear algebra*, Canadian Math. Bull **9** (1966), no. 6, 757–801. ←216

[29] D. M. Kane and J. Nelson, *Sparser johnson-lindenstrauss transforms*, J. ACM **61** (January 2014), no. 1, 4:1–4:23. ←200

[30] E. Liberty, *Accelerated dense random projections*, Ph.D. Thesis, 2009. ←200

[31] E. Liberty, F. Woolfe, P.-G. Martinsson, V. Rokhlin, and M. Tygert, *Randomized algorithms for the low-rank approximation of matrices*, Proc. Natl. Acad. Sci. USA **104** (2007), no. 51, 20167–20172. ←190

[32] M. W Mahoney, *Randomized algorithms for matrices and data*, Foundations and Trends® in Machine Learning **3** (2011), no. 2, 123–224. ←191

[33] P.-G. Martinsson, *Blocked rank-revealing qr factorizations: How randomized sampling can be used to avoid single-vector pivoting*, 2015. ←221, 223

[34] P.-G. Martinsson, *Randomized methods for matrix computations and analysis of high dimensional data* (2016), available at arXiv:1607.01649. ←203, 207

[35] P.-G. Martinsson, G. Quintana Orti, and N. Heavner, *randUTV: A blocked randomized algorithm for computing a rank-revealing UTV factorization* (2017), available at arXiv:1703.00998. ←190, 227

[36] P.-G. Martinsson, G. Quintana-Ortí, N. Heavner, and R. van de Geijn, *Householder qr factorization with randomization for column pivoting (HQRRP)*, SIAM Journal on Scientific Computing **39** (2017), no. 2, C96–C115. ←190, 210, 221, 222, 223

[37] P.-G. Martinsson, V. Rokhlin, and M. Tygert, *On interpolation and integration in finite-dimensional spaces of bounded functions*, Comm. Appl. Math. Comput. Sci (2006), 133–142. ←222

[38] P.-G. Martinsson and S. Voronin, *A randomized blocked algorithm for efficiently computing rank-revealing factorizations of matrices.*, 2015. To appear in *SIAM Journal on Scientific Computation*, arXiv:1503.07157. ←190, 210

[39] P.-G. Martinsson, V. Rokhlin, and M. Tygert, *A randomized algorithm for the approximation of matrices*, Technical Report Yale CS research report YALEU/DCS/RR-1361, Yale University, Computer Science Department, 2006. ←189, 190

[40] P.-G. Martinsson, V. Rokhlin, and M. Tygert, *A randomized algorithm for the decomposition of matrices*, Appl. Comput. Harmon. Anal. **30** (2011), no. 1, 47–68. MR2737933 (2011i:65066) ←190

[41] N. Mitrovic, M. T. Asif, U. Rasheed, J. Dauwels, and P. Jaillet, *Cur decomposition for compression and compressed sensing of large-scale traffic data*, Intelligent transportation systems-(itsc), 2013 16th international ieee conference on, 2013, pp. 1475–1480. ←212

[42] C. Musco and C. Musco, *Randomized block krylov methods for stronger and faster approximate singular value decomposition*, Proceedings of the 28th international conference on neural information processing systems, 2015, pp. 1396–1404. ←205

[43] F. Pourkamali-Anaraki and S. Becker, *Randomized clustered nystrom for large-scale kernel machines*, 2016. arXiv:1612.06470. ←206

[44] V. Rokhlin, A. Szlam, and M. Tygert, *A randomized algorithm for principal component analysis*, SIAM Journal on Matrix Analysis and Applications **31** (2009), no. 3, 1100–1124. ←205

[45] T. Sarlos, *Improved approximation algorithms for large matrices via random projections*, 2006 47th annual ieee symposium on foundations of computer science (focs'06), 2006, pp. 143–152. ←190, 191, 200

[46] D. C Sorensen and M. Embree, *A deim induced cur factorization*, SIAM Journal on Scientific Computing **38** (2016), no. 3, A1454–A1482. ←213

[47] G. W. Stewart, *On the early history of the singular value decomposition*, SIAM Rev. **35** (1993), no. 4, 551–566. MR1247916 (94f:15001) ←194

[48] G. W. Stewart, *Matrix algorithms volume 1: Basic decompositions*, SIAM, 1998. ←218, 223, 224

[49] S. Voronin and P.-G. Martinsson, *A CUR factorization algorithm based on the interpolative decomposition* (2014), available at arXiv:1412.8447. ←190, 207, 213

[50] S. Wang and Z. Zhang, *Improving CUR matrix decomposition and the Nyström approximation via adaptive sampling*, J. Mach. Learn. Res. **14** (2013), 2729–2769. MR3121656 ←207, 212

[51] R. Witten and E. Candes, *Randomized algorithms for low-rank matrix factorizations: sharp performance bounds*, Algorithmica **72** (2015), no. 1, 264–281. ←201

[52] D. P. Woodruff, *Sketching as a tool for numerical linear algebra*, Foundations and Trends in Theoretical Computer Science **10** (2014), no. 1Ű2, 1–157. ←191

[53] F. Woolfe, E. Liberty, V. Rokhlin, and M. Tygert, *A fast randomized algorithm for the approximation of matrices*, Applied and Computational Harmonic Analysis **25** (2008), no. 3, 335–366. ←200, 219

Mathematical Institute, Andrew Wiles Building, University of Oxford, Oxford, OX2 6GG, United Kingdom

Email address: martinsson@maths.ox.ac.uk

IAS/Park City Mathematics Series
Volume 25, Pages 231–271
https://doi.org/10.1090/pcms/025/00833

Four Lectures on Probabilistic Methods for Data Science

Roman Vershynin

Abstract. Methods of high-dimensional probability play a central role in applications for statistics, signal processing, theoretical computer science and related fields. These lectures present a sample of particularly useful tools of high-dimensional probability, focusing on the classical and matrix Bernstein's inequality and the uniform matrix deviation inequality. We illustrate these tools with applications for dimension reduction, network analysis, covariance estimation, matrix completion and sparse signal recovery. The lectures are geared towards beginning graduate students who have taken a rigorous course in probability but may not have any experience in data science applications.

Contents

Partially supported by NSF Grant DMS 1265782 and U.S. Air Force Grant FA9550-14-1-0009.

1. Lecture 1: Concentration of sums of independent random variables

These lectures present a sample of modern methods of high dimensional probability and illustrate these methods with applications in data science. This sample is not comprehensive by any means, but it could serve as a point of entry into a branch of modern probability that is motivated by a variety of data-related problems.

To get the most out of these lectures, you should have taken a graduate course in probability, have a good command of linear algebra (including the singular value decomposition) and be familiar with very basic concepts of functional analysis (familiarity with L^p norms should be enough).

All of the material of these lectures is covered more systematically, at a slower pace, and with a wider range of applications, in my forthcoming textbook [60]. You may also be interested in two similar tutorials: [58] is focused on random matrices, and a more advanced text [59] discusses high-dimensional inference problems.

It should be possible to use these lectures for a self-study or group study. You will find here many places where you are invited to do some work (marked in the text e.g. by "check this!"), and you are encouraged to do it to get a better grasp of the material. Each lecture ends with a section called "Notes" where you will find bibliographic references of the results just discussed, as well as various improvements and extensions.

We are now ready to start.

Probabilistic reasoning has a major impact on modern data science. There are roughly two ways in which this happens.

- *Randomized algorithms*, which perform some operations at random, have long been developed in computer science and remain very popular. Randomized algorithms are among the most effective methods – and sometimes the only known ones – for many data problems.
- *Random models of data* form the usual premise of statistical analysis. Even when the data at hand is deterministic, it is often helpful to think of it as a random sample drawn from some unknown distribution ("population").

In these lectures, we will encounter both randomized algorithms and random models of data.

1.1. Sub-gaussian distributions Before we start discussing probabilistic methods, we will introduce an important class of probability distributions that forms a natural "habitat" for random variables in many theoretical and applied problems. These are *sub-gaussian* distributions. As the name suggests, we will be looking at an extension of the most fundamental distribution in probability theory – the gaussian, or normal, distribution $N(\mu, \sigma)$.

It is a good exercise to check that the the following basic properties of the standard normal random variable $X \sim N(0,1)$:

Tails: $\mathbb{P}\left\{|X| \geqslant t\right\} \leqslant 2\exp(-t^2/2)$ for all $t \geqslant 0$.
Moments: $\|X\|_p := (\mathbb{E}|X|^p)^{1/p} = O(\sqrt{p})$ as $p \to \infty$.
MGF of square: [1] $\mathbb{E}\exp(cX^2) \leqslant 2$ for some $c > 0$.
MGF: $\mathbb{E}\exp(\lambda X) = \exp(\lambda^2)$ for all $\lambda \in \mathbb{R}$.

All these properties tell the same story from four different perspectives. It is not very difficult to show (although we will not do it here) that for any random variable X, not necessarily Gaussian, these four properties are essentially equivalent.

Proposition 1.1.1 (Sub-gaussian properties)**.** *For a random variable* X*, the following properties are equivalent.*[2]

Tails: $\mathbb{P}\left\{|X| \geqslant t\right\} \leqslant 2\exp(-t^2/K_1^2)$ *for all* $t \geqslant 0$.
Moments: $\|X\|_p \leqslant K_2\sqrt{p}$ *for all* $p \geqslant 1$.
MGF of square: $\mathbb{E}\exp(X^2/K_3^2) \leqslant 2$.

Moreover, if $\mathbb{E}X = 0$ *then these properties are also equivalent to the following one:*

MGF: $\mathbb{E}\exp(\lambda X) \leqslant \exp(\lambda^2 K_4^2)$ *for all* $\lambda \in \mathbb{R}$.

Random variables that satisfy one of the first three properties (and thus all of them) are called *sub-gaussian*. The best K_3 is called the *sub-gaussian norm* of X, and is usually denoted $\|X\|_{\psi_2}$, that is

$$\|X\|_{\psi_2} := \inf\left\{t > 0 : \ \mathbb{E}\exp(X^2/t^2) \leqslant 2\right\}.$$

One can check that $\|\cdot\|_{\psi_2}$ indeed defines a norm; it is an example of the general concept of the *Orlicz norm*. Proposition 1.1.1 states that the numbers K_i in all four properties are equivalent to $\|X\|_{\psi_2}$ up to absolute constant factors.

Example 1.1.2. As already noted, the standard normal random variable $X \sim N(0,1)$ is sub-gaussian. Similarly, arbitrary *normal* random variables $X \sim N(\mu, \sigma)$ are sub-gaussian. Another example is a *Bernoulli* random variable X that takes values 0 and 1 with probabilities 1/2 each. More generally, any *bounded* random variable X is sub-gaussian. On the contrary, Poisson, exponential, Pareto and Cauchy distributions are not sub-gaussian. (Verify all these claims; this is not difficult.)

[1]MGF stands for moment generation function.

[2]The parameters $K_i > 0$ appearing in these properties can be different. However, they may differ from each other by at most an absolute constant factor. This means that there exists an absolute constant C such that property 1 implies property 2 with parameter $K_2 \leqslant CK_1$, and similarly for every other pair or properties.

1.2. Hoeffding's inequality You may remember from a basic course in probability that the normal distribution $N(\mu, \sigma)$ has a remarkable property: the sum of independent normal random variables is also normal. Here is a version of this property for sub-gaussian distributions.

Proposition 1.2.1 (Sums of sub-gaussians). *Let $X_1, \dots, X_N$ be independent, mean zero, sub-gaussian random variables. Then $\sum_{i=1}^N X_i$ is a sub-gaussian, and*

$$\Big\| \sum_{i=1}^N X_i \Big\|_{\psi_2}^2 \leqslant C \sum_{i=1}^N \|X_i\|_{\psi_2}^2$$

where C is an absolute constant.[3]

Proof. Let us bound the moment generating function of the sum for any $\lambda \in \mathbb{R}$:

$$\begin{aligned}
\mathbb{E} \exp \Big(\lambda \sum_{i=1}^N X_i\Big) &= \prod_{i=1}^N \mathbb{E} \exp(\lambda X_i) \quad \text{(using independence)} \\
&\leqslant \prod_{i=1}^N \exp(C\lambda^2 \|X_i\|_{\psi_2}^2) \quad \text{(by last property in Proposition 1.1.1)} \\
&= \exp(\lambda^2 K^2) \quad \text{where } K^2 := C \sum_{i=1}^N \|X_i\|_{\psi_2}^2.
\end{aligned}$$

Using again the last property in Proposition 1.1.1, we conclude that the sum $S = \sum_{i=1}^N X_i$ is sub-gaussian, and $\|S\|_{\psi_2} \leqslant C_1 K$ where C_1 is an absolute constant. The proof is complete. □

Let us rewrite Proposition 1.2.1 in a form that is often more useful in applications, namely as a *concentration inequality*. To do this, we simply use the first property in Proposition 1.1.1 for the sum $\sum_{i=1}^N X_i$. We immediately get the following.

Theorem 1.2.2 (General Hoeffding's inequality). *Let $X_1, \dots, X_N$ be independent, mean zero, sub-gaussian random variables. Then, for every $t \geqslant 0$ we have*

$$\mathbb{P}\Big\{ \Big| \sum_{i=1}^N X_i \Big| \geqslant t \Big\} \leqslant 2 \exp\Big(- \frac{ct^2}{\sum_{i=1}^N \|X_i\|_{\psi_2}^2} \Big).$$

Hoeffding's inequality controls how far and with what probability a sum of independent random variables can deviate from its mean, which is zero.

1.3. Sub-exponential distributions Sub-gaussian distributions constitute a sufficiently wide class of distributions. Many results in probability and data science are proved nowadays for sub-gaussian random variables. Still, as we noted, there are some natural random variables that are not sub-gaussian. For example, the

[3] In the future, we will always denote positive absolute constants by C, c, C_1, etc. These numbers do not depend on anything. In most cases, one can get good bounds on these constants from the proof, but the optimal constants for each result are rarely known.

square X^2 of a normal random variable $X \sim N(0,1)$ is not sub-gaussian. (Check!) To cover examples like this, we will introduce the similar but weaker notion of *sub-exponential distributions*.

Proposition 1.3.1 (Sub-exponential properties). *For a random variable* X, *the following properties are equivalent, in the same sense as in Proposition* 1.1.1.

Tails: $\mathbb{P}\{|X| \geqslant t\} \leqslant 2\exp(-t/K_1)$ *for all* $t \geqslant 0$.
Moments: $\|X\|_p \leqslant K_2 p$ *for all* $p \geqslant 1$.
MGF of the square: $\mathbb{E}\exp(|X|/K_3) \leqslant 2$.

Moreover, if $\mathbb{E}X = 0$ *then these properties imply the following one:*

MGF: $\mathbb{E}\exp(\lambda X) \leqslant \exp(\lambda^2 K_4^2)$ *for* $|\lambda| \leqslant 1/K_4$.

Just like we did for sub-gaussian distributions, we call the best K_3 the *sub-exponential norm* of X and denote it by $\|X\|_{\psi_1}$, that is

$$\|X\|_{\psi_1} := \inf\{t > 0 : \mathbb{E}\exp(|X|/t) \leqslant 2\}.$$

All sub-exponential random variables are squares of sub-gaussian random variables. Indeed, inspecting the definitions you will quickly see that

$$\|X^2\|_{\psi_1} = \|X\|_{\psi_2}^2. \tag{1.3.2}$$

(Check!)

1.4. Bernstein's inequality A version of Hoeffding's inequality known as Bernstein's inequality holds for sub-exponential random variables. You may naturally expect to see a sub-exponential tail bound in this result. So it may come as a surprise that Bernstein's inequality actually has a mixture of *two* tails – sub-gaussian and sub-exponential. Let us state and prove the inequality first, and then we will comment on the mixture of the two tails.

Theorem 1.4.1 (Bernstein's inequality). *Let* $X_1, \dots, X_N$ *be independent, mean zero, sub-exponential random variables. Then, for every* $t \geqslant 0$ *we have*

$$\mathbb{P}\Big\{\Big|\sum_{i=1}^N X_i\Big| \geqslant t\Big\} \leqslant 2\exp\Big[-c\min\Big(\frac{t^2}{\sum_{i=1}^N \|X_i\|_{\psi_1}^2}, \frac{t}{\max_i \|X_i\|_{\psi_1}}\Big)\Big].$$

Proof. For simplicity, we will assume that $K = 1$ and only prove the one-sided bound (without absolute value); the general case is not much harder. Our approach will be based on bounding the *moment generating function* of the sum $S := \sum_{i=1}^N X_i$. To see how MGF can be helpful here, choose $\lambda \geqslant 0$ and use Markov's inequality to get

$$\mathbb{P}\{S \geqslant t\} = \mathbb{P}\{\exp(\lambda S) \geqslant \exp(\lambda t)\} \leqslant e^{-\lambda t}\,\mathbb{E}\exp(\lambda S). \tag{1.4.2}$$

Recall that $S = \sum_{i=1}^N X_i$ and use independence to express the right side of (1.4.2) as

$$e^{-\lambda t}\prod_{i=1}^N \mathbb{E}\exp(\lambda X_i).$$

(Check!) It remains to bound the MGF of each term X_i, and this is a much simpler task. If we choose λ small enough so that

$$0 < \lambda \leqslant \frac{c}{\max_i \|X_i\|_{\psi_1}}, \tag{1.4.3}$$

then we can use the last property in Proposition 1.3.1 to get

$$\mathbb{E}\exp(\lambda X_i) \leqslant \exp\left(C\lambda^2 \|X_i\|_{\psi_1}^2\right).$$

Substitute into (1.4.2) and conclude that

$$\mathbb{P}\{S \geqslant t\} \leqslant \exp\left(-\lambda t + C\lambda^2\sigma^2\right)$$

where $\sigma^2 = \sum_{i=1}^N \|X_i\|_{\psi_1}^2$. The left side does not depend on λ while the right side does. So we can choose λ that minimizes the right side subject to the constraint (1.4.3). When this is done carefully, we obtain the tail bound stated in Bernstein's inequality. (Do this!) □

Now, why does Bernstein's inequality have a mixture of two tails? The sub-exponential tail should of course be there. Indeed, even if the entire sum consisted of a single term X_i, the best bound we could hope for would be of the form $\exp(-ct/\|X_i\|_{\psi_1})$. The sub-gaussian term could be explained by the central limit theorem, which states that the sum should becomes approximately *normal* as the number of terms N increases to infinity.

Remark 1.4.4 (Bernstein's inequality for bounded random variables). Suppose the random variables X_i are *uniformly bounded*, which is a stronger assumption than being sub-gaussian. Then there is a useful version of Bernstein's inequality, which unlike Theorem 1.4.1 is sensitive to the *variances* of X_i's. It states that if $K > 0$ is such that $|X_i| \leqslant K$ almost surely for all i, then, for every $t \geqslant 0$, we have

$$\mathbb{P}\left\{\left|\sum_{i=1}^N X_i\right| \geqslant t\right\} \leqslant 2\exp\left(-\frac{t^2/2}{\sigma^2 + CKt}\right). \tag{1.4.5}$$

Here $\sigma^2 = \sum_{i=1}^N \mathbb{E} X_i^2$ is the variance of the sum. This version of Bernstein's inequality can be proved in essentially the same way as Theorem 1.4.1. We will not do it here, but a stronger Theorem 2.2.1, which is valid for matrix-valued random variables X_i, will be proved in Lecture 2.

To compare this with Theorem 1.4.1, note that $\sigma^2 + CKt \leqslant 2\max(\sigma^2, CKt)$. So we can state the probability bound (1.4.5) as

$$2\exp\left[-c\min\left(\frac{t^2}{\sigma^2}, \frac{t}{K}\right)\right].$$

Just as before, here we also have a mixture of two tails, sub-gaussian and sub-exponential. The sub-gaussian tail is a bit sharper than in Theorem 1.4.1, since it depends on the *variances* rather than sub-gaussian norms of X_i. The sub-exponential tail, on the other hand, is weaker, since it depends on the sup-norms rather than the sub-exponential norms of X_i.

1.5. Sub-gaussian random vectors The concept of sub-gaussian distributions can be extended to higher dimensions. Consider a random vector X taking values in $\mathbb{R}^n$. We call X a *sub-gaussian random vector* if all one-dimensional *marginals* of X, i.e., the random variables $\langle X, x\rangle$ for $x \in \mathbb{R}^n$, are sub-gaussian. The sub-gaussian norm of X is defined as

$$\|X\|_{\psi_2} := \sup_{x \in S^{n-1}} \| \langle X, x \rangle \|_{\psi_2}$$

where S^{n-1} denotes the unit Euclidean sphere in $\mathbb{R}^n$.

Example 1.5.1. Examples of sub-gaussian random distributions in $\mathbb{R}^n$ include the standard normal distribution $N(0, I_n)$ (why?), the uniform distribution on the centered Euclidean sphere of radius $\sqrt{n}$, the uniform distribution on the cube $\{-1,1\}^n$, and many others. The last example can be generalized: a random vector $X = (X_1, \dots, X_n)$ with independent and sub-gaussian coordinates is sub-gaussian, with $\|X\|_{\psi_2} \leqslant C \max_i \|X_i\|_{\psi_2}$.

1.6. Johnson-Lindenstrauss Lemma Concentration inequalities like Hoeffding's and Bernstein's are successfully used in the analysis of algorithms. Let us give one example for the problem of *dimension reduction*. Suppose we have some data that is represented as a set of N points in $\mathbb{R}^n$. (Think, for example, of n gene expressions of N patients.)

We would like to compress the data by representing it in a lower dimensional space $\mathbb{R}^m$ instead of $\mathbb{R}^n$ with $m \ll n$. By how much can we reduce the dimension without loosing the important features of the data?

The basic result in this direction is the Johnson-Lindenstrauss Lemma. It states that a remarkably simple dimension reduction method works – a *random linear map* from $\mathbb{R}^n$ to $\mathbb{R}^m$ with

$$m \sim \log N,$$

see Figure 1.6.3. The logarithmic function grows very slowly, so we can usually reduce the dimension dramatically.

What exactly is a random linear map? Several models are possible to use. Here we will model such a map using a *Gaussian random matrix* – an $m \times n$ matrix A with independent $N(0,1)$ entries. More generally, we can consider an $m \times n$ matrix A whose rows are independent, mean zero, isotropic[4] and sub-gaussian random vectors in $\mathbb{R}^n$. For example, the entries of A can be independent *Rademacher* entries – those taking values ± 1 with equal probabilities.

Theorem 1.6.1 (Johnson-Lindenstrauss Lemma). *Let $\mathcal{X}$ be a set of* N *points in* $\mathbb{R}^n$ *and* $\varepsilon \in (0,1)$. *Consider an* $m \times n$ *matrix* A *whose rows are independent, mean zero, isotropic and sub-gaussian random vectors in* $\mathbb{R}^n$. *Rescale* A *by defining the "Gaussian*

[4]A random vector $X \in \mathbb{R}^n$ is called *isotropic* if $\mathbb{E}\, XX^T = I_n$.

random projection"[5]

$$P := \frac{1}{\sqrt{m}} A.$$

Assume that

$$m \geqslant C\varepsilon^{-2} \log N,$$

where C *is an appropriately large constant that depends only on the sub-gaussian norms of the vectors* X_i. *Then, with high probability (say, 0.99), the map* P *preserves the distances between all points in* $\mathcal{X}$ *with error* ε, *that is*

$$(1.6.2) \qquad (1-\varepsilon)\|x-y\|_2 \leqslant \|Px - Py\|_2 \leqslant (1+\varepsilon)\|x-y\|_2 \quad \textit{for all } x, y \in \mathcal{X}.$$

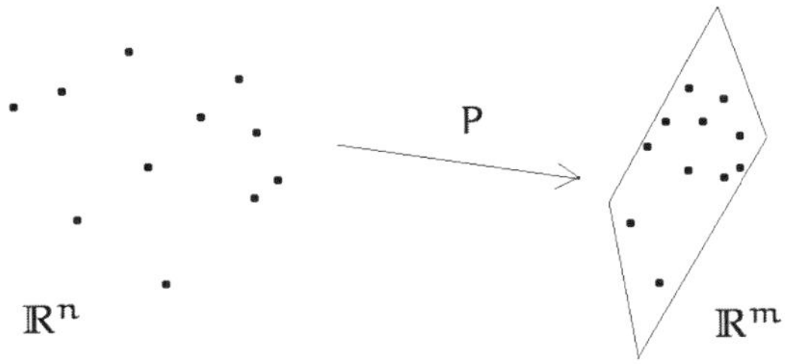

Figure 1.6.3. Johnson-Lindenstrauss Lemma states that a random projection of N data points from dimension n to dimension $m \sim \log N$ approximately preserves the distances between the points.

Proof. Let us take a closer look at the desired conclusion (1.6.2). By linearity, $Px - Py = P(x-y)$. So, dividing the inequality by $\|x-y\|_2$, we can rewrite (1.6.2) in the following way:

$$(1.6.4) \qquad 1-\varepsilon \leqslant \|Pz\|_2 \leqslant 1+\varepsilon \quad \text{for all } z \in T$$

where

$$T := \left\{ \frac{x-y}{\|x-y\|_2} : x, y \in \mathcal{X} \text{ distinct points} \right\}.$$

It will be convenient to square the inequality (1.6.4). Using that $1+\varepsilon \leqslant (1+\varepsilon)^2$ and $1-\varepsilon \geqslant (1-\varepsilon)^2$, we see that it is enough to show that

$$(1.6.5) \qquad 1-\varepsilon \leqslant \|Pz\|_2^2 \leqslant 1+\varepsilon \quad \text{for all } z \in T.$$

By construction, the coordinates of the vector $Pz = \frac{1}{\sqrt{m}} Az$ are $\frac{1}{\sqrt{m}} \langle X_i, z \rangle$. Thus we can restate (1.6.5) as

$$(1.6.6) \qquad \left| \frac{1}{m} \sum_{i=1}^{m} \langle X_i, z \rangle^2 - 1 \right| \leqslant \varepsilon \quad \text{for all } z \in T.$$

Results like (1.6.6) are often proved by combining *concentration* and a *union bound*. In order to use concentration, we first fix $z \in T$. By assumption, the random variables $\langle X_i, z \rangle^2 - 1$ are independent; they have zero mean (use isotropy

[5]Strictly speaking, this P is not a projection since it maps $\mathbb{R}^n$ to a different space $\mathbb{R}^m$.

to check this!), and they are sub-exponential (use (1.3.2) to check this). Then Bernstein's inequality (Theorem 1.4.1) gives

$$\mathbb{P}\left\{\left|\frac{1}{m}\sum_{i=1}^{m}\langle X_i, z\rangle^2 - 1\right| > \varepsilon\right\} \leqslant 2\exp(-c\varepsilon^2 m).$$

(Check!)

Finally, we can unfix z by taking a union bound over all possible $z \in T$:

$$\mathbb{P}\left\{\max_{z\in T}\left|\frac{1}{m}\sum_{i=1}^{m}\langle X_i, z\rangle^2 - 1\right| > \varepsilon\right\} \leqslant \sum_{z\in T}\mathbb{P}\left\{\left|\frac{1}{m}\sum_{i=1}^{m}\langle X_i, z\rangle^2 - 1\right| > \varepsilon\right\}$$

$$\leqslant |T| \cdot 2\exp(-c\varepsilon^2 m). \tag{1.6.7}$$

By definition of T, we have $|T| \leqslant N^2$. So, if we choose $m \geqslant C\varepsilon^{-2}\log N$ with appropriately large constant C, we can make (1.6.7) bounded by 0.01. The proof is complete. □

1.7. Notes The material presented in Sections 1.1–1.5 is basic and can be found e.g. in [58] and [60] with all the proofs. Bernstein's and Hoeffding's inequalities that we covered here are two basic examples of *concentration inequalities*. There are many other useful concentration inequalities for sums of independent random variables (e.g. Chernoff's and Bennett's) and for more general objects. The textbook [60] is an elementary introduction into concentration; the books [10, 38, 39] offer more comprehensive and more advanced accounts of this area.

The original version of Johnson-Lindenstrauss Lemma was proved in [31]. The version we gave here, Theorem 1.6.1, was stated with probability of success 0.99, but an inspection of the proof gives probability $1 - 2\exp(-c\varepsilon^2 m)$ which is much better for large m. A great variety of ramifications and applications of Johnson-Lindenstrauss lemma are known, see e.g. [2, 4, 7, 10, 34, 42].

2. Lecture 2: Concentration of sums of independent random matrices

In the previous lecture we proved Bernstein's inequality, which quantifies how a sum of independent random variables concentrates about its mean. We will now study an extension of Bernstein's inequality to higher dimensions, which holds for sums of independent *random matrices*.

2.1. Matrix calculus The key idea of developing a matrix Bernstein's inequality will be to use *matrix calculus*, which allows us to operate with matrices as with scalars – adding and multiplying them of course, but also comparing matrices and applying functions to matrices. Let us explain this.

We can compare matrices to each other using the notion of being *positive semi-definite*. Let us focus here on $n \times n$ symmetric matrices. If $A - B$ is a positive semidefinite matrix,[6] which we denote $A - B \succeq 0$, then we say that $A \succeq B$ (and, of course, $B \preceq A$). This defines a *partial order* on the set of $n \times n$ symmetric matrices. The term "partial" indicates that, unlike the real numbers, there exist $n \times n$ symmetric matrices A and B that can not be compared. (Give an example where neither $A \preceq B$ nor $B \succeq A$!)

Next, let us guess how to measure the *magnitude* of a matrix A. The magnitude of a scalar $a \in \mathbb{R}$ is measured by the absolute value $|a|$; it is the smallest non-negative number t such that

$$-t \leqslant a \leqslant t.$$

Extending this reasoning to matrices, we can measure the magnitude of an $n \times n$ symmetric matrix A by the smallest non-negative number t such that[7]

$$-tI_n \preceq A \preceq tI_n.$$

The smallest t is called the *operator norm* of A and is denoted $\|A\|$. Diagonalizing A, we can see that

$$\|A\| = \max\{|\lambda| : \lambda \text{ is an eigenvalue of } A\}. \tag{2.1.1}$$

With a little more work (do it!), we can see that $\|A\|$ is the norm of A acting as a linear operator on the Euclidean space $(\mathbb{R}^n, \|\cdot\|_2)$; this is why $\|A\|$ is called the operator norm. Thus $\|A\|$ is the smallest non-negative number M such that

$$\|Ax\|_2 \leqslant M\|x\|_2 \quad \text{for all } x \in \mathbb{R}^n.$$

Finally, we will need to be able to take *functions of matrices*. Let $f : \mathbb{R} \to \mathbb{R}$ be a function and X be an $n \times n$ symmetric matrix. We can define $f(X)$ in two equivalent ways. The spectral theorem allows us to represent X as

$$X = \sum_{i=1}^{n} \lambda_i u_i u_i^{\mathsf{T}}$$

where λ_i are the eigenvalues of X and u_i are the corresponding eigenvectors. Then we can simply define

$$f(X) := \sum_{i=1}^{n} f(\lambda_i) u_i u_i^{\mathsf{T}}.$$

Note that $f(X)$ has the same eigenvectors as X, but the eigenvalues change under the action of f. An equivalent way to define $f(X)$ is using power series. Suppose the function f has a convergent power series expansion about some point $x_0 \in \mathbb{R}$, i.e.

$$f(x) = \sum_{k=1}^{\infty} a_k (x - x_0)^k.$$

[6]Recall that a symmetric real $n \times n$ matrix M is called positive semidefinite if $x^{\mathsf{T}} M x \geqslant 0$ for any vector $x \in \mathbb{R}^n$.

[7]Here and later, I_n denotes the $n \times n$ identity matrix.

Then one can check that the following matrix series converges[8] and defines $f(X)$:

$$f(X) = \sum_{k=1}^{\infty} a_k (X - X_0)^k.$$

(Check!)

2.2. Matrix Bernstein's inequality We are now ready to state and prove a remarkable generalization of Bernstein's inequality for random matrices.

Theorem 2.2.1 (Matrix Bernstein's inequality). *Let $X_1, \dots, X_N$ be independent, mean zero, $n \times n$ symmetric random matrices, such that $\|X_i\| \leqslant K$ almost surely for all i. Then, for every $t \geqslant 0$ we have*

$$\mathbb{P}\Big\{\Big\|\sum_{i=1}^{N} X_i\Big\| \geqslant t\Big\} \leqslant 2n \cdot \exp\Big(-\frac{t^2/2}{\sigma^2 + Kt/3}\Big).$$

Here $\sigma^2 = \Big\|\sum_{i=1}^N \mathbb{E}\, X_i^2\Big\|$ is the norm of the "matrix variance" of the sum.

The scalar case, where $n = 1$, is the classical Bernstein's inequality we stated in (1.4.5). A remarkable feature of matrix Bernstein's inequality, which makes it especially powerful, is that it *does not require any independence of the entries* (or the rows or columns) of X_i; all is needed is that the random matrices X_i be independent *from each other.*

In the rest of this section we will prove matrix Bernstein's inequality, and give a few applications in this and next lecture.

Our proof will be based on bounding the *moment generating function* (MGF) $\mathbb{E}\exp(\lambda S)$ of the sum $S = \sum_{i=1}^N X_i$. Note that to exponentiate the matrix λS in order to define the matrix MGF, we rely on the matrix calculus that we introduced in Section 2.1.

If the terms X_i were *scalars*, independence would yield the classical fact that the MGF of a product is the product of MGF's, i.e.

$$\mathbb{E}\exp(\lambda S) = \mathbb{E}\prod_{i=1}^{N}\exp(\lambda X_i) = \prod_{i=1}^{N}\mathbb{E}\exp(\lambda X_i). \tag{2.2.2}$$

But for *matrices*, this reasoning breaks down badly, for in general

$$e^{X+Y} \neq e^X e^Y$$

even for 2×2 symmetric matrices X and Y. (Give a counterexample!)

Fortunately, there are some trace inequalities that can often serve as proxies for the missing equality $e^{X+Y} = e^X e^Y$. One of such proxies is the *Golden-Thompson inequality*, which states that

$$\operatorname{tr}(e^{X+Y}) \leqslant \operatorname{tr}(e^X e^Y) \tag{2.2.3}$$

[8]The convergence holds in any given metric on the set of matrices, for example in the metric given by the operator norm. In this series, the terms $(X - X_0)^k$ are defined by the usual matrix product.

for any $n \times n$ symmetric matrices X and Y. Another result, which we will actually use in the proof of matrix Bernstein's inequality, is *Lieb's inequality.*

Theorem 2.2.4 (Lieb's inequality). *Let H be an $n \times n$ symmetric matrix. Then the function*

$$f(X) = \operatorname{tr}\exp(H + \log X)$$

is concave[9] *on the space on $n \times n$ symmetric matrices.*

Note that in the scalar case, where $n = 1$, the function f in Lieb's inequality is linear and the result is trivial.

To use Lieb's inequality in a probabilistic context, we will combine it with the classical Jensen's inequality. It states that for any concave function f and a random matrix X, one has[10]

$$\mathbb{E} f(X) \leqslant f(\mathbb{E} X). \tag{2.2.5}$$

Using this for the function f in Lieb's inequality, we get

$$\mathbb{E} \operatorname{tr}\exp(H + \log X) \leqslant \operatorname{tr}\exp(H + \log \mathbb{E} X).$$

And changing variables to $X = e^Z$, we get the following:

Lemma 2.2.6 (Lieb's inequality for random matrices). *Let H be a fixed $n \times n$ symmetric matrix and Z be an $n \times n$ symmetric random matrix. Then*

$$\mathbb{E} \operatorname{tr}\exp(H + Z) \leqslant \operatorname{tr}\exp(H + \log \mathbb{E} e^Z).$$

Lieb's inequality is a perfect tool for bounding the MGF of a sum of independent random variables $S = \sum_{i=1}^N X_i$. To do this, let us condition on the random variables $X_1, \dots, X_{N-1}$. Apply Lemma 2.2.6 for the fixed matrix $H := \sum_{i=1}^{N-1} \lambda X_i$ and the random matrix $Z := \lambda X_N$, and afterwards take the expectation with respect to $X_1, \dots, X_{N-1}$. By the law of total expectation, we get

$$\mathbb{E} \operatorname{tr}\exp(\lambda S) \leqslant \mathbb{E} \operatorname{tr}\exp\Big[\sum_{i=1}^{N-1} \lambda X_i + \log \mathbb{E} e^{\lambda X_N}\Big].$$

Next, apply Lemma 2.2.6 in a similar manner for $H := \sum_{i=1}^{N-2} \lambda X_i + \log \mathbb{E} e^{\lambda X_N}$ and $Z := \lambda X_{N-1}$, and so on. After N times, we obtain:

Lemma 2.2.7 (MGF of a sum of independent random matrices). *Let $X_1, \dots, X_N$ be independent $n \times n$ symmetric random matrices. Then the sum $S = \sum_{i=1}^N X_i$ satisfies*

$$\mathbb{E} \operatorname{tr}\exp(\lambda S) \leqslant \operatorname{tr}\exp\Big[\sum_{i=1}^{N} \log \mathbb{E} e^{\lambda X_i}\Big].$$

[9]Formally, *concavity* of f means that $f(\lambda X + (1-\lambda)Y) \geqslant \lambda f(X) + (1-\lambda) f(Y)$ for all symmetric matrices X and Y and all $\lambda \in [0,1]$.

[10]Jensen's inequality is usually stated for a *convex* function g and a *scalar* random variable X, and it reads $g(\mathbb{E} X) \leqslant \mathbb{E} g(X)$. From this, inequality (2.2.5) for concave functions and random matrices easily follows (Check!).

Think of this inequality is a matrix version of the scalar identity (2.2.2). The main difference is that it bounds the trace of the MGF[11] rather the MGF itself.

You may recall from a course in probability theory that the quantity $\log \mathbb{E}\, e^{\lambda X_i}$ that appears in this bound is called the *cumulant generating function* of X_i.

Lemma 2.2.7 reduces the complexity of our task significantly, for it is much easier to bound the cumulant generating function of each *single* random variable X_i than to say something about their sum. Here is a simple bound.

Lemma 2.2.8 (Moment generating function). *Let X be an $n \times n$ symmetric random matrix. Assume that $\mathbb{E}\, X = 0$ and $\|X\| \leqslant K$ almost surely. Then, for all $0 < \lambda < 3/K$ we have*

$$\mathbb{E} \exp(\lambda X) \preceq \exp\left(g(\lambda)\, \mathbb{E}\, X^2\right) \quad \textit{where} \quad g(\lambda) = \frac{\lambda^2/2}{1-\lambda K/3}.$$

Proof. First, check that the following scalar inequality holds for $0 < \lambda < 3/K$ and $|x| \leqslant K$:

$$e^{\lambda x} \leqslant 1 + \lambda x + g(\lambda) x^2.$$

Then extend it to matrices using matrix calculus: if $0 < \lambda < 3/K$ and $\|X\| \leqslant K$ then

$$e^{\lambda X} \preceq I + \lambda X + g(\lambda) X^2.$$

(Do these two steps carefully!) Finally, take the expectation and recall that $\mathbb{E}\, X = 0$ to obtain

$$\mathbb{E}\, e^{\lambda X} \preceq I + g(\lambda)\, \mathbb{E}\, X^2 \preceq \exp\left(g(\lambda)\, \mathbb{E}\, X^2\right).$$

In the last inequality, we use the matrix version of the scalar inequality $1 + z \leqslant e^z$ that holds for all $z \in \mathbb{R}$. The lemma is proved. □

Proof of Matrix Bernstein's inequality. We would like to bound the operator norm of the random matrix $S = \sum_{i=1}^N X_i$, which, as we know from (2.1.1), is the largest eigenvalue of S *by magnitude*. For simplicity of exposition, let us drop the absolute value from (2.1.1) and just bound the maximal eigenvalue of S, which we denote $\lambda_{\max}(S)$. (Once this is done, we can repeat the argument for $-S$ to reinstate the absolute value. Do this!) So, we are to bound

$$\begin{aligned}
\mathbb{P}\left\{\lambda_{\max}(S) \geqslant t\right\} &= \mathbb{P}\left\{e^{\lambda \cdot \lambda_{\max}(S)} \geqslant e^{\lambda t}\right\} \quad \text{(multiply by } \lambda > 0 \text{ and exponentiate)} \\
&\leqslant e^{-\lambda t}\; \mathbb{E}\, e^{\lambda \cdot \lambda_{\max}(S)} \quad \text{(by Markov's inequality)} \\
&= e^{-\lambda t}\; \mathbb{E}\, \lambda_{\max}(e^{\lambda S}) \quad \text{(check!)} \\
&\leqslant e^{-\lambda t}\; \mathbb{E} \operatorname{tr} e^{\lambda S} \quad \text{(max of eigenvalues is bounded by the sum)} \\
&\leqslant e^{-\lambda t} \operatorname{tr} \exp\left[\sum_{i=1}^N \log \mathbb{E}\, e^{\lambda X_i}\right] \quad \text{(use Lemma 2.2.7)} \\
&\leqslant \operatorname{tr} \exp\left[-\lambda t + g(\lambda) Z\right] \quad \text{(by Lemma 2.2.8)}
\end{aligned}$$

[11]Note that the order of expectation and trace can be swapped using linearity.

where

$$Z := \sum_{i=1}^{N} \mathbb{E} X_i^2.$$

It remains to optimize this bound in λ. The minimum value is attained for $\lambda = t/(\sigma^2 + Kt/3)$. (Check!) With this value of λ, we conclude

$$\mathbb{P}\left\{\lambda_{\max}(S) \geqslant t\right\} \leqslant n \cdot \exp\Big(-\frac{t^2/2}{\sigma^2 + Kt/3}\Big).$$

This completes the proof of Theorem 2.2.1. □

Bernstein's inequality gives a powerful *tail bound* for $\|\sum_{i=1}^N X_i\|$. This easily implies a useful bound on the *expectation*:

Corollary 2.2.9 (Expected norm of sum of random matrices). *Let $X_1, \ldots, X_N$ be independent, mean zero, $n \times n$ symmetric random matrices, such that $\|X_i\| \leqslant K$ almost surely for all i. Then*

$$\mathbb{E}\Big\|\sum_{i=1}^{N} X_i\Big\| \lesssim \sigma\sqrt{\log n} + K \log n$$

where $\sigma = \|\sum_{i=1}^N \mathbb{E} X_i^2\|^{1/2}$.

Proof. The link from tail bounds to expectation is provided by the basic identity

$$\mathbb{E} Z = \int_0^\infty \mathbb{P}\left\{Z > t\right\} dt \tag{2.2.10}$$

which is valid for any non-negative random variable Z. (Check it!) Integrating the tail bound given by matrix Bernstein's inequality, you will arrive at the expectation bound we claimed. (Check!) □

Notice that the bound in this corollary has mild, logarithmic, dependence on the ambient dimension n. As we will see shortly, this can be an important feature in some applications.

2.3. Community recovery in networks Matrix Bernstein's inequality has many applications. The one we are going to discuss first is for the analysis of *networks*. A network can be mathematically represented by a graph, a set of n vertices with edges connecting some of them. For simplicity, we will consider undirected graphs where the edges do not have arrows. Real world networks often tend to have clusters, or *communities* – subsets of vertices that are connected by unusually many edges. (Think, for example, about a friendship network where communities form around some common interests.) An important problem in data science is to recover communities from a given network.

We are going to explain one of the simplest methods for community recovery, which is called *spectral clustering*. But before we introduce it, we will first of all place a probabilistic model on the networks we consider. In other words, it will be convenient for us to view networks as *random graphs* whose edges are formed at random. Although not all real-world networks are truly random, this simplistic model can motivate us to develop algorithms that may empirically succeed also for real-world networks.

The basic probabilistic model of random graphs is the *Erdös-Rényi model.*

Definition 2.3.1 (Erdös-Rényi model). *Consider a set of* n *vertices and connect every pair of vertices independently and with fixed probability* p. *The resulting random graph is said to follow the* Erdös-Rényi model $G(n, p)$.

The Erdös-Rényi random model is very simple. But it is not a good choice if we want to model a network with communities, for every pair of vertices has the same chance to be connected. So let us introduce a natural generalization of the Erdös-Rényi random model that does allow for community structure:

Definition 2.3.2 (Stochastic block model). *Partition a set of* n *vertices into two subsets ("communities") with* $n/2$ *vertices each, and connect every pair of vertices independently with probability* p *if they belong to the same community and* $q < p$ *if not. The resulting random graph is said to follow the* stochastic block model $G(n, p, q)$.

Figure 2.3.3 illustrates a simulation of a stochastic block model.

Figure 2.3.3. A network generated according to the stochastic block model $G(n, p, q)$ with $n = 200$ nodes and connection probabilities $p = 1/20$ and $q = 1/200$.

Suppose we are shown one instance of a random graph generated according to a stochastic block model $G(n, p, q)$. How can we find which vertices belong to which community?

The *spectral clustering* algorithm we are going to explain will do precisely this. It will be based on the spectrum of the *adjacency matrix* A of the graph, which is the $n \times n$ symmetric matrix whose entries A_{ij} equal 1 if the vertices i and j are connected by an edge, and 0 otherwise.[12]

The adjacency matrix A is a random matrix. Let us compute its expectation first. This is easy, since the entires of A are Bernoulli random variables. If i and j

[12]For convenience, we call the vertices of the graph $1, 2, \dots, n$.

belong to the same community then $\mathbb{E} A_{ij} = p$ and otherwise $\mathbb{E} A_{ij} = q$. Thus A has block structure: for example, if $n = 4$ then A looks like this:

$$\mathbb{E} A = \left[\begin{array}{cc|cc} p & p & q & q \\ p & p & q & q \\ \hline q & q & p & p \\ q & q & p & p \end{array}\right]$$

(For illustration purposes, we grouped the vertices from each community together.)

You will easily check that A has rank 2, and the non-zero eigenvalues and the corresponding eigenvectors are

$$\lambda_1(\mathbb{E} A) = \left(\frac{p+q}{2}\right)n, \ v_1(\mathbb{E} A) = \left[\begin{array}{c} 1 \\ 1 \\ \hline 1 \\ 1 \end{array}\right]; \ \lambda_2(\mathbb{E} A) = \left(\frac{p-q}{2}\right)n, \ v_2(\mathbb{E} A) = \left[\begin{array}{c} 1 \\ 1 \\ \hline -1 \\ -1 \end{array}\right].$$

(Check!)

The eigenvalues and eigenvectors of $\mathbb{E} A$ tell us a lot about the community structure of the underlying graph. Indeed, the first (larger) eigenvalue,

$$d := \left(\frac{p+q}{2}\right)n,$$

is the *expected degree* of any vertex of the graph.[13] The second eigenvalue tells us whether there is any community structure at all (which happens when $p \neq q$ and thus $\lambda_2(\mathbb{E} A) \neq 0$). The first eigenvector v_1 is not informative of the structure of the network at all. It is the second eigenvector v_2 that tells us exactly how to separate the vertices into the two communities: the signs of the coefficients of v_2 can be used for this purpose.

Thus if we know $\mathbb{E} A$, we can recover the community structure of the network from the signs of the second eigenvector. The problem is that we do not know $\mathbb{E} A$. Instead, we know the adjacency matrix A. If, by some chance, A is not far from $\mathbb{E} A$, we may hope to use the A to approximately recover the community structure. So is it true that $A \approx \mathbb{E} A$? The answer is yes, and we can prove it using matrix Bernstein's inequality.

Theorem 2.3.4 (Concentration of the stochastic block model). *Let A be the adjacency matrix of a $G(n, p, q)$ random graph. Then*

$$\mathbb{E} \|A - \mathbb{E} A\| \lesssim \sqrt{d \log n} + \log n.$$

Here $d = (p+q)n/2$ is the expected degree.

[13]The degree of the vertex is the number of edges connected to it.

Proof. Let us sketch the argument. To use matrix Bernstein's inequality, let us break A into a sum of independent random matrices

$$A = \sum_{i,j:\, i \leqslant j} X_{ij},$$

where each matrix X_{ij} contains a pair of symmetric entries of A, or one diagonal entry.[14] Matrix Bernstein's inequality obviously applies for the sum

$$A - \mathbb{E}\, A = \sum_{i \leqslant j} (X_{ij} - \mathbb{E}\, X_{ij}).$$

Corollary 2.2.9 gives[15]

$$\mathbb{E}\, \|A - \mathbb{E}\, A\| \lesssim \sigma \sqrt{\log n} + K \log n \tag{2.3.5}$$

where $\sigma^2 = \left\| \sum_{i \leqslant j} \mathbb{E}(X_{ij} - \mathbb{E}\, X_{ij})^2 \right\|$ and $K = \max_{ij} \|X_{ij} - \mathbb{E}\, X_{ij}\|$. It is a good exercise to check that

$$\sigma^2 \lesssim d \quad \text{and} \quad K \leqslant 2.$$

(Do it!) Substituting into (2.3.5), we complete the proof. □

How useful is Theorem 2.3.4 for community recovery? Suppose that the network is not too sparse, namely

$$d \gg \log n.$$

Then

$$\|A - \mathbb{E}\, A\| \lesssim \sqrt{d \log n} \quad \text{while} \quad \|\mathbb{E}\, A\| = \lambda_1(\mathbb{E}\, A) = d,$$

which implies that

$$\|A - \mathbb{E}\, A\| \ll \|\mathbb{E}\, A\|.$$

In other words, A nicely approximates $\mathbb{E}\, A$: the relative error or approximation is small in the operator norm.

At this point one can apply classical results from the *perturbation theory* for matrices, which state that since A and $\mathbb{E}\, A$ are close, their eigenvalues and eigenvectors must also be close. The relevant perturbation results are *Weyl's inequality* for eigenvalues and *Davis-Kahan's inequality* for eigenvectors, which we will not reproduce here. Heuristically, what they give us is

$$v_2(A) \approx v_2(\mathbb{E}\, A) = \begin{bmatrix} 1 \\ 1 \\ \hline -1 \\ -1 \end{bmatrix}.$$

[14] Precisely, if $i \neq j$, then X_{ij} has all zero entries except the (i,j) and (j,i) entries that can potentially equal 1. If $i = j$, the only non-zero entry of X_{ij} is the (i,i).

[15] We will liberally use the notation $\lesssim$ to hide constant factors appearing in the inequalities. Thus, $a \lesssim b$ means that $a \leqslant Cb$ for some constant C.

Then we should expect that most of the coefficients of $v_2(A)$ are positive on one community and negative on the other. So we can use $v_2(A)$ to approximately recover the communities. This method is called *spectral clustering*:

Spectral Clustering Algorithm. *Compute $v_2(A)$, the eigenvector corresponding to the second largest eigenvalue of the adjacency matrix A of the network. Use the signs of the coefficients of $v_2(A)$ to predict the community membership of the vertices.*

We saw that spectral clustering should perform well for the stochastic block model $G(n, p, q)$ if it is not too sparse, namely if the expected degrees satisfy $d = (p + q)n/2 \gg \log n$.

A more careful analysis along these lines, which you should be able to do yourself with some work, leads to the following more rigorous result.

Theorem 2.3.6 (Guarantees of spectral clustering). *Consider a random graph generated according to the stochastic block model $G(n, p, q)$ with $p > q$, and set $a = pn$, $b = qn$. Suppose that*

$$(a - b)^2 \gg \log(n)(a + b). \tag{2.3.7}$$

Then, with high probability, the spectral clustering algorithm recovers the communities up to $o(n)$ misclassified vertices.

Note that condition (2.3.7) implies that the expected degrees are not too small, namely $d = (a + b)/2 \gg \log(n)$ (check!). It also ensures that a and b are sufficiently different: recall that if $a = b$ the network is Erdös-Rényi graph without any community structure.

2.4. Notes The idea to extend concentration inequalities like Bernstein's to matrices goes back to R. Ahlswede and A. Winter [3]. They used Golden-Thompson inequality (2.2.3) and proved a slightly weaker form of matrix Bernstein's inequality than we gave in Section 2.2. R. Oliveira [48, 49] found a way to improve this argument and gave a result similar to Theorem 2.2.1. The version of matrix Bernstein's inequality we gave here (Theorem 2.2.1) and a proof based on Lieb's inequality is due to J. Tropp [52].

The survey [53] contains a comprehensive introduction of matrix calculus, a proof of Lieb's inequality (Theorem 2.2.4), a detailed proof of matrix Bernstein's inequality (Theorem 2.2.1) and a variety of applications. A proof of Golden-Thompson inequality (2.2.3) can be found in [8, Theorem 9.3.7].

In Section 2.3 we scratched the surface of an interdisciplinary area of *network analysis*. For a systematic introduction into networks, refer to the book [47]. Stochastic block models (Definition 2.3.2) were introduced in [33]. The community recovery problem in stochastic block models, sometimes also called community detection problem, has been in the spotlight in the last few years. A vast and still growing body of literature exists on algorithms and theoretical results for community recovery, see the book [47], the survey [22], papers such as [9,29,30,32,37,46,61] and the references therein.

A concentration result similar to Theorem 2.3.4 can be found in [48]; the argument there is also based on matrix concentration. This theorem is not quite optimal. For dense networks, where the expected degree d satisfies $d \gtrsim \log n$, the concentration inequality in Theorem 2.3.4 can be improved to

$$\mathbb{E}\,\|A - \mathbb{E}\,A\| \lesssim \sqrt{d}. \tag{2.4.1}$$

This improved bound goes back to the original paper [21] which studies the simpler Erdös-Rényi model but the results extend to stochastic block models [17]; it can also be deduced from [6,32,37].

If the network is relatively dense, i.e. $d \gtrsim \log n$, one can improve the guarantee (2.3.7) of spectral clustering in Theorem 2.3.6 to

$$(a-b)^2 \gg (a+b).$$

All one has to do is use the improved concentration inequality (2.4.1) instead of Theorem 2.3.4. Furthermore, in this case there exist algorithms that can recover the communities *exactly*, i.e. without any misclassified vertices, and with high probability, see e.g. [1,17,32,43].

For sparser networks, where $d \ll \log n$ and possibly even $d = O(1)$, relatively few algorithms were known until recently, but now there exist many approaches that provably recover communities in sparse stochastic block models, see, for example, [9,17,29,30,37,46,61].

3. Lecture 3: Covariance estimation and matrix completion

In the last lecture, we proved matrix Bernstein's inequality and gave an application for network analysis. We will spend this lecture discussing a couple of other interesting applications of matrix Bernstein's inequality. In Section 3.1 we will work on *covariance estimation*, a basic problem in high-dimensional statistics. In Section 3.2, we will derive a useful bound on norms of random matrices, which unlike Bernstein's inequality does not require any boundedness assumptions on the distribution. We will apply this bound in Section 3.3 for a problem of *matrix completion*, where we are shown a small sample of the entries of a matrix and asked to guess the missing entries.

3.1. Covariance estimation Covariance estimation is a problem of fundamental importance in high-dimensional statistics. Suppose we have a sample of data points $X_1, \ldots, X_N$ in $\mathbb{R}^n$. It is often reasonable to assume that these points are independently sampled from the same probability distribution (or "population") which is unknown. We would like to learn something useful about this distribution.

Denote by X a random vector that has this (unknown) distribution. The most basic parameter of the distribution is the *mean* $\mathbb{E}\,X$. One can estimate $\mathbb{E}\,X$ from the sample by computing the *sample mean* $\frac{1}{N}\sum_{i=1}^N X_i$. The law of large numbers guarantees that the estimate becomes tight as the sample size N grows to infinity.

In other words,

$$\frac{1}{N}\sum_{i=1}^{N} X_i \to \mathbb{E}\,X \quad \text{as } N \to \infty.$$

The next most basic parameter of the distribution is the *covariance matrix*

$$\Sigma := \mathbb{E}(X - \mathbb{E}\,X)(X - \mathbb{E}\,X)^\mathsf{T}.$$

This is a higher-dimensional version of the usual notion of *variance* of a random variable Z, which is

$$\operatorname{Var}(Z) = \mathbb{E}(Z - \mathbb{E}\,Z)^2.$$

The eigenvectors of the covariance matrix of Σ are called the *principal components*. Principal components that correspond to large eigenvalues of Σ are the directions in which the distribution of X is most extended, see Figure 3.1.1. These are often the most interesting directions in the data. Practitioners often visualize the high-dimensional data by projecting it onto the span of a few (maybe two or three) of such principal components; the projection may reveal some hidden structure of the data. This method is called *Principal Component Analysis* (PCA).

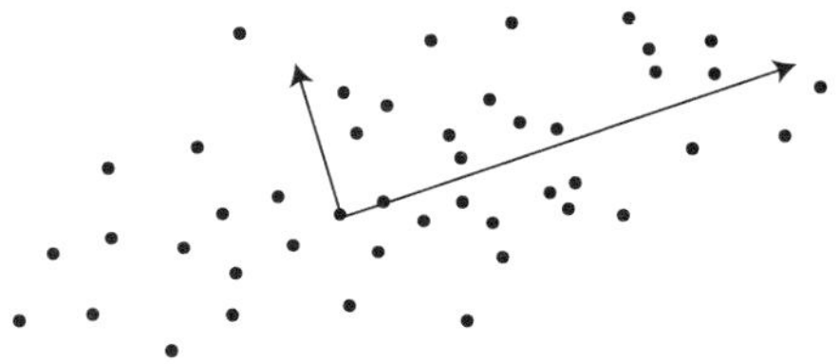

Figure 3.1.1. Data points $X_1, \dots, X_N$ sampled from a distribution in $\mathbb{R}^n$ and the principal components of the covariance matrix.

One can estimate the covariance matrix Σ from the sample by computing the *sample covariance*

$$\Sigma_N := \frac{1}{N}\sum_{i=1}^{N} (X_i - \mathbb{E}\,X_i)(X_i - \mathbb{E}\,X_i)^\mathsf{T}.$$

Again, the law of large numbers guarantees that the estimate becomes tight as the sample size N grows to infinity, i.e.

$$\Sigma_N \to \Sigma \quad \text{as } N \to \infty.$$

But how large should the sample size N be for covariance estimation? Generally, one can not have $N < n$ for dimension reasons. (Why?) We are going to show that

$$N \sim n \log n$$

is enough. In other words, covariance estimation is possible with just *logarithmic oversampling*.

For simplicity, we shall state the covariance estimation bound for *mean zero* distributions. (If the mean is not zero, we can estimate it from the sample and subtract. Check that the mean can be accurately estimated from a sample of size $N = O(n)$.)

Theorem 3.1.2 (Covariance estimation). *Let X be a random vector in $\mathbb{R}^n$ with covariance matrix Σ. Suppose that*

$$\|X\|_2^2 \lesssim \mathbb{E}\,\|X\|_2^2 = \operatorname{tr}\Sigma \quad \textit{almost surely.} \tag{3.1.3}$$

Then, for every $N \geqslant 1$, we have

$$\mathbb{E}\,\|\Sigma_N - \Sigma\| \lesssim \|\Sigma\| \Big(\sqrt{\frac{n \log n}{N}} + \frac{n \log n}{N} \Big).$$

Before we pass to the proof, let us note that Theorem 3.1.2 yields the covariance estimation result we promised. Let $\varepsilon \in (0,1)$. If we take a sample of size

$$N \sim \varepsilon^{-2} n \log n,$$

then we are guaranteed covariance estimation with a good relative error:

$$\mathbb{E}\,\|\Sigma_N - \Sigma\| \leqslant \varepsilon \|\Sigma\|.$$

Proof. Apply matrix Bernstein's inequality (Corollary 2.2.9) for the sum of independent random matrices $X_i X_i^\mathsf{T} - \Sigma$ and get

$$\mathbb{E}\,\|\Sigma_N - \Sigma\| = \frac{1}{N}\,\mathbb{E}\,\Big\| \sum_{i=1}^N (X_i X_i^\mathsf{T} - \Sigma) \Big\| \lesssim \frac{1}{N} (\sigma \sqrt{\log n} + K \log n) \tag{3.1.4}$$

where

$$\sigma^2 = \Big\| \sum_{i=1}^N \mathbb{E}(X_i X_i^\mathsf{T} - \Sigma)^2 \Big\| = N \|\, \mathbb{E}(XX^\mathsf{T} - \Sigma)^2 \|$$

and K is chosen so that

$$\|XX^\mathsf{T} - \Sigma\| \leqslant K \quad \text{almost surely.}$$

It remains to bound σ and K. Let us start with σ. We have

$$\begin{aligned}
\mathbb{E}(XX^\mathsf{T} - \Sigma)^2 &= \mathbb{E}\,\|X\|_2^2 XX^\mathsf{T} - \Sigma^2 \quad \text{(check by expanding the square)} \\
&\precsim \operatorname{tr}(\Sigma) \cdot \mathbb{E}\, XX^\mathsf{T} \quad \text{(drop } \Sigma^2 \text{ and use (3.1.3))} \\
&= \operatorname{tr}(\Sigma) \cdot \Sigma.
\end{aligned}$$

Thus

$$\sigma^2 \lesssim N \operatorname{tr}(\Sigma) \|\Sigma\|.$$

Next, to bound K, we have

$$\begin{aligned}
\|XX^\mathsf{T} - \Sigma\| &\leqslant \|X\|_2^2 + \|\Sigma\| \quad \text{(by triangle inequality)} \\
&\lesssim \operatorname{tr}\Sigma + \|\Sigma\| \quad \text{(using (3.1.3))} \\
&\leqslant 2 \operatorname{tr}\Sigma =: K.
\end{aligned}$$

Substitute the bounds on σ and K into (3.1.4) and get

$$\mathbb{E}\,\|\Sigma_N - \Sigma\| \lesssim \frac{1}{N} \big(\sqrt{N \operatorname{tr}(\Sigma) \|\Sigma\| \log n} + \operatorname{tr}(\Sigma) \log n \big)$$

Complete the proof by using $\operatorname{tr}\Sigma \leqslant n\|\Sigma\|$ (check this!) to simplify the bound. □

Remark 3.1.5 (Low-dimensional distributions). Far fewer samples are needed for covariance estimation for low-dimensional, or approximately low-dimensional, distributions. To measure approximate low-dimensionality we can use the notion of the *stable rank* of Σ^2. The stable rank of a matrix A is defined as the square of the ratio of the Frobenius to operator norms:[16]

$$r(A) := \frac{\|A\|_F^2}{\|A\|^2}.$$

The stable rank is always bounded by the usual, linear algebraic rank,

$$r(A) \leqslant \operatorname{rank}(A),$$

and it can be much smaller. (Check both claims.)

Our proof of Theorem 3.1.2 actually gives

$$\mathbb{E}\,\|\Sigma_N - \Sigma\| \leqslant \|\Sigma\| \left(\sqrt{\frac{r\log n}{N}} + \frac{r\log n}{N}\right).$$

where

$$r = r(\Sigma^{1/2}) = \frac{\operatorname{tr}\Sigma}{\|\Sigma\|}.$$

(Check this!) Therefore, covariance estimation is possible with

$$N \sim r\log n$$

samples.

Remark 3.1.6 (The boundedness condition). It is a good exercise to check that if we remove the boundedness condition (3.1.3), a nontrivial covariance estimation is impossible in general. (Show this!) But how do we know whether the boundedness condition holds for data at hand? We may not, but we can enforce this condition by *truncation*. All we have to do is to discard 1% of data points with largest norms. (Check this accurately, assuming that such truncation does not change the covariance significantly.)

3.2. Norms of random matrices We have worked a lot with the *operator norm* of matrices, denoted $\|A\|$. One may ask if is there exists a formula that expresses $\|A\|$ in terms of the entires A_{ij}. Unfortunately, there is no such formula. The operator norm is a more difficult quantity in this respect than the *Frobenius norm*, which as we know can be easily expressed in terms of entries: $\|A\|_F = (\sum_{i,j} A_{ij}^2)^{1/2}$.

If we can not express $\|A\|$ in terms of the entires, can we at least get a good estimate? Let us consider $n \times n$ symmetric matrices for simplicity. In one direction,

[16]The Frobenius norm of an $n \times m$ matrix, sometimes also called the *Hilbert-Schmidt norm*, is defined as $\|A\|_F = (\sum_{i=1}^n \sum_{j=1}^m A_{ij}^2)^{1/2}$. Equivalently, for an $n \times n$ symmetric matrix, we have $\|A\|_F = (\sum_{i=1}^n \lambda_i(A)^2)^{1/2}$, where $\lambda_i(A)$ are the eigenvalues of A. Thus the stable rank of A can be expressed as $r(A) = \sum_{i=1}^n \lambda_i(A)^2 / \max_i \lambda_i(A)^2$.

$\|A\|$ is always bounded *below* by the largest Euclidean norm of the rows A_i:

$$\|A\| \geqslant \max_i \|A_i\|_2 = \max_i \Big(\sum_j A_{ij}^2 \Big)^{1/2}. \tag{3.2.1}$$

(Check!) Unfortunately, this bound is sometimes very loose, and the best possible upper bound is

$$\|A\| \leqslant \sqrt{n} \cdot \max_i \|A_i\|_2. \tag{3.2.2}$$

(Show this bound, and give an example where it is sharp.)

Fortunately, for *random* matrices with independent entries the bound (3.2.2) can be improved to the point where the upper and lower bounds almost match.

Theorem 3.2.3 (Norms of random matrices without boundedness assumptions). *Let A be an $n \times n$ symmetric random matrix whose entries on and above the diagonal are independent, mean zero random variables. Then*

$$\mathbb{E} \max_i \|A_i\|_2 \leqslant \mathbb{E}\,\|A\| \leqslant C \log n \cdot \mathbb{E} \max_i \|A_i\|_2,$$

where A_i denote the rows of A.

In words, the operator norm of a random matrix is almost determined by the norm of the rows.

Our proof of this result will be based on matrix Bernstein's inequality – more precisely, Corollary 2.2.9. There is one surprising point. How can we use matrix Bernstein's inequality, which applies only for bounded distributions, to prove a result like Theorem 3.2.3 that does not have any boundedness assumptions? We will do this using a trick based on conditioning and symmetrization. Let us introduce this technique first.

Lemma 3.2.4 (Symmetrization). *Let $X_1, \dots, X_N$ be independent, mean zero random vectors in a normed space and $\varepsilon_1, \dots, \varepsilon_N$ be independent Rademacher random variables.*[17] *Then*

$$\frac{1}{2} \mathbb{E} \Big\| \sum_{i=1}^N \varepsilon_i X_i \Big\| \leqslant \mathbb{E} \Big\| \sum_{i=1}^N X_i \Big\| \leqslant 2\, \mathbb{E} \Big\| \sum_{i=1}^N \varepsilon_i X_i \Big\|.$$

Proof. To prove the upper bound, let (X_i') be an independent copy of the random vectors (X_i), i.e. just different random vectors with the same joint distribution as (X_i) and independent from (X_i). Then

$$\begin{aligned}
\mathbb{E} \Big\| \sum_i X_i \Big\| &= \mathbb{E} \Big\| \sum_i X_i - \mathbb{E} \Big(\sum_i X_i' \Big) \Big\| \quad \text{(since } \mathbb{E} \sum_i X_i' = 0 \text{ by assumption)} \\
&\leqslant \mathbb{E} \Big\| \sum_i X_i - \sum_i X_i' \Big\| \quad \text{(by Jensen's inequality)} \\
&= \mathbb{E} \Big\| \sum_i (X_i - X_i') \Big\|.
\end{aligned}$$

[17]This means that random variables ε_i take values ± 1 with probability $1/2$ each. We require that all random variables we consider here, i.e. $\{X_i, \varepsilon_i : i = 1, \dots, N\}$ are jointly independent.

The distribution of the random vectors $Y_i := X_i - X_i'$ is *symmetric*, which means that the distributions of Y_i and $-Y_i'$ are the same. (Why?) Thus the distribution of the random vectors Y_i and $\varepsilon_i Y_i$ is also the same, for all we do is change the signs of these vectors at random and independently of the values of the vectors. Summarizing, we can replace $X_i - X_i'$ in the sum above with $\varepsilon_i(X_i - X_i')$. Thus

$$\begin{aligned}\mathbb{E}\Big\|\sum_i X_i\Big\| &\leqslant \mathbb{E}\Big\|\sum_i \varepsilon_i(X_i - X_i')\Big\| \\ &\leqslant \mathbb{E}\Big\|\sum_i \varepsilon_i X_i\Big\| + \mathbb{E}\Big\|\sum_i \varepsilon_i X_i'\Big\| \quad \text{(using triangle inequality)} \\ &= 2\,\mathbb{E}\Big\|\sum_i \varepsilon_i X_i\Big\| \quad \text{(the two sums have the same distribution).}\end{aligned}$$

This proves the upper bound in the symmetrization inequality. The lower bound can be proved by a similar argument. (Do this!) □

Proof of Theorem 3.2.3. We already mentioned in (3.2.1) that the bound in Theorem 3.2.3 is trivial. The proof of the upper bound will be based on matrix Bernstein's inequality.

First, we decompose A in the same way as we did in the proof of Theorem 2.3.4. Thus we represent A as a sum of independent, mean zero, symmetric random matrices Z_{ij} each of which contains a pair of symmetric entries of A (or one diagonal entry):

$$A = \sum_{i,j:\, i \leqslant j} Z_{ij}.$$

Apply the symmetrization inequality (Lemma 3.2.4) for the random matrices Z_{ij} and get

$$\mathbb{E}\,\|A\| = \mathbb{E}\Big\|\sum_{i\leqslant j} Z_{ij}\Big\| \leqslant 2\,\mathbb{E}\Big\|\sum_{i\leqslant j} X_{ij}\Big\| \tag{3.2.5}$$

where we set

$$X_{ij} := \varepsilon_{ij} Z_{ij}$$

and ε_{ij} are independent Rademacher random variables.

Now we condition on A. The random variables Z_{ij} become *fixed values* and all randomness remains in the Rademacher random variables ε_{ij}. Note that X_{ij} are (conditionally) *bounded* almost surely, and this is exactly what we have lacked to apply matrix Bernstein's inequality. Now we can do it. Corollary 2.2.9 gives[18]

$$\mathbb{E}_\varepsilon \Big\|\sum_{i\leqslant j} X_{ij}\Big\| \lesssim \sigma\sqrt{\log n} + K \log n, \tag{3.2.6}$$

where $\sigma^2 = \big\|\sum_{i\leqslant j} \mathbb{E}_\varepsilon X_{ij}^2\big\|$ and $K = \max_{i\leqslant j}\|X_{ij}\|$.

[18]We stick a subscript ε to the expected value to remember that this is a conditional expectation, i.e. we average only with respect to ε_i.

A good exercise is to check that

$$\sigma \lesssim \max_i \|A_i\|_2 \quad \text{and} \quad K \lesssim \max_i \|A_i\|_2.$$

(Do it!) Substituting into (3.2.6), we get

$$\mathbb{E}_\varepsilon \Big\| \sum_{i \leqslant j} X_{ij} \Big\| \lesssim \log n \cdot \max_i \|A_i\|_2.$$

Finally, we unfix A by taking expectation of both sides of this inequality with respect to A and using the law of total expectation. The proof is complete. □

We stated Theorem 3.2.3 for symmetric matrices, but it is simple to extend it to general $m \times n$ random matrices A. The bound in this case becomes

$$\mathbb{E}\,\|A\| \leqslant C \log(m+n) \cdot \big(\mathbb{E} \max_i \|A_i\|_2 + \mathbb{E} \max_j \|A^j\|_2 \big) \tag{3.2.7}$$

where A_i and A^j denote the rows and columns of A. To see this, apply Theorem 3.2.3 to the $(m+n) \times (m+n)$ symmetric random matrix

$$\begin{bmatrix} 0 & A \\ A^\mathsf{T} & 0 \end{bmatrix}.$$

(Do this!)

3.3. Matrix completion Consider a fixed, unknown $n \times n$ matrix X. Suppose we are shown m randomly chosen entries of X. Can we guess all the missing entries? This important problem is called *matrix completion*. We will analyze it using the bounds on the norms on random matrices we just obtained.

Obviously, there is no way to guess the missing entries unless we know something extra about the matrix X. So let us assume that X has *low rank*:

$$\operatorname{rank}(X) =: r \ll n.$$

The number of degrees of freedom of an $n \times n$ matrix with rank r is $O(rn)$. (Why?) So we may hope that

$$m \sim rn \tag{3.3.1}$$

observed entries of X will be enough to determine X completely. But how?

Here we will analyze what is probably the simplest method for matrix completion. Take the matrix Y that consists of the observed entries of X while all unobserved entries are set to zero. Unlike X, the matrix Y may not have small rank. Compute the best rank r approximation[19] of Y. The result, as we will show, will be a good approximation to X.

But before we show this, let us define sampling of entries more rigorously. Assume each entry of X is shown or hidden independently of others with fixed

[19]The best rank r approximation of an $n \times n$ matrix A is a matrix B of rank r that minimizes the operator norm $\|A - B\|$ or, alternatively, the Frobenius norm $\|A - B\|_F$ (the minimizer turns out to be the same). One can compute B by truncating the singular value decomposition $A = \sum_{i=1}^n s_i u_i v_i^\mathsf{T}$ of A as follows: $B = \sum_{i=1}^r s_i u_i v_i^\mathsf{T}$, where we assume that the singular values s_i are arranged in non-increasing order.

probability p. Which entries are shown is decided by independent Bernoulli random variables

$$\delta_{ij} \sim \text{Ber}(p) \quad \text{with} \quad p := \frac{m}{n^2}$$

which are often called *selectors* in this context. The value of p is chosen so that among n^2 entries of X, the expected number of selected (known) entries is m. Define the $n \times n$ matrix Y with entries

$$Y_{ij} := \delta_{ij} X_{ij}.$$

We can assume that we are shown Y, for it is a matrix that contains the observed entries of X while all unobserved entries are replaced with zeros. The following result shows how to estimate X based on Y.

Theorem 3.3.2 (Matrix completion). *Let $\hat{X}$ be a best rank r approximation to $p^{-1}Y$. Then*

$$\mathbb{E}\, \frac{1}{n}\|\hat{X} - X\|_F \leqslant C \log(n) \sqrt{\frac{rn}{m}}\, \|X\|_\infty, \tag{3.3.3}$$

Here $\|X\|_\infty = \max_{i,j} |X_{ij}|$ denotes the maximum magnitude of the entries of X.

Before we prove this result, let us understand what this bound says about the quality of matrix completion. The recovery error is measured in the Frobenius norm, and the left side of (3.3.3) is

$$\frac{1}{n}\|\hat{X} - X\|_F = \Big(\frac{1}{n^2} \sum_{i,j=1}^{n} |\hat{X}_{ij} - X_{ij}|^2 \Big)^{1/2}.$$

Thus Theorem 3.3.2 controls *the average error per entry* in the mean-squared sense. To make the error small, let us assume that we have a sample of size

$$m \gg rn \log^2 n,$$

which is slightly larger than the ideal size we discussed in (3.3.1). This makes $C \log(n) \sqrt{rn/m} = o(1)$ and forces the recovery error to be bounded by $o(1)\|X\|_\infty$. Summarizing, Theorem 3.3.2 says that *the expected average error per entry is much smaller than the maximal magnitude of the entries of* X. This is true for a sample of almost optimal size m. The smaller the rank r of the matrix X, the fewer entries of X we need to see in order to do matrix completion.

Proof of Theorem 3.3.2. **Step 1: The error in the operator norm.** Let us first bound the recovery error in the *operator* norm. Decompose the error into two parts using triangle inequality:

$$\|\hat{X} - X\| \leqslant \|\hat{X} - p^{-1}Y\| + \|p^{-1}Y - X\|.$$

Recall that $\hat{X}$ is a best approximation to $p^{-1}Y$. Then the first part of the error is smaller than the second part, i.e. $\|\hat{X} - p^{-1}Y\| \leqslant \|p^{-1}Y - X\|$, and we have

$$\|\hat{X} - X\| \leqslant 2\|p^{-1}Y - X\| = \frac{2}{p}\|Y - pX\|. \tag{3.3.4}$$

The entries of the matrix $Y - pX$,

$$(Y - pX)_{ij} = (\delta_{ij} - p)X_{ij},$$

are independent and mean zero random variables. Thus we can apply the bound (3.2.7) on the norms of random matrices and get

$$\mathbb{E}\,\|Y - pX\| \leqslant C \log n \cdot \big(\mathbb{E} \max_{i \in [n]} \|(Y - pX)_i\|_2 + \mathbb{E} \max_{j \in [n]} \|(Y - pX)^j\|_2 \big). \tag{3.3.5}$$

All that remains is to bound the norms of the rows and columns of $Y - pX$. This is not difficult if we note that they can be expressed as sums of independent random variables:

$$\|(Y - pX)_i\|_2^2 = \sum_{j=1}^{n} (\delta_{ij} - p)^2 X_{ij}^2 \leqslant \sum_{j=1}^{n} (\delta_{ij} - p)^2 \cdot \|X\|_\infty^2,$$

and similarly for columns.

Taking expectation and noting that $\mathbb{E}(\delta_{ij} - p)^2 = \mathrm{Var}(\delta_{ij}) = p(1-p)$, we get[20]

$$\mathbb{E}\,\|(Y - pX)_i\|_2 \leqslant (\mathbb{E}\,\|(Y - pX)_i\|_2^2)^{1/2} \leqslant \sqrt{pn}\,\|X\|_\infty. \tag{3.3.6}$$

This is a good bound, but we need something stronger in (3.3.5). Since the maximum appears inside the expectation, we need a *uniform* bound, which will say that all rows are bounded simultaneously with high probability.

Such uniform bounds are usually proved by applying concentration inequalities followed by a union bound. Bernstein's inequality (1.4.5) yields

$$\mathbb{P}\left\{ \sum_{j=1}^{n} (\delta_{ij} - p)^2 > tpn \right\} \leqslant \exp(-ctpn) \quad \text{for } t \geqslant 3.$$

(Check!) This probability can be further bounded by n^{-ct} using the assumption that $m = pn^2 \geqslant n \log n$. A union bound over n rows leads to

$$\mathbb{P}\left\{ \max_{i \in [n]} \sum_{j=1}^{n} (\delta_{ij} - p)^2 > tpn \right\} \leqslant n \cdot n^{-ct} \quad \text{for } t \geqslant 3.$$

Integrating this tail, we conclude using (2.2.10) that

$$\mathbb{E} \max_{i \in [n]} \sum_{j=1}^{n} (\delta_{ij} - p)^2 \lesssim pn.$$

(Check!) And this yields the desired bound on the rows,

$$\mathbb{E} \max_{i \in [n]} \|(Y - pX)_i\|_2 \lesssim \sqrt{pn},$$

which is an improvement of (3.3.6) we wanted. We can do similarly for the columns. Substituting into (3.3.5), this gives

$$\mathbb{E}\,\|Y - pX\| \lesssim \log(n) \sqrt{pn}\,\|X\|_\infty.$$

[20]The first bound below that compares the L^1 and L^2 averages follows from Hölder's inequality.

Then, by (3.3.4), we get

$$\mathbb{E}\,\|\hat{X}-X\| \lesssim \log(n)\sqrt{\frac{n}{p}}\,\|X\|_\infty. \tag{3.3.7}$$

Step 2: Passing to Frobenius norm. Now we will need to pass from the operator to Frobenius norm. This is where we will use for the first (and only) time the rank of X. We know that $\operatorname{rank}(X) \leqslant r$ by assumption and $\operatorname{rank}(\hat{X}) \leqslant r$ by construction, so $\operatorname{rank}(\hat{X}-X) \leqslant 2r$. There is a simple relationship between the operator and Frobenius norms:

$$\|\hat{X}-X\|_F \leqslant \sqrt{2r}\|\hat{X}-X\|.$$

(Check it!) Take expectation of both sides and use (3.3.7); we get

$$\mathbb{E}\,\|\hat{X}-X\|_F \leqslant \sqrt{2r}\,\mathbb{E}\,\|\hat{X}-X\| \lesssim \log(n)\sqrt{\frac{rn}{p}}\,\|X\|_\infty.$$

Dividing both sides by n, we can rewrite this bound as

$$\mathbb{E}\,\frac{1}{n}\|\hat{X}-X\|_F \lesssim \log(n)\sqrt{\frac{rn}{pn^2}}\,\|X\|_\infty.$$

This yields (3.3.3) since $pn^2 = m$ by definition of the sampling probability p. □

3.4. Notes Theorem 3.1.2 on covariance estimation is a version of [58, Corollary 5.52], see also [36]. The logarithmic factor is in general necessary. This theorem is a general-purpose result. If one knows some additional structural information about the covariance matrix (such as sparsity), then fewer samples may be needed, see e.g. [12, 16, 40].

A version of Theorem 3.2.3 was proved in [51] in a more technical way. Although the logarithmic factor in Theorem 3.2.3 can not be completely removed in general, it can be improved. Our argument actually gives

$$\mathbb{E}\,\|A\| \leqslant C\sqrt{\log n}\cdot\mathbb{E}\max_i\|A_i\|_2 + C\log n\cdot\mathbb{E}\max_{ij}|A_{ij}|,$$

Using different methods, one can save an extra $\sqrt{\log n}$ factor and show that

$$\mathbb{E}\,\|A\| \leqslant C\,\mathbb{E}\max_i\|A_i\|_2 + C\sqrt{\log n}\cdot\mathbb{E}\max_{ij}|A_{ij}|$$

(see [6]) and

$$\mathbb{E}\,\|A\| \leqslant C\sqrt{\log n\cdot\log\log n}\cdot\mathbb{E}\max_i\|A_i\|_2,$$

see [55]. (The results in [6, 55] are stated for Gaussian random matrices; the two bounds above can be deduced by using conditioning and symmetrization.) The surveys [6, 58] and the textbook [60] present several other useful techniques to bound the operator norm of random matrices.

The matrix completion problem, which we discussed in Section 3.3, has attracted a lot of recent attention. E. Candes and B. Recht [14] showed that one can often achieve *exact* matrix completion, thus computing the precise values of all missing values of a matrix, from $m \sim rn\log^2(n)$ randomly sampled entries. For exact matrix completion, one needs an extra *incoherence* assumption that is not present in Theorem 3.3.2. This assumption basically excludes matrices that are simultaneously sparse and low rank (such as a matrix whose all but one entries are zero – it would be extremely hard to complete it, since sampling will likely miss the non-zero entry). Many further results on exact matrix completion are known, e.g. [15, 18, 28, 56].

Theorem 3.3.2 with a simple proof is borrowed from [50]; see also the tutorial [59]. This result only guarantees approximate matrix completion, but it does not have any incoherence assumptions on the matrix.

4. Lecture 4: Matrix deviation inequality

In this last lecture we will study a new uniform deviation inequality for random matrices. This result will be a far reaching generalization of the Johnson-Lindenstrauss Lemma we proved in Lecture 1.

Consider the same setup as in Theorem 1.6.1, where A is an $m \times n$ random matrix whose rows are independent, mean zero, isotropic and sub-gaussian random vectors in $\mathbb{R}^n$. (If you find it helpful to think in terms of concrete examples, let the entries of A be independent $N(0,1)$ random variables.) Like in the Johnson-Lindenstrauss Lemma, we will be looking at A as a linear transformation from $\mathbb{R}^n$ to $\mathbb{R}^m$, and we will be interested in what A does to points in some set in $\mathbb{R}^n$. This time, however, we will allow for *infinite* sets $T \subset \mathbb{R}^n$.

Let us start by analyzing what A does to a single fixed vector $x \in \mathbb{R}^n$. We have

$$\begin{aligned}
\mathbb{E}\,\|Ax\|_2^2 &= \mathbb{E}\sum_{j=1}^m \langle A_j, x\rangle^2 \quad \text{(where } A_j^\mathsf{T} \text{ denote the rows of } A\text{)}\\
&= \sum_{j=1}^m \mathbb{E}\,\langle A_j, x\rangle^2 \quad \text{(by linearity)}\\
&= m\|x\|_2^2 \quad \text{(using isotropy of } A_j\text{)}.
\end{aligned}$$

Further, if we assume that concentration about the mean holds here (and in fact, it does), we should expect that

$$\|Ax\|_2 \approx \sqrt{m}\,\|x\|_2 \tag{4.0.1}$$

with high probability.

Similarly to Johnson-Lindenstrauss Lemma, our next goal is to make (4.0.1) hold simultaneously over all vectors x in some fixed set $T \subset \mathbb{R}^n$. Precisely, we may ask – how large is the average *uniform deviation*:

$$\mathbb{E}\sup_{x\in T}\Big|\|Ax\|_2 - \sqrt{m}\,\|x\|\Big| \;? \tag{4.0.2}$$

This quantity should clearly depend on some notion of the size of T: the larger T, the larger should the uniform deviation be. So, how can we quantify the size of T for this problem? In the next section we will do precisely this – introduce a convenient, geometric measure of the sizes of sets in $\mathbb{R}^n$, which is called *Gaussian width*.

4.1. Gaussian width

Definition 4.1.1. *Let* $T \subset \mathbb{R}^n$ *be a bounded set, and* g *be a standard normal random vector in* $\mathbb{R}^n$, *i.e.* $g \sim N(0, I_n)$. *Then the quantities*

$$w(T) := \mathbb{E} \sup_{x \in T} \langle g, x \rangle \quad \textit{and} \quad \gamma(T) := \mathbb{E} \sup_{x \in T} |\langle g, x \rangle|$$

are called the Gaussian width *of* T *and the* Gaussian complexity *of* T, *respectively.*

Gaussian width and Gaussian complexity are closely related. Indeed,[21]

$$2w(T) = w(T-T) = \mathbb{E} \sup_{x,y \in T} \langle g, x-y \rangle = \mathbb{E} \sup_{x,y \in T} |\langle g, x-y \rangle| = \gamma(T-T). \tag{4.1.2}$$

(Check these identities!)

Gaussian width has a natural geometric interpretation. Suppose g is a unit vector in $\mathbb{R}^n$. Then a moment's thought reveals that $\sup_{x,y \in T} \langle g, x-y \rangle$ is simply the *width of* T *in the direction of* g, i.e. the distance between the two hyperplanes with normal g that touch T on both sides as shown in Figure 4.1.3. Then $2w(T)$ can be obtained by averaging the width of T over all directions g in $\mathbb{R}^n$.

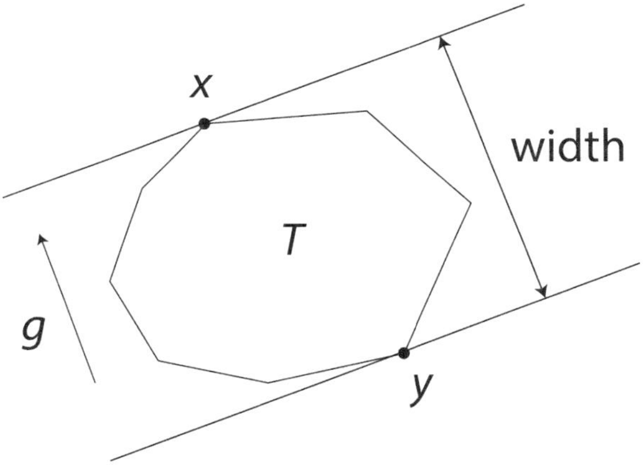

Figure 4.1.3. The width of a set T in the direction of g.

This reasoning is valid except where we assumed that g is a unit vector. Instead, for $g \sim N(0, I_n)$ we have $\mathbb{E}\|g\|_2^2 = n$ and

$$\|g\|_2 \approx \sqrt{n} \quad \text{with high probability.}$$

(Check both these claims using Bernstein's inequality.) Thus, we need to scale by the factor $\sqrt{n}$. Ultimately, the geometric interpretation of the Gaussian width becomes the following: $w(T)$ *is approximately* $\sqrt{n}/2$ *larger than the usual, geometric width of* T *averaged over all directions.*

[21] The set $T - T$ is defined as $\{x - y : x, y \in T\}$. More generally, given two sets A and B in the same vector space, the *Minkowski sum* of A and B is defined as $A + B = \{a + b : a \in A, b \in B\}$.

A good exercise is to compute the Gaussian width and complexity for some simple sets, such as the unit balls of the ℓ_p norms in $\mathbb{R}^n$, which we denote by $B_p^n = \{x \in \mathbb{R}^n : \|x\|_p \leqslant 1\}$. In particular, we have

$$\gamma(B_2^n) \sim \sqrt{n}, \quad \gamma(B_1^n) \sim \sqrt{\log n}. \tag{4.1.4}$$

For any finite set $T \subset B_2^n$, we have

$$\gamma(T) \lesssim \sqrt{\log |T|}. \tag{4.1.5}$$

The same holds for Gaussian width $w(T)$. (Check these facts!)

A look a these examples reveals that the Gaussian width captures some non-obvious geometric qualities of sets. Of course, the fact that the Gaussian width of the unit Euclidean ball B_2^n is or order $\sqrt{n}$ is not surprising: the usual, geometric width in all directions is 2 and the Gaussian width is about $\sqrt{n}$ times that. But it may be surprising that the Gaussian width of the ℓ_1 ball B_1^n is much smaller, and so is the width of any finite set T (unless the set has exponentially large cardinality). As we will see later, Gaussian width nicely captures the geometric size of "the bulk" of a set.

4.2. Matrix deviation inequality Now we are ready to answer the question we asked in the beginning of this lecture: what is the magnitude of the uniform deviation (4.0.2)? The answer is surprisingly simple: it is bounded by the Gaussian complexity of T. The proof is not too simple however, and we will skip it (see the notes after this lecture for references).

Theorem 4.2.1 (Matrix deviation inequality). *Let A be an $m \times n$ matrix whose rows A_i are independent, isotropic and sub-gaussian random vectors in $\mathbb{R}^n$. Let $T \subset \mathbb{R}^n$ be a fixed bounded set. Then*

$$\mathbb{E} \sup_{x \in T} \Big| \|Ax\|_2 - \sqrt{m}\|x\|_2 \Big| \leqslant CK^2 \gamma(T)$$

where $K = \max_i \|A_i\|_{\psi_2}$ is the maximal sub-gaussian norm[22] of the rows of A.

Remark 4.2.2 (Tail bound). It is often useful to have results that hold *with high probability* rather than in expectation. There exists a high-probability version of the matrix deviation inequality, and it states the following. Let $u \geqslant 0$. Then the event

$$\sup_{x \in T} \Big| \|Ax\|_2 - \sqrt{m}\|x\|_2 \Big| \leqslant CK^2 \left[\gamma(T) + u \cdot \mathrm{rad}(T)\right] \tag{4.2.3}$$

holds with probability at least $1 - 2\exp(-u^2)$. Here $\mathrm{rad}(T)$ is the *radius* of T, defined as

$$\mathrm{rad}(T) := \sup_{x \in T} \|x\|_2.$$

[22]A definition of the sub-gaussian norm of a random vector was given in Section 1.5. For example, if A is a Gaussian random matrix with independent $N(0,1)$ entries, then K is an absolute constant.

Since $\mathrm{rad}(T) \lesssim \gamma(T)$ (check!) we can continue the bound (4.2.3) by

$$\lesssim K^2 u \gamma(T)$$

for all $u \geqslant 1$. This is a weaker but still a useful inequality. For example, we can use it to bound all higher moments of the deviation:

$$\left(\mathbb{E} \sup_{x \in T} \Big| \|Ax\|_2 - \sqrt{m}\|x\|_2 \Big|^p \right)^{1/p} \leqslant C_p K^2 \gamma(T) \tag{4.2.4}$$

where $C_p \leqslant C\sqrt{p}$ for $p \geqslant 1$. (Check this using Proposition 1.1.1.)

Remark 4.2.5 (Deviation of squares). It is sometimes helpful to bound the deviation of the *square* $\|Ax\|_2^2$ rather than $\|Ax\|_2$ itself. We can easily deduce the deviation of squares by using the identity $a^2 - b^2 = (a-b)^2 + 2b(a-b)$ for $a = \|Ax\|_2$ and $b = \sqrt{m}\|x\|_2$. Doing this, we conclude that

$$\mathbb{E} \sup_{x \in T} \Big| \|Ax\|_2^2 - m\|x\|_2^2 \Big| \leqslant CK^4 \gamma(T)^2 + CK^2 \sqrt{m}\, \mathrm{rad}(T) \gamma(T). \tag{4.2.6}$$

(Do this calculation using (4.2.4) for $p = 2$.) We will use this bound in Section 4.4.

Matrix deviation inequality has many consequences. We will explore some of them now.

4.3. Deriving Johnson-Lindenstrauss Lemma We started this lecture by promising a result that is more general than Johnson-Lindenstrauss Lemma. So let us show how to quickly derive Johnson-Lindenstrauss from the matrix deviation inequality. Theorem 1.6.1 from Theorem 4.2.1.

Assume we are in the situation of the Johnson-Lindenstrauss Lemma (Theorem 1.6.1). Given a set $\mathcal{X} \subset \mathbb{R}$, consider the normalized difference set

$$T := \left\{ \frac{x-y}{\|x-y\|_2} : x, y \in \mathcal{X} \text{ distinct vectors} \right\}.$$

Then T is a finite subset of the unit sphere of $\mathbb{R}^n$, and thus (4.1.5) gives

$$\gamma(T) \lesssim \sqrt{\log |T|} \leqslant \sqrt{\log |\mathcal{X}|^2} \lesssim \sqrt{\log |\mathcal{X}|}.$$

Matrix deviation inequality (Theorem 4.2.1) then yields

$$\sup_{x,y \in \mathcal{X}} \left| \frac{\|A(x-y)\|_2}{\|x-y\|_2} - \sqrt{m} \right| \lesssim \sqrt{\log N} \leqslant \varepsilon\sqrt{m}.$$

with high probability, say 0.99. (To pass from expectation to high probability, we can use Markov's inequality. To get the last bound, we use the assumption on m in Johnson-Lindenstrauss Lemma.)

Multiplying both sides by $\|x-y\|_2/\sqrt{m}$, we can write the last bound as follows. With probability at least 0.99, we have

$$(1-\varepsilon)\|x-y\|_2 \leqslant \frac{1}{\sqrt{m}}\|Ax - Ay\|_2 \leqslant (1+\varepsilon)\|x-y\|_2 \quad \text{for all } x, y \in \mathcal{X}.$$

This is exactly the consequence of Johnson-Lindenstrauss lemma.

The argument based on matrix deviation inequality, which we just gave, can be easily extended for infinite sets. It allows one to state a version of Johnson-Lindenstrauss lemma for general, possibly infinite, sets, which depends on the Gaussian complexity of T rather than cardinality. (Try to do this!)

4.4. Covariance estimation In Section 3.1, we introduced the problem of covariance estimation, and we showed that

$$N \sim n \log n$$

samples are enough to estimate the covariance matrix of a general distribution in $\mathbb{R}^n$. We will now show how to do better if the distribution is sub-gaussian — see Section 1.5 for the definition of sub-gaussian random vectors — when we can get rid of the logarithmic oversampling and the boundedness condition (3.1.3).

Theorem 4.4.1 (Covariance estimation for sub-gaussian distributions)**.** *Let X be a random vector in $\mathbb{R}^n$ with covariance matrix Σ. Suppose X is sub-gaussian, and more specifically*

$$\| \langle X, x\rangle \|_{\psi_2} \lesssim \| \langle X, x\rangle \|_{L^2} = \|\Sigma^{1/2}x\|_2 \quad \text{for any } x \in \mathbb{R}^n. \tag{4.4.2}$$

Then, for every $N \geqslant 1$, we have

$$\mathbb{E}\, \|\Sigma_N - \Sigma\| \lesssim \|\Sigma\| \left(\sqrt{\frac{n}{N}} + \frac{n}{N}\right).$$

This result implies that if, for $\varepsilon \in (0,1)$, we take a sample of size

$$N \sim \varepsilon^{-2} n,$$

then we are guaranteed covariance estimation with a good relative error:

$$\mathbb{E}\, \|\Sigma_N - \Sigma\| \leqslant \varepsilon \|\Sigma\|.$$

Proof. Since we are going to use Theorem 4.2.1, we will need to first bring the random vectors $X, X_1, \dots, X_N$ to the isotropic position. This can be done by a suitable linear transformation. You will easily check that there exists an *isotropic* random vector Z such that

$$X = \Sigma^{1/2} Z.$$

(For example, Σ has full rank, set $Z := \Sigma^{-1/2} X$. Check the general case.) Similarly, we can find independent and isotropic random vectors Z_i such that

$$X_i = \Sigma^{1/2} Z_i, \quad i = 1, \dots, N.$$

The sub-gaussian assumption (4.4.2) then implies that

$$\|Z\|_{\psi_2} \lesssim 1.$$

(Check!) Then

$$\|\Sigma_N - \Sigma\| = \|\Sigma^{1/2} R_N \Sigma^{1/2}\| \quad \text{where} \quad R_N := \frac{1}{N}\sum_{i=1}^N Z_i Z_i^T - I_n.$$

The operator norm of a symmetric $n \times n$ matrix A can be computed by maximizing the quadratic form over the unit sphere: $\|A\| = \max_{x \in S^{n-1}} |\langle Ax, x\rangle|$. (To see this, recall that the operator norm is the biggest eigenvalue of A in magnitude.) Then

$$\|\Sigma_N - \Sigma\| = \max_{x \in S^{n-1}} \left\langle \Sigma^{1/2} R_N \Sigma^{1/2} x, x \right\rangle = \max_{x \in T} \langle R_N x, x\rangle$$

where T is the ellipsoid

$$T := \Sigma^{1/2} S^{n-1}.$$

Recalling the definition of R_N, we can rewrite this as

$$\|\Sigma_N - \Sigma\| = \max_{x \in T} \left| \frac{1}{N} \sum_{i=1}^{N} \langle Z_i, x\rangle^2 - \|x\|_2^2 \right| = \frac{1}{N} \max_{x \in T} \left| \|Ax\|_2^2 - N\|x\|_2^2 \right|.$$

Now apply the matrix deviation inequality for squares (4.2.6) to conclude that

$$\|\Sigma_N - \Sigma\| \lesssim \frac{1}{N} \left(\gamma(T)^2 + \sqrt{N}\, \mathrm{rad}(T) \gamma(T) \right).$$

(Do this calculation!) The radius and Gaussian width of the ellipsoid T are easy to compute:

$$\mathrm{rad}(T) = \|\Sigma\|^{1/2} \quad \text{and} \quad \gamma(T) \leqslant (\operatorname{tr} \Sigma)^{1/2}.$$

Substituting, we get

$$\|\Sigma_N - \Sigma\| \lesssim \frac{1}{N} \left(\operatorname{tr} \Sigma + \sqrt{N \|\Sigma\| \operatorname{tr} \Sigma} \right).$$

To complete the proof, use that $\operatorname{tr} \Sigma \leqslant n\|\Sigma\|$ (check!) and simplify the bound. □

Remark 4.4.3 (Low-dimensional distributions). Similarly to Section 3.1, we can show that much fewer samples are needed for covariance estimation of *low-dimensional* sub-gaussian distributions. Indeed, the proof of Theorem 4.4.1 actually yields

$$\mathbb{E}\, \|\Sigma_N - \Sigma\| \leqslant \|\Sigma\| \left(\sqrt{\frac{r}{N}} + \frac{r}{N} \right) \tag{4.4.4}$$

where

$$r = r(\Sigma^{1/2}) = \frac{\operatorname{tr} \Sigma}{\|\Sigma\|}$$

is the *stable rank* of $\Sigma^{1/2}$. This means that covariance estimation is possible with

$$N \sim r$$

samples.

4.5. Underdetermined linear equations We will give one more application of the matrix deviation inequality – this time, to the area of high dimensional inference. Suppose we need to solve a severely underdetermined system of linear equations: say, we have m equations in $n \gg m$ variables. Let us write it in the matrix form as

$$y = Ax$$

where A is a given $m \times n$ matrix, $y \in \mathbb{R}^m$ is a given vector and $x \in \mathbb{R}^n$ is an unknown vector. We would like to compute x from A and y.

When the linear system is underdetermined, we can not find x with any accuracy, unless we know something extra about x. So, let us assume that we do have some a-priori information. We can describe this situation mathematically by assuming that

$$x \in K$$

where $K \subset \mathbb{R}^n$ is some known set in $\mathbb{R}^n$ that describes anything that we know about x a-priori. (Admittedly, we are operating on a high level of generality here. If you need a concrete example, we will consider one in Section 4.6.)

Summarizing, here is the problem we are trying to solve. Determine a solution $x = x(A, y, K)$ to the underdetermined linear equation $y = Ax$ as accurately as possible, assuming that $x \in K$.

A variety of approaches to this and similar problems were proposed during the last decade; see the notes after this lecture for pointers to some literature. The one we will describe here is based on optimization. To do this, it will be convenient to convert the *set* K into a *function* on $\mathbb{R}^n$ which is called the *Minkowski functional* of K. This is basically a function whose level sets are multiples of K. To define it formally, assume that K is *star-shaped*, which means that together with any point x, the set K must contain the entire interval that connects x with the origin; see Figure 4.5.1 for illustration. The Minkowski functional of K is defined as

$$\|x\|_K := \inf\{t > 0 : \; x/t \in K\}, \quad x \in \mathbb{R}^n.$$

If the set K is convex and symmetric about the origin, $\|x\|_K$ is actually a *norm* on $\mathbb{R}^n$. (Check this!)

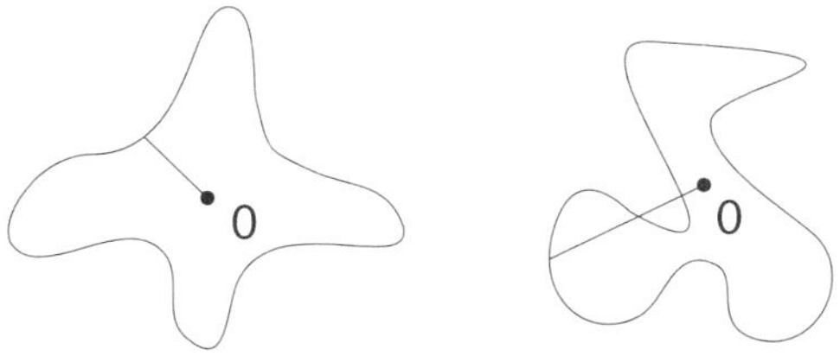

Figure 4.5.1. The set on the left (whose boundary is shown) is star-shaped, the set on the right is not.

Now we propose the following way to solve the recovery problem: solve the optimization program

$$\min \|x'\|_K \quad \text{subject to} \quad y = Ax'. \tag{4.5.2}$$

Note that this is a very natural program: it looks at all solutions to the equation $y = Ax'$ and tries to "shrink" the solution x' toward K. (This is what minimization of Minkowski functional is about.)

Also note that if K is convex, this is a *convex optimization* program, and thus can be solved effectively by one of the many available numeric algorithms.

The main question we should now be asking is – would the solution to this program approximate the original vector x? The following result bounds the approximation error for a *probabilistic model* of linear equations. Assume that A is a random matrix as in Theorem 4.2.1, i.e. A is an $m \times n$ matrix whose rows A_i are independent, isotropic and sub-gaussian random vectors in $\mathbb{R}^n$.

Theorem 4.5.3 (Recovery by optimization). *The solution $\hat{x}$ of the optimization program* (4.5.2) *satisfies*[23]

$$\mathbb{E}\,\|\hat{x} - x\|_2 \lesssim \frac{w(K)}{\sqrt{m}},$$

where $w(K)$ is the Gaussian width of K.

Proof. Both the original vector x and the solution $\hat{x}$ are feasible vectors for the optimization program (4.5.2). Then

$$\begin{aligned} \|\hat{x}\|_K &\leqslant \|x\|_K \quad \text{(since } \hat{x} \text{ minimizes the Minkowski functional)} \\ &\leqslant 1 \quad \text{(since } x \in K\text{)}. \end{aligned}$$

Thus both $\hat{x}, x \in K$.

We also know that $A\hat{x} = Ax = y$, which yields

$$A(\hat{x} - x) = 0. \tag{4.5.4}$$

Let us apply matrix deviation inequality (Theorem 4.2.1) for $T := K - K$. It gives

$$\mathbb{E} \sup_{u,v \in K} \Big| \|A(u-v)\|_2 - \sqrt{m}\,\|u - v\|_2 \Big| \lesssim \gamma(K - K) = 2w(K),$$

where we used (4.1.2) in the last identity. Substitute $u = \hat{x}$ and $v = x$ here. We may do this since, as we noted above, both these vectors belong to K. But then the term $\|A(u-v)\|_2$ will equal zero by (4.5.4). It disappears from the bound, and we get

$$\mathbb{E}\,\sqrt{m}\,\|\hat{x} - x\|_2 \lesssim w(K).$$

Dividing both sides by $\sqrt{m}$ we complete the proof. □

Theorem 4.5.3 says that a signal $x \in K$ can be efficiently recovered from

$$m \sim w(K)^2$$

random linear measurements.

4.6. Sparse recovery Let us illustrate Theorem 4.5.3 with an important specific example of the feasible set K. Suppose we know that the signal x is *sparse*, which means that only a few coordinates of x are nonzero. As before, our task is to recover x from the random linear measurements given by the vector

$$y = Ax,$$

[23]Here and in other similar results, the notation $\lesssim$ will hide possible dependence on the sub-gaussian norms of the rows of A.

where A is an $m \times n$ random matrix. This is a basic example of *sparse recovery problems*, which are ubiquitous in various disciplines.

The number of nonzero coefficients of a vector $x \in \mathbb{R}^n$, or the sparsity of x, is often denoted $\|x\|_0$. This is reminiscent of the notation for the ℓ_p norm $\|x\|_p = (\sum_{i=1}^n |x_i|^p)^{1/p}$, and for a reason. You can quickly check that

$$\|x\|_0 = \lim_{p \to 0} \|x\|_p \tag{4.6.1}$$

(Do this!) Keep in mind that neither $\|x\|_0$ nor $\|x\|_p$ for $0 < p < 1$ are actually *norms* on $\mathbb{R}^n$, since they fail triangle inequality. (Give an example.)

Let us go back to the sparse recovery problem. Our first attempt to recover x is to try the following optimization problem:

$$\min \|x'\|_0 \quad \text{subject to} \quad y = Ax'. \tag{4.6.2}$$

This is sensible because this program selects the sparsest feasible solution. But there is an implementation caveat: the function $f(x) = \|x\|_0$ is highly non-convex and even discontinuous. There is simply no known algorithm to solve the optimization problem (4.6.2) efficiently.

To overcome this difficulty, let us turn to the relation (4.6.1) for an inspiration. What if we replace $\|x\|_0$ in the optimization problem (4.6.2) by $\|x\|_p$ with $p > 0$? The smallest p for which $f(x) = \|x\|_p$ is a genuine norm (and thus a convex function on $\mathbb{R}^n$) is $p = 1$. So let us try

$$\min \|x'\|_1 \quad \text{subject to} \quad y = Ax'. \tag{4.6.3}$$

This is a *convexification* of the non-convex program (4.6.2), and a variety of numeric convex optimization methods are available to solve it efficiently.

We will now show that ℓ_1 minimization works nicely for sparse recovery. As before, we assume that A is a random matrix as in Theorem 4.2.1.

Theorem 4.6.4 (Sparse recovery by optimization). *If an unknown vector $x \in \mathbb{R}^n$ has at most s non-zero coordinates, i.e. $\|x\|_0 \leqslant s$, then the solution $\hat{x}$ of the optimization program* (4.6.3) *satisfies*

$$\mathbb{E} \|\hat{x} - x\|_2 \lesssim \sqrt{\frac{s \log n}{m}} \|x\|_2.$$

Proof. Since $\|x\|_0 \leqslant s$, Cauchy-Schwarz inequality shows that

$$\|x\|_1 \leqslant \sqrt{s}\, \|x\|_2. \tag{4.6.5}$$

(Check!) Denote the unit ball of the ℓ_1 norm in $\mathbb{R}^n$ by B_1^n, i.e. $B_1^n := \{x \in \mathbb{R}^n : \|x\|_1 \leqslant 1\}$. Then we can rewrite (4.6.5) as the inclusion

$$x \in \sqrt{s}\, \|x\|_2 \cdot B_1^n := K.$$

Apply Theorem 4.5.3 for this set K. We noted the Gaussian width of B_1^n in (4.1.4), so

$$w(K) = \sqrt{s}\, \|x\|_2 \cdot w(B_1^n) \leqslant \sqrt{s}\, \|x\|_2 \cdot \gamma(B_1^n) \leqslant \sqrt{s}\, \|x\|_2 \cdot \sqrt{\log n}.$$

Substitute this in Theorem 4.5.3 and complete the proof. □

Theorem 4.6.4 says that an s-sparse signal $x \in \mathbb{R}^n$ can be efficiently recovered from

$$m \sim s \log n$$

random linear measurements.

4.7. Notes For a more thorough introduction to Gaussian width and its role in high-dimensional estimation, refer to the tutorial [59] and the textbook [60]; see also [5]. Related to Gaussian complexity is the notion of *Rademacher complexity* of T, obtained by replacing the coordinates of g by independent Rademacher (i.e. ± 1 symmetric) random variables. Rademacher complexity of classes of functions plays an important role in statistical learning theory, see e.g. [44]

Matrix deviation inequality (Theorem 4.2.1) is borrowed from [41]. In the special case where A is a Gaussian random matrix, this result follows from the work of G. Schechtman [57] and could be traced back to results of Gordon [24–27].

In the general case of sub-gaussian distributions, earlier variants of Theorem 4.2.1 were proved by B. Klartag and S. Mendelson [35], S. Mendelson, A. Pajor and N. Tomczak-Jaegermann [45] and S. Dirksen [20].

Theorem 4.4.1 for covariance estimation can be proved alternatively using more elementary tools (Bernstein's inequality and ε-nets), see [58]. However, no known elementary approach exists for the *low-rank* covariance estimation discussed in Remark 4.4.3. The bound (4.4.4) was proved by V. Koltchinskii and K. Lounici [36] by a different method.

In Section 4.5, we scratched the surface of a recently developed area of *sparse signal recovery*, which is also called *compressed sensing*. Our presentation there essentially follows the tutorial [59]. Theorem 4.6.4 can be improved: if we take

$$m \gtrsim s \log(n/s)$$

measurements, then with high probability the optimization program (4.6.3) recovers the unknown signal x *exactly*, i.e.

$$\hat{x} = x.$$

First results of this kind were proved by J. Romberg, E. Candes and T. Tao [13] and a great number of further developments followed; refer e.g. to the book [23] and the chapter in [19] for an introduction into this research area.

Acknowledgement

I am grateful to the referees who made a number of useful suggestions, which led to better presentation of the material in this chapter.

References

[1] E. Abbe, A. S. Bandeira, G. Hall. *Exact recovery in the stochastic block model*, IEEE Transactions on Information Theory 62 (2016), 471–487. MR3447993 249

[2] D. Achlioptas, *Database-friendly random projections: Johnson-Lindenstrauss with binary coins*, Journal of Computer and System Sciences, 66 (2003), 671–687. MR2005771 239

[3] R. Ahlswede, A. Winter, *Strong converse for identification via quantum channels,* IEEE Trans. Inf. Theory 48 (2002), 569–579. MR1889969 248

[4] N. Ailon, B. Chazelle, *Approximate nearest neighbors and the fast Johnson-Lindenstrauss transform,* Proceedings of the 38th Annual ACM Symposium on Theory of Computing. New York: ACM Press, 2006. pp. 557–563. MR2277181 239

[5] D. Amelunxen, M. Lotz, M. McCoy, J. Tropp, *Living on the edge: phase transitions in convex programs with random data,* Inf. Inference 3 (2014), 224–294. MR3311453 268

[6] A. Bandeira, R. van Handel, *Sharp nonasymptotic bounds on the norm of random matrices with independent entries*, Ann. Probab. 44 (2016), 2479–2506. MR3531673 249, 258

[7] R. Baraniuk, M. Davenport, R. DeVore, M. Wakin, *A simple proof of the restricted isometry property for random matrices,* Constructive Approximation, 28 (2008), 253–263. MR2453366 239

[8] R. Bhatia, *Matrix Analysis.* Graduate Texts in Mathematics, vol. 169. Springer, Berlin, 1997. MR1477662 248

[9] C. Bordenave, M. Lelarge, L. Massoulie, *Non-backtracking spectrum of random graphs: community detection and non-regular Ramanujan graphs,* Annals of Probability, to appear. MR3758726 248, 249

[10] S. Boucheron, G. Lugosi, P. Massart, *Concentration inequalities. A nonasymptotic theory of independence.* With a foreword by Michel Ledoux. Oxford University Press, Oxford, 2013. MR3185193 239

[11] O. Bousquet1, S. Boucheron, G. Lugosi, *Introduction to statistical learning theory,* in: Advanced Lectures on Machine Learning, Lecture Notes in Computer Science 3176, pp.169–207, Springer Verlag 2004.

[12] T. Cai, R. Zhao, H. Zhou, *Estimating structured high-dimensional covariance and precision matrices: optimal rates and adaptive estimation,* Electron. J. Stat. 10 (2016), 1–59. MR3466172 258

[13] E. Candes, J. Romberg, T. Tao, *Robust uncertainty principles: exact signal reconstruction from highly incomplete frequency information,* IEEE Trans. Inform. Theory 52 (2006), 489–509. MR2236170 268

[14] E. Candes, B. Recht, *Exact Matrix Completion via Convex Optimization,* Foundations of Computational Mathematics 9 (2009), 717–772. MR2565240 258

[15] E. Candes, T. Tao, *The power of convex relaxation: near-optimal matrix completion,* IEEE Trans. Inform. Theory 56 (2010), 2053–2080. MR2723472 259

[16] R. Chen, A. Gittens, J. Tropp, *The masked sample covariance estimator: an analysis using matrix concentration inequalities,* Inf. Inference 1 (2012), 2–20. MR3311439 258

[17] P. Chin, A. Rao, and V. Vu, *Stochastic block model and community detection in the sparse graphs: A spectral algorithm with optimal rate of recovery*, preprint, 2015. 249

[18] M. Davenport, Y. Plan, E. van den Berg, M. Wootters, *1-bit matrix completion,* Inf. Inference 3 (2014), 189–223. MR3311452 259

[19] M. Davenport, M. Duarte, Yonina C. Eldar, Gitta Kutyniok, *Introduction to compressed sensing,* in: *Compressed sensing.* Edited by Yonina C. Eldar and Gitta Kutyniok. Cambridge University Press, Cambridge, 2012. MR2963166 268

[20] S. Dirksen, *Tail bounds via generic chaining,* Electron. J. Probab. 20 (2015), 1–29. MR3354613 268

[21] U. Feige, E. Ofek, *Spectral techniques applied to sparse random graphs,* Random Structures Algorithms 27 (2005), 251–275. MR2155709 249

[22] S. Fortunato, Santo; D. Hric, *Community detection in networks: A user guide.* Phys. Rep. 659 (2016), 1–44. MR3566093 248

[23] S. Foucart, H. Rauhut, *A mathematical introduction to compressive sensing.* Applied and Numerical Harmonic Analysis. Birkhäuser/Springer, New York, 2013. MR3100033 268

[24] Y. Gordon, *Some inequalities for Gaussian processes and applications,* Israel J. Math. 50 (1985), 265–289. MR800188 268

[25] Y. Gordon, *Elliptically contoured distributions,* Prob. Th. Rel. Fields 76 (1987), 429–438. MR917672 268

[26] Y. Gordon, *On Milman's inequality and random subspaces which escape through a mesh in* $\mathbb{R}^n$, Geometric aspects of functional analysis (1986/87), Lecture Notes in Math., vol. 1317, pp. 84–106. MR950977 268

[27] Y. Gordon, *Majorization of Gaussian processes and geometric applications,* Prob. Th. Rel. Fields 91 (1992), 251–267. MR1147616 268

[28] D. Gross, *Recovering low-rank matrices from few coefficients in any basis*, IEEE Trans. Inform. Theory 57 (2011), 1548–1566. MR2815834 259

[29] O. Guedon, R. Vershynin, *Community detection in sparse networks via Grothendieck's inequality*, Probability Theory and Related Fields 165 (2016), 1025–1049. MR3520025 248, 249

[30] A. Javanmard, A. Montanari, F. Ricci-Tersenghi, *Phase transitions in semidefinite relaxations*, PNAS, April 19, 2016, vol. 113, no.16, E2218–E2223. MR3494080 248, 249

[31] W. B. Johnson, J. Lindenstrauss, *Extensions of Lipschitz mappings into a Hilbert space*. In Beals, Richard; Beck, Anatole; Bellow, Alexandra; et al. Conference in modern analysis and probability (New Haven, Conn., 1982). Contemporary Mathematics. 26. Providence, RI: American Mathematical Society, 1984. pp. 189–206. MR737400 239

[32] B. Hajek, Y. Wu, J. Xu, *Achieving exact cluster recovery threshold via semidefinite programming*, IEEE Transactions on Information Theory 62 (2016), 2788–2797. MR3493879 248, 249

[33] P. W. Holland, K. B. Laskey, S. Leinhardt, *Stochastic blockmodels: first steps*, Social Networks 5 (1983), 109–137. MR718088 248

[34] D. Kane, J. Nelson, *Sparser Johnson-Lindenstrauss Transforms*, Journal of the ACM 61 (2014): 1. MR3167920 239

[35] B. Klartag, S. Mendelson, *Empirical processes and random projections*, J. Funct. Anal. 225 (2005), 229–245. MR2149924 268

[36] V. Koltchinskii, K. Lounici, *Concentration inequalities and moment bounds for sample covariance operators*, Bernoulli 23 (2017), 110–133. MR3556768 258, 268

[37] C. Le, E. Levina, R. Vershynin, *Concentration and regularization of random graphs*, Random Structures and Algorithms, to appear. MR3689343 248, 249

[38] M. Ledoux, *The concentration of measure phenomenon*. American Mathematical Society, Providence, RI, 2001. MR1849347 239

[39] M. Ledoux, M. Talagrand, *Probability in Banach spaces. Isoperimetry and processes*. Springer-Verlag, Berlin, 1991. MR1102015 239

[40] E. Levina, R. Vershynin, *Partial estimation of covariance matrices*, Probability Theory and Related Fields 153 (2012), 405–419. MR2948681 258

[41] C. Liaw, A. Mehrabian, Y. Plan, R. Vershynin, *A simple tool for bounding the deviation of random matrices on geometric sets*, Geometric Aspects of Functional Analysis, Lecture Notes in Mathematics, Springer, Berlin, to appear. MR3645128 268

[42] J. Matoušek, *Lectures on discrete geometry*. Graduate Texts in Mathematics, 212. Springer-Verlag, New York, 2002. MR1899299 239

[43] F. McSherry, *Spectral partitioning of random graphs*, Proc. 42nd FOCS (2001), 529–537. MR1948742 249

[44] S. Mendelson, S. Mendelson, *A few notes on statistical learning theory*, in: Advanced Lectures on Machine Learning, S. Mendelson, A. J. Smola (Eds.) LNAI 2600, pp. 1–40, 2003. 268

[45] S. Mendelson, A. Pajor, N. Tomczak-Jaegermann, *Reconstruction and subgaussian operators in asymptotic geometric analysis*. Geom. Funct. Anal. 17 (2007), 1248–1282. MR2373017 268

[46] E. Mossel, J. Neeman, A. Sly, *Belief propagation, robust reconstruction and optimal recovery of block models*. Ann. Appl. Probab. 26 (2016), 2211–2256. MR3543895 248, 249

[47] M. E. Newman, *Networks. An introduction*. Oxford University Press, Oxford, 2010. MR2676073 248

[48] R. I. Oliveira, *Concentration of the adjacency matrix and of the Laplacian in random graphs with independent edges*, unpublished (2010), arXiv:0911.0600. 248, 249

[49] R. I. Oliveira, *Sums of random Hermitian matrices and an inequality by Rudelson*, Electron. Commun. Probab. 15 (2010), 203–212. MR2653725 248

[50] Y. Plan, R. Vershynin, E. Yudovina, *High-dimensional estimation with geometric constraints*, Information and Inference 0 (2016), 1–40. MR3636866 259

[51] S. Riemer, C. Schütt, *On the expectation of the norm of random matrices with non-identically distributed entries*, Electron. J. Probab. 18 (2013), no. 29, 13 pp. MR3035757 258

[52] J. Tropp, *User-friendly tail bounds for sums of random matrices*. Found. Comput. Math. 12 (2012), 389–434. MR2946459 248

[53] J. Tropp, *An introduction to matrix concentration inequalities*. Found. Trends Mach. Learning 8 (2015), 10-230. 248

[54] R. van Handel, *Structured random matrices*. in: IMA Volume "Discrete Structures: Analysis and Applications", Springer, to appear.

[55] R. van Handel, *On the spectral norm of Gaussian random matrices,* Trans. Amer. Math. Soc., to appear. MR3695857 258

[56] B. Recht, *A simpler approach to matrix completion,* J. Mach. Learn. Res. 12 (2011), 3413–3430. MR2877360 259

[57] G. Schechtman, *Two observations regarding embedding subsets of Euclidean spaces in normed spaces,* Adv. Math. 200 (2006), 125–135. MR2199631 268

[58] R. Vershynin, *Introduction to the non-asymptotic analysis of random matrices.* Compressed sensing, 210–268, Cambridge University Press, Cambridge, 2012. MR2963170 232, 239, 258, 268

[59] R. Vershynin, *Estimation in high dimensions: a geometric perspective.* Sampling Theory, a Renaissance, 3–66, Birkhauser Basel, 2015. MR3467418 232, 259, 268

[60] R. Vershynin, *High-Dimensional Probability. An Introduction with Applications in Data Science.* Cambridge University Press, to appear. 232, 239, 258, 268

[61] H. Zhou, A. Zhang, *Minimax Rates of Community Detection in Stochastic Block Models,* Annals of Statistics, to appear. MR3546450 248, 249

Department of Mathematics, University of Michigan, 530 Church Street, Ann Arbor, MI 48109, U.S.A.

Email address: romanv@umich.edu

IAS/Park City Mathematics Series
Volume 25, Pages 273–325
https://doi.org/10.1090/pcms/025/00834

Homological Algebra and Data

Robert Ghrist

Contents

Introduction and Motivation

These lectures are meant as an introduction to the methods and perspectives of Applied Topology for students and researchers in areas including but not limited to data science, neuroscience, complex systems, and statistics. Though the tools

2010 *Mathematics Subject Classification.* Primary 55-01; Secondary 18G35, 55N30.
Key words and phrases. cohomology, complexes, homology, persistence, sheaves.
RG supported by the Office of the Assistant Secretary of Defense Research & Engineering through ONR N00014-16-1-2010.

are mathematical in nature, this article will treat the formalities with a light touch and heavy references, in order to make the subject more accessible to practitioners. See the concluding section for a roadmap for finding more details. The material is written for beginning graduate students in any of the applied mathematical sciences (though some mathematical maturity is helpful).

What is Homology?

Homology is an algebraic compression scheme that excises all but the essential topological features from a particular class of data structures arising naturally from topological spaces. Homology therefore pairs with topology. Topology is the mathematics of abstract space and transformations between them. The notion of a space, X, requires only a set together with a notion of nearness, expressed as a system of subsets comprising the "open" neighborhoods satisfying certain consistency conditions. Metrics are permissible but not required. So many familiar notions in applied mathematics – networks, graphs, data sets, signals, imagery, and more – are interpretable as topological spaces, often with useful auxiliary structures. Furthermore, manipulations of such objects, whether as comparison, inference, or metadata, are expressible in the language of mappings, or continuous relationships between spaces. Topology concerns the fundamental notions of equivalence up to the loose nearness of what makes a space. Thus, connectivity and holes are significant; bends and corners less so. Topological invariants of spaces and mappings between them record the essential qualitative features, insensitive to coordinate changes and deformations.

Homology is the simplest, general, computable invariant of topological data. In its most primal manifestation, the homology of a space X returns a sequence of vector spaces $H_\bullet(X)$, the dimensions of which count various types of *linearly independent* holes in X. Homology is inherently linear-algebraic, but transcends linear algebra, serving as the inspiration for *homological algebra*. It is this algebraic engine that powers the subject.

When is Homology Useful?

Homological methods are, almost by definition, robust, relying on neither precise coordinates nor careful estimates for efficacy. As such, they are most useful in settings where geometric precision fails. With great robustness comes both great flexibility and great weakness. Topological data analysis is more fundamental than revolutionary: such methods are not intended to supplant analytic, probabilistic, or spectral techniques. They can however reveal a deeper basis for why some data sets and systems behave the way they do. It is unwise to wield topological techniques in isolation, assuming that the weapons of unfamiliar "higher" mathematics are clad in incorruptible silver.

Scheme

There is far too much material in the subject of algebraic topology to be surveyed here. Existing applications alone span an enormous range of principles and techniques, and the subject of applications of homology and homological algebra is in its infancy still. As such, these notes are selective to a degree that suggests caprice. For deeper coverage of the areas touched on here, complete with illustrations, see [51]. For alternate ranges and perspectives, there are now a number of excellent sources, including [40, 62, 76]. These notes will deemphasize formalities and ultimate formulations, focusing instead on principles, with examples and exercises. The reader should not infer that the theorems or theoretic minutiae are anything less than critical in practice.

These notes err on the side of simplicity. The many included exercises are not of the typical lemma-lemma-theorem form appropriate for a mathematics course; rather, they are meant to ground the student in examples. There is an additional layer of unstated problems for the interested reader: these notes are devoid of figures. The student apt with a pen should endeavor to create cartoons to accompany the various definitions and examples presented here, with the aim of minimality and clarity of encapsulation. The author's attempt at such can be found in [51].

Lecture 1: Complexes and Homology

This lecture will introduce the initial objects and themes of applied algebraic topology. There is little novel here: all definitions are standard and found in standard texts. The quick skip over formalities, combined with a linear-algebraic sensibility, allows for a rapid ascent to the interesting relationships to be found in homology and homological algebra.

Spaces

A *space* is a set X together with a compendium of all subsets in X deemed "open," which subcollection must of necessity satisfy a list of intuitively obvious properties. The interested reader should consult any point-set topology book (such as [70]) briefly. All the familiar spaces of elementary calculus – surfaces, level sets of functions, Euclidean spaces – are indeed topological spaces and just the beginning of the interesting spaces studied in manifold theory, algebraic geometry, differential geometry, and more. These tend to be frustratingly indiscrete. Applications involving computation prompt an emphasis on those spaces that are easily digitized. Such are usually called *complexes*, often with an adjectival prefix. Several are outlined below.

Simplicial Complexes Consider a set X of discrete objects. A k-*simplex* in X is an unordered collection σ of $k+1$ distinct elements of X. Though the definition is combinatorial, for X a set of points in a Euclidean space [*viz.* point-cloud data set] one visualizes a simplex as the geometric convex hull of the $k+1$ points, a "filled-in" clique: thus, 0-simplices are points, 1-simplices are edges, 2-simplices are filled-in triangles, etc. A *complex* is a collection of multiple *simplices*.[1] In particular, a *simplicial complex* on X is a collection of simplices in X that is *downward closed*, in the sense that every subset of every simplex is also a simplex in the complex. One says that X *contains all its faces*. Greek letters (especially σ and τ) will be used to denote simplices in what follows.

Exercise 1.1. Recall that a collection of random variables $\mathcal{X} = \{X_i\}_1^k$ on a fixed domain are statistically independent if their probability densities f_{X_i} are jointly multiplicative (that is, the probability density $f_{\mathcal{X}}$ of the combined random variable $(X_1, \ldots, X_k)$ satisfies $f_{\mathcal{X}} = \prod_i f_{X_i}$). Given a set of n random variables on a fixed domain, explain how one can build a simplicial complex using statistical independence to define simplices. What is the maximal dimension of this *independence complex*? What does the number of connected components of the independence complex tell you? Is it possible to have all edges present and no higher-dimensional faces?

Exercise 1.2. Not all interesting simplicial complexes are simple to visualize. Consider a finite-dimensional real vector space V and consider V to be the vertex set of a simplicial complex defined as follows: a k-simplex consists of $k+1$ linearly independent members of V. Is the resulting independence complex finite? Finite-dimensional? What does the dimension of this complex tell you?

Simplicial complexes as described are purely combinatorial objects, like the graphs they subsume. As a graph, one topologizes a simplicial complex as a quotient space built from topological simplices. The *standard* k-*simplex* is the following incarnation of its Platonic ideal:

$$\Delta^k = \left\{ x \in [0,1]^{k+1} : \sum_{i=0}^{k} x_i = 1 \right\}. \tag{1.3}$$

One topologizes an abstract simplicial complex into a space X by taking one formal copy of Δ^k for each k-simplex of X, then identifying these together along faces inductively. The way to do this formally is to take a disjoint union of standard simplices, one for each simplex in the complex; then identify or "glue" using an equivalence relation. Specifically, define the k-*skeleton* of X, $k \in \mathbb{N}$, to be the quotient space:

$$X^{(k)} = \left(X^{(k-1)} \bigcup \coprod_{\sigma:\dim \sigma = k} \Delta^k \right) \Big/ \sim, \tag{1.4}$$

where $\sim$ is the equivalence relation that identifies faces of Δ^k with the corresponding combinatorial faces of σ in $X^{(j)}$ for $j < k$. Thus, for example, $X^{(0)}$ is a discrete

[1] The etymology of both words is salient.

collection of points (the vertices) and $X^{(1)}$ is the abstract space obtained by gluing edges to the vertices using information stored in the 1-simplices.

Exercise 1.5. How many total k-simplices are there in the closed n-simplex for $k < n$?

Vietoris-Rips Complexes A data set in the form of a finite metric space (X, d) gives rise to a family of simplicial complexes in the following manner. The *Vietoris-Rips complex* (or VR-complex) of (X, d) at scale $\epsilon > 0$ is the simplicial complex $VR_\epsilon(X)$ whose simplices are precisely those collections of points with pairwise distance $\leqslant \epsilon$. Otherwise said, one connects points that are sufficiently close, filling in sufficiently small holes, with sufficiency specified by ϵ.

These VR complexes have been used as a way of associating a simplicial complex to point cloud data sets. One obvious difficulty, however, lies in the choice of ϵ: too small, and nothing is connected; too large, and everything is connected. The question of which ϵ to use has no easy answer. However, the perspectives of algebraic topology offer a modified question. *How to integrate structures across all ϵ values?* This will be considered in Lecture 2 of this series.

Flag/clique complexes The VR complex is a particular instance of the following construct. Given a graph (network) X, the *flag complex* or *clique complex* of X is the maximal simplicial complex X that has the graph as its 1-skeleton: $X^{(1)} = X$. What this means in practice is that whenever you "see" the skeletal frame of a simplex in X, you fill it and all its faces in with simplices. Flag complexes are advantageous as data structures for spaces, in that you do not need to input/store all of the simplices in a simplicial complex: the 1-skeleton consisting of vertices and edges suffices to define the rest of the complex.

Exercise 1.6. Consider a combinatorial simplicial complex X on a vertex set of size n. As a function of this n, how difficult is it to store in memory enough information about X to reconstruct the list of its simplices? (There are several ways to approach this: see Exercise 1.5 for one approach.) Does this worst-case complexity improve if you know that X is a flag complex?

Nerve Complexes This is a particular example of a *nerve complex* associated to a collection of subsets.

Let $\mathcal{U} = \{U_\alpha\}$ be a collection of open subsets of a topological space X. The *nerve* of $\mathcal{U}$, $\mathcal{N}(\mathcal{U})$, is the simplicial complex defined by the intersection lattice of $\mathcal{U}$. The k-simplices of $\mathcal{N}(\mathcal{U})$ correspond to nonempty intersections of $k+1$ distinct elements of $\mathcal{U}$. Thus, vertices of the nerve correspond to elements of $\mathcal{U}$; edges correspond to pairs in $\mathcal{U}$ which intersect nontrivially. This definition respects faces: the faces of a k-simplex are obtained by removing corresponding elements of $\mathcal{U}$, leaving the resulting intersection still nonempty.

Exercise 1.7. Compute all possible nerves of four bounded convex subsets in the Euclidean plane. What is and is not possible? Now, repeat, but with two *nonconvex* subsets of Euclidean $\mathbb{R}^3$.

Dowker Complexes There is a matrix version of the nerve construction that is particularly relevant to applications, going back (at least) to the 1952 paper of Dowker [39]. For simplicity, let X and Y be finite sets with $\mathcal{R} \subset X \times Y$ representing the ones in a binary matrix (also denoted $\mathcal{R}$) whose columns are indexed by X and whose rows are indexed by Y. The *Dowker complex* of $\mathcal{R}$ on X is the simplicial complex on the vertex set X defined by the rows of the matrix $\mathcal{R}$. That is, each row of $\mathcal{R}$ determines a subset of X: use this to generate a simplex and all its faces. Doing so for all the rows gives the Dowker complex on X. There is a *dual Dowker complex* on Y whose simplices on the vertex set Y are determined by the ones in *columns* of $\mathcal{R}$.

Exercise 1.8. Compute the Dowker complex and the dual Dowker complex of the following relation $\mathcal{R}$:

$$\mathcal{R} = \begin{bmatrix} 1 & 0 & 0 & 0 & 1 & 1 & 0 & 0 \\ 0 & 1 & 1 & 0 & 0 & 0 & 1 & 0 \\ 0 & 1 & 0 & 0 & 1 & 1 & 0 & 1 \\ 1 & 0 & 1 & 0 & 1 & 0 & 0 & 1 \\ 1 & 0 & 1 & 0 & 0 & 1 & 1 & 0 \end{bmatrix}. \tag{1.9}$$

Dowker complexes have been used in a variety of social science contexts (where X and Y represent *agents* and *attributes* respectively) [7]. More recent applications of these complexes have arisen in settings ranging from social networks [89] to sensor networks [54]. The various flavors of *witness complexes* in the literature on topological data analysis [37,57] are special cases of Dowker complexes.

Cell Complexes There are other ways to build spaces out of simple pieces. These, too, are called complexes, though not simplicial, as they are not necessarily built from simplicies. They are best described as *cell complexes*, being built from cells of various dimensions sporting a variety of possible auxiliary structures.

A *cubical complex* is a cell complex built from cubes of various dimensions, the formal definition mimicking Equation (1.4): see [51,62]. These often arise as the natural model for pixel or voxel data in imagery and time series. Cubical complexes have found other uses in modelling spaces of phylogenetic trees [17, 77] and robot configuration spaces [1,53,55].

There are much more general cellular complexes built from simple pieces with far less rigidity in the attachments between pieces. Perhaps the most general useful model of a cell complex is the *CW complex* used frequently in algebraic topology. The idea of a CW complex is this: one begins with a disjoint union of points $X^{(0)}$ as the 0-skeleton. One then inductively defines the n-skeleton of

X, $X^{(n)}$ as the $(n-1)$-skeleton along with a collection of closed n-dimensional balls, $\mathbb{D}^n$, each glued to $X^{(n-1)}$ via attaching maps on the boundary spheres $\partial\mathbb{D}^n \to X^{(n-1)}$. In dimension one, [finite] CW complexes, simplicial complexes, and cubical complexes are identical and equivalent to [finite] graphs.[2] In higher dimensions, these types of cell complexes diverge in expressivity and ease of use.

Spaces and Equivalence

Many of the spaces of interest in topological data analysis are finite metric spaces [point clouds] and simplicial approximations and generalizations of these. However, certain spaces familiar from basic calculus are relevant. We have already referenced $\mathbb{D}^n$, the closed unit n-dimensional ball in Euclidean $\mathbb{R}^n$. Its boundary defines the standard sphere $\mathbb{S}^{n-1}$ of dimension $n-1$. The 1-sphere $\mathbb{S}^1$ is also the 1-torus, where, by n-torus is meant the [Cartesian] product $\mathbb{T}^n = (\mathbb{S}^1)^n$ of n circles. The 2-sphere $\mathbb{S}^2$ and 2-torus $\mathbb{T}^2$ are compact, orientable surfaces of *genus* 0 and 1 respectively. For any genus $g \in \mathbb{N}$, there is a compact orientable surface Σ_g with that genus: for $g > 1$ these look like g 2-tori merged together so as to have the appearance of having g holes. All orientable genus g surfaces are *"topologically equivalent"*, though this is as yet imprecise.

One soon runs into difficulty with descriptive language for spaces and equivalences, whether via coordinates or visual features. Another language is needed. Many of the core results of topology concern equivalence, detection, and resolution: are two spaces or maps between spaces qualitatively the same? This presumes a notion of equivalence, of which there are many. In what follows, *map* always means a *continuous* function between spaces.

Homeomorphism and Homotopy A *homeomorphism* is a map $f\colon X \to Y$ with continuous inverse. This is the strongest form of topological equivalence, distinguishing spaces of different (finite) dimensions or different essential features (e.g., genus of surfaces) and also distinguishing an open from a closed interval. The more loose and useful equivalence is that generated by homotopy. A *homotopy* between maps, $f_0 \simeq f_1\colon X \to Y$ is a continuous 1-parameter family of maps $f_t\colon X \to Y$. A *homotopy equivalence* is a map $f\colon X \to Y$ with a homotopy inverse, $g\colon Y \to X$ satisfying $f \circ g \simeq \mathrm{Id}_Y$ and $g \circ f \simeq \mathrm{Id}_X$. One says that such an X and Y are *homotopic*. This is the core equivalence relation among spaces in topology.

Exercise 1.10. A space is *contractible* if it is homotopic to a point. (1) Show explicitly that $\mathbb{D}^n$ is contractible. (2) Show that $\mathbb{D}^3$ with a point in the interior removed is homotopic to $\mathbb{S}^2$. (3) Argue that the twice-punctured plane is homotopic to a "figure-eight." It's not so easy to do this with explicit maps and coordinates, is it?

[2]With the exception of loop edges, which are generally not under the aegis of a graph, but are permissible in CW complexes.

Many of the core results in topology are stated in the language of homotopy (and are not true when *homotopy* is replaced with the more restrictive *homeomorphism*). For example:

Theorem 1.11. *If $\mathcal{U}$ is a finite collection of open contractible subsets of X with all non-empty intersections of subcollections of $\mathcal{U}$ contractible, then $\mathcal{N}(\mathcal{U})$ is homotopic to the union $\cup_\alpha U_\alpha$.*

Theorem 1.12. *Given any binary relation $\mathcal{R} \subset X \times Y$, the Dowker and dual Dowker complexes are homotopic.*

Homotopy invariants — functions that assign equivalent values to homotopic inputs — are central both to topology and its applications to data (as noise perturbs spaces in an often non-homeomorphic but homotopic manner). Invariants of finite simplicial and cell complexes invite a computational perspective, since one has the hope of finite inputs and felicitous data structures.

Euler Characteristic The simplest nontrivial topological invariant of finite cell complexes dates back to Euler. It is elementary, combinatorial, and sublime. The *Euler characteristic* of a finite cell complex X is:

$$\chi(X) = \sum_\sigma (-1)^{\dim \sigma}, \tag{1.13}$$

where the sum is over all cells σ of X.

Exercise 1.14. Compute explicitly the Euler characteristics of the following cell complexes: (1) the decompositions of the 2-sphere, $\mathbb{S}^2$, defined by the boundaries of the five regular Platonic solids; (2) the CW complex having one 0-cell and one 2-cell disc whose boundary is attached directly to the 0-cell ("collapse the boundary circle to a point"); and (3) the thickened 2-sphere $\mathbb{S}^2 \times [0,1]$. How did you put a cell structure on this last 3-dimensional space?

Completion of this exercise suggests the following result:

Theorem 1.15. *Euler characteristic is a homotopy invariant among finite cell complexes.*

That this is so would seem to require a great deal of combinatorics to prove. The modern proof transcends combinatorics, making the problem hopelessly uncomputable before pulling back to the finite world, as will be seen in Lecture 3.

Exercise 1.16. Prove that the Euler characteristic distinguishes [connected] trees from [connected] graphs with cycles. What happens if the connectivity requirement is dropped?

Euler characteristic is a wonderfully useful invariant, with modern applications ranging from robotics [42, 47] and AI [68] to sensor networks [9–11] to Gaussian random fields [4,6]. In the end, however, it is a numerical invariant, and has a limited resolution. The path to improving the resolution of this invariant is to enrich the underlying algebra that the Eulerian ± 1 obscures in Equation (1.13).

Lifting to Linear Algebra One of the core themes of this lecture series is the lifting of cell complexes to algebraic complexes on which the tools of homological algebra can be brought to bear. This is not a novel idea: most applied mathematicians learn, e.g., to use the adjacency matrix of a graph as a means of harnessing linear-algebraic ideas to understand networks. What is novel is the use of higher-dimensional structure and the richer algebra this entails.

Homological algebra is often done with modules over a commutative ring. For clarity of exposition, let us restrict to the nearly trivial setting of finite-dimensional vector spaces over a field $\mathbb{F}$, typically either $\mathbb{R}$ or, when orientations are bothersome, $\mathbb{F}_2$, the binary field.

Given a cell complex, one lifts the topological cells to algebraic objects by using them as bases for vector spaces. One remembers the dimensions of the cells by using a sequence of vector spaces, with dimension as a *grading* that indexes the vector spaces. Consider the following sequence $\mathcal{C} = (C_k)$ of vector spaces, where the grading k is in $\mathbb{N}$.

$$\cdots \qquad C_k \qquad C_{k-1} \qquad \cdots \qquad C_1 \qquad C_0\ . \tag{1.17}$$

In algebraic topology, one often uses a "star" or "dot" to denote a grading: this chapter will use a dot, as in $\mathcal{C} = (C_\bullet)$. For a finite (and thus finite-dimensional) cell complex, the sequence becomes all zeros eventually. Such a sequence does not obviously offer an algebraic advantage over the original space; indeed, much of the information on how cells are glued together has been lost. However, it is easy to "lift" the Euler characteristic to this class of algebraic objects. For $\mathcal{C}$ a sequence of finite-dimensional vector spaces with finitely many nonzero terms, define:

$$\chi(\mathcal{C}) = \sum_k (-1)^k \dim C_k. \tag{1.18}$$

Chain Complexes Recall that basic linear algebra does not focus overmuch on vector spaces and bases; it is in linear transformations that power resides. Augmenting a sequence of vector spaces with a matching sequence of linear transformations adds in the assembly instructions and permits a fuller algebraic representation of a topological complex. Given a simplicial[3] complex X, fix a field $\mathbb{F}$ and let $\mathcal{C} = (C_k, \partial_k)$ denote the following sequence of $\mathbb{F}$-vector spaces and linear transformations.

$$\cdots \longrightarrow C_k \xrightarrow{\partial_k} C_{k-1} \xrightarrow{\partial_{k-1}} \cdots \xrightarrow{\partial_2} C_1 \xrightarrow{\partial_1} C_0 \xrightarrow{\partial_0} 0\ . \tag{1.19}$$

Each C_k has as basis the k-simplices of X. Each ∂_k restricted to a k-simplex basis element sends it to a linear combination of those basis elements in C_{k-1} determined by the $k+1$ faces of the k-simplex. This is simplest in the case $\mathbb{F} = \mathbb{F}_2$,

[3]Cell complexes in full generality can be used with more work put into the definitions of the linear transformations: see [58].

in which case orientations can be ignored; otherwise, one must affix an orientation to each simplex and proceed accordingly: see [51, 58] for details on how this is performed.

The chain complex is the primal algebraic object in homological algebra. It is rightly seen as the higher-dimensional analogue of a graph together with its adjacency matrix. The chain complex will often be written as $\mathcal{C} = (C_\bullet, \partial)$, with the grading implied in the subscript dot. The boundary operator, $\partial = \partial_\bullet$, can be thought of either as a sequence of linear transformations, or as a single operator acting on the direct sum of the C_k.

Homology Homological algebra begins with the following suspiciously simple statement about simplicial complexes.

Lemma 1.20. *The boundary of a boundary is null:*

$$\partial^2 = \partial_{k-1} \circ \partial_k = 0, \tag{1.21}$$

for all k.

Proof. For simplicity, consider the case of an abstract simplicial complex on a vertex set $V = \{v_i\}$ with chain complex having $\mathbb{F}_2$ coefficients. The *face map* D_i acts on a simplex by removing the ith vertex v_i from the simplex's list, if present; else, return zero. The graded boundary operator $\partial\colon C_\bullet \to C_\bullet$ is thus a formal sum of face maps $\partial = \bigoplus_i D_i$. It suffices to show that $\partial^2 = 0$ on each basis simplex σ. Computing the composition in terms of face maps, one obtains:

$$\partial^2 \sigma = \sum_{i \neq j} D_j D_i \sigma. \tag{1.22}$$

Each $(k-2)$-face of the k-simplex σ is represented exactly twice in the image of $D_j D_i$ over all $i \neq j$. Thanks to $\mathbb{F}_2$ coefficients, the sum over this pair is zero. □

Inspired by what happens with simplicial complexes, one defines an algebraic *complex* to be any sequence $\mathcal{C} = (C_\bullet, \partial)$ of vector spaces and linear transformations with the property that $\partial^2 = 0$. Going two steps along the sequence is the zero-map.

Exercise 1.23. Show that for any algebraic complex $\mathcal{C}$, $\mathsf{im}\ \partial_{k+1}$ is a subspace of $\mathsf{ker}\ \partial_k$ for all k.

The homology of an algebraic complex $\mathcal{C}$, $H_\bullet(\mathcal{C})$, is a complex of vector spaces defined as follows. The k-*cycles* of $\mathcal{C}$ are elements of C_k with "zero boundary", denoted $Z_k = \mathsf{ker}\ \partial_k$. The k-*boundaries* of $\mathcal{C}$ are elements of C_k that are the boundary of something in C_{k+1}, and are denoted $B_k = \mathsf{im}\ \partial_{k+1}$. The *homology* of $\mathcal{C}$ is the complex $H_\bullet(\mathcal{C})$ of quotient vector spaces $H_k(\mathcal{C})$, for $k \in \mathbb{N}$, given by:

$$\begin{aligned} H_k(\mathcal{C}) &= Z_k/B_k \\ &= \mathsf{ker}\ \partial_k / \mathsf{im}\ \partial_{k+1} \\ &= \text{cycles/boundaries}. \end{aligned} \tag{1.24}$$

Homology inherits the grading of the complex $\mathcal{C}$ and has trivial (zero) linear transformations connecting the individual vector spaces. Elements of $H_\bullet(\mathcal{C})$ are *homology classes* and are denoted $[\alpha] \in H_k$, where $\alpha \in Z_k$ is a k-cycle and $[\cdot]$ denotes the equivalence class modulo elements of B_k.

Exercise 1.25. If $\mathcal{C}$ has boundary maps that are all zero, what can you say about $H_\bullet(\mathcal{C})$? What if all the boundary maps (except at the ends) are isomorphisms (injective and surjective)?

Homology of Simplicial Complexes For X a simplicial complex, the chain complex of $\mathbb{F}_2$-vector spaces generated by simplices of X and the boundary attaching maps is a particularly simple algebraic complex associated to X. The homology of this complex is usually denoted $H_\bullet(X)$ or perhaps $H_\bullet(X; \mathbb{F}_2)$ when the binary coefficients are to be emphasized.

Exercise 1.26. Show that if X is a connected simplicial complex, then $H_0(X) = \mathbb{F}_2$. Argue that $\dim H_0(X)$ equals the number of connected components of X.

Exercise 1.27. Let X be a one-dimensional simplicial complex with five vertices and eight edges that looks like $\otimes$. Show that $\dim H_1(X) = 4$.

One of the important aspects of homology is that it allows one to speak of cycles that are *linearly independent*. It is true that for a graph, $\dim H_1$ is the number of *independent* cycles in the graph. One might have guessed that the number of cycles in the previous exercise is five, not four; however, the fifth can be always be expressed as a linear combination of the other four basis cycles.

Graphs have nearly trivial homology, since there are no simplices of higher dimension. Still, one gets from graphs the [correct] intuition that H_0 counts connected components and H_1 counts loops. Higher-dimensional homology measures higher-dimensional "holes" as detectable by cycles.

Exercise 1.28. Compute explicitly the $\mathbb{F}_2$-homology of the cell decompositions of the 2-sphere, S^2, defined by the boundaries of the five regular Platonic solids. When doing so, recall that ∂_2 takes the various types of 2-cells (triangles, squares, pentagons) to the formal sum of their boundary edges, using addition with $\mathbb{F}_2$ coefficients.

These examples testify as to one of the most important features of homology.

Theorem 1.29. *Homology is a homotopy invariant.*

As stated, the above theorem would seem to apply only to cell complexes. However, as we will detail in Lecture 3, we can define homology for any topological space independent of cell structure; to this, as well, the above theorem applies. Thus, we can talk of the homology of a space independent of any cell structure or concrete representation: homotopy type is all that matters. It therefore makes sense to explore some basic examples. The following are the homologies of the

n-dimensional sphere, $\mathbb{S}^n$; the n-dimensional torus, $\mathbb{T}^n$; and the oriented surface Σ_g of genus g.

(1.30)
$$\dim H_k(\mathbb{S}^n) = \begin{cases} 1 & k = n, 0 \\ 0 & k \neq n, 0 \end{cases},$$

(1.31)
$$\dim H_k(\mathbb{T}^n) = \binom{n}{k},$$

(1.32)
$$\dim H_k(\Sigma_g) = \begin{cases} 0 & k > 2 \\ 1 & k = 2 \\ 2g & k = 1 \\ 1 & k = 0 \end{cases}.$$

Betti Numbers We see in the above examples that the *dimensions* of the homology are the most notable features. In the history of algebraic topology, these dimensions of the homology groups — called *Betti numbers* $\beta_k = \dim H_k$ — were the first invariants investigated. They are just the beginning of the many connections to other topological invariants. For example, we will explain the following in Lecture 3:

Theorem 1.33. *The Euler characteristic of a finite cell complex is the alternating sum of its Betti numbers:* $\chi(X) = \sum_k (-1)^k \beta_k$.

For the moment, we will focus on applications of Betti numbers as a topological statistic. In Lecture 2 and following, however, we will go beyond Betti numbers and consider the richer internal structure of homologies.

Application: Neuroscience

Each of these lectures ends with a sketch of some application(s): this first sketch will focus on the use of Betti numbers. Perhaps the best-to-date example of the use of homology in data analysis is the following recent work of Giusti, Pastalkova, Curto, & Itskov [56] on network inference in neuroscience using parametrized Betti numbers as a statistic.

Consider the challenge of inferring how a collection of neurons is wired together. Because of the structure of a neuron (in particular the length of the axon), mere physical proximity does not characterize the wiring structure: neurons which are far apart may in fact be "wired" together. Experimentalists can measure the responses of individual neurons and their firing sequences as a response to stimuli. By comparing time-series data from neural probes, the correlations of neuron activity can be estimated, resulting in a correlation matrix with entries, say, between zero and one, referencing the estimated correlation between neurons, with the diagonal, of course, consisting of ones. By thresholding the correlation matrix at some value, one can estimate the "wiring network" of how neurons are connected.

Unfortunately, things are more complicated than this simple scenario suggests. First, again, the problem of which threshold to choose is present. Worse, the correlation matrix is not the truth, but an experimentally measured estimation that relies on how the experiment was performed (Where were the probes inserted? How was the spike train data handled?). Repeating an experiment may lead to a very different correlation matrix – a difference not accountable by a linear transformation. This means, in particular, that methods based on spectral properties such as PCA are misleading [56].

What content does the experimentally-measured correlation matrix hold? The entries satisfy an *order principle*: if neurons A and B seem more correlated than C and D, then, in truth, they are. In other words, repeated experiments lead to a nonlinear, but order-preserving, homeomorphism of the correlation axis. It is precisely this nonlinear coordinate-free nature of the problem that prompts a topological approach.

The approach is this. Given a correlation matrix $\mathcal{R}$, let $1 \geqslant \epsilon \geqslant 0$ be a decreasing threshold parameter, and, for each ϵ, let $\mathcal{R}_\epsilon$ be the binary matrix generated from $\mathcal{R}$ with ones wherever the correlation exceeds ϵ. Let X_ϵ be the Dowker complex of $\mathcal{R}_\epsilon$ (or dual; the same, by symmetry). Then consider the kth Betti number distribution $\beta_k : [1, 0] \to \mathbb{N}$. These distributions are unique under change of correlation axis coordinates up to order-preserving homeomorphisms of the domain.

What do these distributions look like? For $\epsilon \to 1$, the complex is an isolated set of points, and for $\epsilon \to 0$ it is one large connected simplex: all the interesting homology lies in the middle. It is known that when this sequence of simplicial complexes is obtained by sampling points from a probability distribution, the Betti distributions β_k for $k > 0$ are unimodal. Furthermore, it is known that homological peaks are ordered by dimension [5]: the peak ϵ value for β_1 precedes that of β_2, etc. Thus, what is readily available as a signature for the network is the ordering of the heights of the peaks of the β_k distributions. The surprise is that this peak height data gives information about the distribution from which the points were sampled.

In particular, one can distinguish between networks that are wired *randomly* versus those that are wired *geometrically*. This is motivated by the neuroscience applications. It has been known since the Nobel prize-winning work of O'Keefe et al. that certain neurons in the visual cortex of rats act as *place cells*, encoding the geometry of a learned domain (e.g., a maze) by how the neurons are wired [74], in manner not unlike that of a nerve complex [35]. Other neural networks are known to be wired together randomly, such as the olfactory system of a fly [29]. Giusti et al., relying on theorems about Betti number distributions for random geometric complexes by Kahle [63], show that one can differentiate between geometrically-wired and randomly wired networks by looking at the peak signatures of β_1, β_2, and β_3 and whether the peaks increase [random] or decrease [geometric]. Follow-on work gives novel signature types [87]. The use of these methods is revolutionary, since actual physical experiments to rigorously determine neuron wiring are prohibitively difficult and expensive, whereas computing homology is, in principle, simple. Lecture 3 will explore this issue of computation more.

Lecture 2: Persistence

We have covered the basic definitions of simplicial and algebraic complexes and their homological invariants. Our goal is to pass from the mechanics of invariants to the principles that animate the subject, culminating in a deeper understanding of how data can be qualitatively compressed and analyzed. In this lecture, we will begin that process, using the following principles as a guide:

(1) A simplicial [or cell] complex is the right type of discrete data structure for capturing the significant features of a space.
(2) A chain complex is a linear-algebraic representation of this data structure – an algebraic set of assembly instructions.
(3) To prove theorems about how cell complexes behave under deformation, study instead deformations of chain complexes.
(4) Homology is the optimal compression of a chain complex down to its qualitative features.

Towards Functoriality

Our chief end is this: homology is *functorial*. This means that one can talk not only about homology of a complex, but also of the homology of a map between complexes. To study continuous maps between spaces algebraically, one translates the concept to chain complexes. Assume that X and Y are simplicial complexes and $f\colon X \to Y$ is a *simplicial map* – a continuous map taking simplices to simplices.[4] This does not imply that the simplices map homeomorphically to simplices of the same dimension.

In the same way that X and Y lift to algebraic chain complexes $C_\bullet(X)$ and $C_\bullet(Y)$, the map f lifts to a graded sequence of linear transformations $f_\bullet\colon C_\bullet(X) \to C_\bullet(Y)$, generated by basis n-simplices of X being sent to basis n-simplices of Y, where, if an n-simplex of X is sent by f to a simplex of dimension less than n, then the algebraic effect is to send the basis chain in $C_n(X)$ to $0 \in C_n(Y)$. The continuity of the map f induces a *chain map* $f_\bullet$ that fits together with the boundary maps of $C_\bullet(X)$ and $C_\bullet(Y)$ to form the following diagram of vector spaces and linear transformations:

$$\begin{array}{ccccccccc} \cdots \longrightarrow & C_{n+1}(X) & \xrightarrow{\partial} & C_n(X) & \xrightarrow{\partial} & C_{n-1}(X) & \xrightarrow{\partial} & \cdots . \\ & \downarrow{\scriptstyle f_{n+1}} & & \downarrow{\scriptstyle f_n} & & \downarrow{\scriptstyle f_{n-1}} & & \\ \cdots \longrightarrow & C_{n+1}(Y) & \xrightarrow{\partial} & C_n(Y) & \xrightarrow{\partial} & C_{n-1}(Y) & \xrightarrow{\partial} & \cdots \end{array} \tag{2.1}$$

[4]For cell complexes, one makes the obvious adjustments.

Commutative Diagrams Equation (2.1) is important – it is our first example of what is known as a *commutative diagram*. These are the gears for algebraic engines of inference. In this example, commutativity means precisely that the chain maps respect the boundary operation, $f_\bullet \partial = \partial f_\bullet$. *This is what continuity means for linear transformations of complexes.* There is no need for simplices or cells to be explicit. One defines a *chain map* to be any sequence of linear transformations $f_\bullet \colon \mathcal{C} \to \mathcal{C}'$ on algebraic complexes making the diagram commutative.

Exercise 2.2. Show that any chain map $f_\bullet \colon \mathcal{C} \to \mathcal{C}'$ takes cycles to cycles and boundaries to boundaries.

Induced Homomorphisms Because of this commutativity, a chain map $f_\bullet$ acts not only on chains but on cycles and boundaries as well. This makes well-defined the *induced homomorphism* $H(f)\colon H_\bullet(\mathcal{C}) \to H_\bullet(\mathcal{C}')$ on homology. For α a cycle in $\mathcal{C}$ with homology class $[\alpha]$, one may thus define $H(f)[\alpha] = [f_\bullet \alpha] = [f \circ \alpha]$. This is well-defined: if $[\alpha] = [\alpha']$, then, as chains, $\alpha' = \alpha + \partial\beta$ for some β, and,

$$f_\bullet \alpha' = f \circ \alpha' = f \circ (\alpha + \partial\beta) = f \circ \alpha + f \circ \partial\beta = f_\bullet \alpha + \partial(f_\bullet \beta), \tag{2.3}$$

so that $H(f)[\alpha'] = [f_\bullet \alpha'] = [f_\bullet \alpha] = H(f)[\alpha]$ in $H_\bullet(\mathcal{C}')$.

The term *homomorphism* is used to accustom the reader to standard terminology. Of course, in the present context, an induced homomorphism is simply a graded linear transformation on homology induced by a chain map.

Exercise 2.4. Consider the disc in $\mathbb{R}^2$ of radius π punctured at the integer points along the x and y axes. Although this space is not a cell complex, let us assume that its homology is well-defined and is "the obvious thing" for H_1, defined by the number of punctures. What are the induced homomorphisms on H_1 of the continuous maps given by (1) rotation by $\pi/2$ counterclockwise; (2) the folding map $x \mapsto |x|$; (3) flipping along the y axis?

Functoriality Homology is functorial, meaning that the homomorphisms induced on homology are an algebraic reflection of the properties of continuous maps between spaces. The following are simple properties of induced homomorphisms, easily shown from the definitions above:

- Given a chain map $f_\bullet \colon \mathcal{C} \to \mathcal{C}'$, $H(f)\colon H_\bullet(\mathcal{C}) \to H_\bullet(\mathcal{C}')$ is a (graded) sequence of linear transformations.
- The identity map $\mathrm{Id}\colon \mathcal{C} \to \mathcal{C}$ induces the identity $\mathrm{Id}\colon H_\bullet(\mathcal{C}) \to H_\bullet(\mathcal{C})$ on homology.
- Given $f_\bullet \colon \mathcal{C} \to \mathcal{C}'$ and $g_\bullet \colon \mathcal{C}' \to \mathcal{C}''$, $H(g \circ f) = H(g) \circ H(f)$.

There is hardly a more important feature of homology than this functoriality.

Exercise 2.5. Show using functoriality that homeomorphisms between spaces induce isomorphisms on homologies.

Exercise 2.6. Can you find explicit counterexamples to the following statements about maps f between simplicial complexes and their induced homomorphisms $H(f)$ (on some grading for homology)?

(1) If f is surjective then $H(f)$ is surjective.
(2) If f is injective then $H(f)$ is injective.
(3) If f is not surjective then $H(f)$ is not surjective.
(4) If f is not injective then $H(f)$ is not injective.
(5) If f is not bijective then $H(f)$ is not bijective.

Functorial Inference It is sometimes the case that what is desired is knowledge of the qualitative features [homology] of an important but unobservable space X; what is observed is an approximation Y to X, of uncertain homological fidelity. One such observation is unhelpful. Two or more homological samplings may lead to increased confidence; however, functoriality can relate observations to truth. Suppose the observed data comprises the homology of a *pair* of spaces Y_1, Y_2, which are related by a map $f\colon Y_1 \to Y_2$ that factors through a map to X, so that $f = f_2 \circ f_1$ with $f_1\colon Y_1 \to X$ and $f_2\colon X \to Y_2$. If the induced homomorphism $H(f)$ is known, then, although $H_\bullet(X)$ is hidden from view, inferences can be made.

$$
\begin{array}{ccc}
H_\bullet Y_1 & \xrightarrow{\quad H(f)\quad} & H_\bullet Y_2 \,. \\
& {\scriptstyle H(f_1)} \searrow \quad \nearrow {\scriptstyle H(f_2)} & \\
& H_\bullet X &
\end{array}
\tag{2.7}
$$

Exercise 2.8. In the above scenario, what can you conclude about $H_\bullet(X)$ if $H(f)$ is an isomorphism? If it is merely injective? Surjective?

The problem of measuring topological features of experimental data by means of sensing is particularly vulnerable to threshold effects. Consider, e.g., an open tank of fluid whose surface waves are experimentally measured and imaged. Perhaps the region of interest is the portion of the fluid surface above the ambient height $h = 0$; the topology of the set $A = \{h \geqslant 0\}$ must be discerned, but can only be approximated by imprecise pixellated images of $\{h \gtrsim 0\}$. One can choose a measurable threshold above and below the zero value to get just such a situation as outlined in Equation (2.7) above. Similar scenarios arise in MRI data, where the structure of a tissue of interest can be imaged as a pair of pixellated approximations, known to over- and under-approximate the truth.

Sequences

Induced homomorphisms in homology are key, as central to homology as the role of linear transformations are in linear algebra. In these lectures, we have seen how, though linear transformations between vector spaces are important, what a great advantage there is in the chaining of linear transformations into sequences and complexes. The advent of induced homomorphisms should prompt the same desire, to chain into sequences, analyze, classify, and infer. This is the plan for the remainder of this lecture as we outline the general notions of persistence, persistent homology, and topological data analysis.

Consider a sequence of inclusions of subcomplexes $\iota\colon X_k \subset X_{k+1}$ of a simplicial complex X for $1 \leqslant k \leqslant N$. These can be arranged into a sequence of spaces with inclusion maps connecting them like so:

$$\varnothing = X_0 \xrightarrow{\iota} X_1 \xrightarrow{\iota} \cdots \xrightarrow{\iota} X_{N-1} \xrightarrow{\iota} X_N \xrightarrow{\iota} X. \tag{2.9}$$

Sequences of spaces are very natural. One motivation comes from a sequence of Vietoris-Rips complexes of a set of data points with an increasing sequence of radii $(\epsilon_i)_{i=1}^N$.

Exercise 2.10. At the end of Lecture 1, we considered a correlation matrix $\mathcal{R}$ on a set V of variables, where correlations are measured from 0 to 1, and used this matrix to look at a sequence of Betti numbers. Explain how to rephrase this as a sequence of homologies with maps (assuming some discretization along the correlation axis). What maps induce the homomorphisms on homologies? What do the homologies look like for very large or very small values of the correlation parameter?

A topological sequence of spaces is converted to an algebraic sequence by passing to homology and using induced homomorphisms:

$$H_\bullet(X_0) \xrightarrow{H(\iota)} H_\bullet(X_1) \xrightarrow{H(\iota)} \cdots \xrightarrow{H(\iota)} H_\bullet(X_{N-1}) \xrightarrow{H(\iota)} H_\bullet(X_N) \xrightarrow{H(\iota)} H_\bullet(X). \tag{2.11}$$

The individual induced homomorphisms on homology encode local topological changes in the X_i; thanks to functoriality, the *sequence* encodes the *global* changes.

Exercise 2.12. Consider a collection of 12 equally-spaced points on a circle — think of tick-marks on a clock. Remove from this all the points corresponding to the prime numbers (2, 3, 5, 7, 11). Use the remaining points on the circle as the basis of a sequence of Vietoris-Rips [VR] complexes based on an increasing sequence $\{\epsilon_i\}$ of distances starting with $\epsilon_0 = 0$. Without worrying about the actual values of the ϵ_i, describe what happens to the sequence of VR complexes. What do you observe? Does H_0 ever increase? Decrease? What about H_1?

What one observes from this example is the evolution of homological features over a sequence: homology classes are born, can merge, split, die, or persist. This evolutionary process as written in the language of sequences is the algebraic means of encoding notions of geometry, significance, and noise.

Persistence Let us formalize some of what we have observed. Consider a sequence of spaces (X_i) and continuous transformations $f_i\colon X_i \to X_{i+1}$, without requiring subcomplexes and inclusions. We again have a sequence of homologies with induced homomorphisms. A homology class in $H_\bullet(X_i)$ is said to *persist* if its image in $H_\bullet(X_{i+1})$ is also nonzero; otherwise it is said to *die*. A homology class in $H_\bullet(X_j)$ is said to be *born* when it is not in the image of $H_\bullet(X_{j-1})$.

One may proceed with this line of argument, at the expense of some sloppiness of language. Does every homology class have an unambiguous birth and death? Can we describe cycles this way, or do we need to work with classes of cycles modulo boundaries? For the sake of precision and clarity, it is best to follow the pattern of these lectures and pass to the context of linear algebra and sequences.

Consider a sequence $V_\bullet$ of finite-dimensional vector spaces, graded over the integers $\mathbb{Z}$, and stitched together with linear transformations like so:

$$V_\bullet = \quad \cdots \longrightarrow V_{i-1} \longrightarrow V_i \longrightarrow V_{i+1} \longrightarrow \cdots . \tag{2.13}$$

These *sequences* are more general than algebraic *complexes*, which must satisfy the restriction of composing two incident linear transformations yielding zero. Two such sequences $V_\bullet$ and $V'_\bullet$ are said to be *isomorphic* if there are isomorphisms $V_k \cong V'_k$ which commute with the linear transformations in $V_\bullet$ and $V'_\bullet$ as in Equation (2.1). The simplest such sequence is an *interval indecomposable* of the form

$$I_\bullet = \quad \cdots \longrightarrow 0 \longrightarrow 0 \longrightarrow \mathbb{F} \xrightarrow{\mathsf{Id}} \mathbb{F} \xrightarrow{\mathsf{Id}} \cdots \xrightarrow{\mathsf{Id}} \mathbb{F} \longrightarrow 0 \longrightarrow 0 \longrightarrow \cdots , \tag{2.14}$$

where the *length* of the interval equals the number of Id maps, so that an interval of length zero consists of $0 \to \mathbb{F} \to 0$ alone. Infinite or bi-infinite intervals are also included as indecomposables.

Representation Theory A very slight amount of representation theory is all that is required to convert a sequence of homologies into a useful data structure for measuring persistence. Consider the following operation: sequences can be formally added by taking the direct sum, $\oplus$, term-by-term and map-by-map. The interval indecomposables are precisely *indecomposable* with respect to $\oplus$ and cannot be expressed as a sum of simpler sequences, even up to isomorphism. The following theorem, though simple, is suitable for our needs.

Theorem 2.15 (Structure Theorem for Sequences). *Any sequence of finite dimensional vector spaces and linear transformations decomposes as a direct sum of interval indecomposables, unique up to reordering.*

What does this mean? It's best to begin with the basics of linear algebra, and then see how that extends to homology.

Exercise 2.16. Any linear transformation $\mathbb{R}^n \xrightarrow{A} \mathbb{R}^m$ extends to a biinfinite sequence with all but two terms zero. How many different isomorphism classes of decompositions into interval indecomposables are there? What types of intervals are present? Can you interpret the numbers of the various types of intervals? What well-known theorem from elementary linear algebra have you recovered?

Barcodes. When we use field coefficients, applying the Structure Theorem to a sequence of homologies gives an immediate clarification of how homology classes evolve. Homology classes correspond to interval indecomposables, and are born, persist, then die at particular (if perhaps infinite) parameter values. This decomposition also impacts how we illustrate evolving homology classes. By drawing pictures of the interval indecomposables over the [discretized] parameter line as horizontal *bars*, we obtain a pictograph that is called a *homology barcode*.

Exercise 2.17. Consider a simple sequence of four vector spaces, each of dimension three. Describe and/or draw pictures of all possible barcodes arising from such a sequence. Up to isomorphism, how many such barcodes are there?

The phenomena of homology class *birth*, *persistence*, and *death* corresponds precisely to the *beginning*, *middle*, and *end* of an interval indecomposable. The barcode is usually presented with horizontal intervals over the parameter line corresponding to interval indecomposables. Note that barcodes, like the homology they illustrate, are graded. There is an H_k-barcode for each $k \geqslant 0$. Since, from the Structure Theorem, the order does not matter, one typically orders the bars in terms of birth time (other orderings are possible).

The barcode provides a simple descriptor for topological significance: the shorter an interval, the more ephemeral the hole; long bars indicate robust topological features with respect to the parameter. This is salient in the context of point clouds $\mathcal{Q}$ and Vietoris-Rips complexes $VR_\epsilon(\mathcal{Q})$ using an increasing sequence $\{\epsilon_i\}$ as parameter. For ϵ too small or too large, the homology of $VR_\epsilon(\mathcal{Q})$ is unhelpful. Instead of trying to choose an *optimal* ϵ, choose them *all*: the barcode reveals significant features.

Exercise 2.18. Persistent homology is useful and powerful in topological data analysis, but sometimes one can get lost in the equivalence relation that comprises homology classes. Often, in applications, one cares less about the homology and more about a particular *cycle* (whose homology class may be too loose to have meaning within one's data). Given a sequence of chain complexes and chain maps, what can be said about *persistent cycles* and *persistent boundaries*? Are these well-defined? Do they have barcodes? How would such structures relate to persistent homology barcodes?

Persistent Homology Let us summarize what we have covered with slightly more formal terminology. A *persistence complex* is a sequence of chain complexes $\mathcal{P} = (\mathcal{C}_i)$, together with chain maps $x\colon \mathcal{C}_i \longrightarrow \mathcal{C}_{i+1}$. For notational simplicity, the index subscripts on the chain maps x are suppressed. Note that each $\mathcal{C}_i = (C_{\bullet,i}, \partial)$ is itself a complex: we have a sequence of sequences. The *persistent homology* of a persistence complex $\mathcal{P}$ is not a simple homology theory, but rather a homology associated to closed intervals in the "parameter domain". Over the interval $[i, j]$, its persistent homology, denoted $H_\bullet(\mathcal{P}[i, j])$, is defined to be the image of the induced homomorphism $H(x^{j-i})\colon H_\bullet(\mathcal{C}_i) \to H_\bullet(\mathcal{C}_j)$ induced by x^{j-i}. That is, one looks at the composition of the chain maps from $\mathcal{C}_i \to \mathcal{C}_j$ and takes the image of the induced homomorphism on homologies. This persistent homology consists of homology classes that *persist*: $\dim H_k(\mathcal{P}[i,j])$ equals the number of intervals in the barcode of $H_k(\mathcal{P})$ containing the parameter interval $[i,j]$.

Exercise 2.19. If, in the indexing for a persistence complex, you have $i < j < k < \ell$, what is the relationship between the various subintervals of $[i,\ell]$ using $\{i,j,k,\ell\}$ as endpoints? Draw the lattice of such intervals under inclusion. What is the relationship between the persistent homologies on these subintervals?

Persistence Diagrams. Barcodes are not the only possible graphical presentation for persistent homology. Since there is a decomposition into homology classes with well-defined initial and terminal parameter values, one can plot each homology class as a point in the plane with axes the parameter line. To each interval indecomposable (homology class) one assigns a single point with coordinates *(birth, death)*. This scatter plot is called the *persistence diagram* and is more practical to plot and interpret than a barcode for very large numbers of homology classes.

Exercise 2.20. In the case of a homology barcode coming from a data set, the "noisy" homology classes are those with the smallest length, with the largest bars holding claim as the "significant" topological features in a data set. What do these noisy and significant bars translate to in the context of a persistence diagram? For a specific example, return to the "clockface" data set of Exercise 2.12, but now consider the set of all even points: 2, 4, 6, 8, 10, and 12. Show that the persistent H_2 contains a "short" bar. Are you surprised at this *artificial bubble* in the VR complex? Does a similar bubble form in the homology when all 12 points are used? In which dimension?

One aspect worth calling out is the notion of persistent homology as a homological data structure *over* the parameter space, in that one associates to each interval $[i,j]$ its persistent homology. This perspective is echoed in the early literature on the subject [32,36,41,91], in which a continuous parameter space was used, with a continuous family of (excursion sets) of spaces X_t, $t \in \mathbb{R}$: in this setting, persistent homology is assigned to an interval $[s,t]$. The discretized parameter interval offers little in the way of restrictions (unless you are working with fractal-like or otherwise degenerate objects) and opens up the simple setting of the Structure Theorem on Sequences as used in this lecture.

Stability

The idea behind the use of barcodes and persistence diagrams in data is grounded in the intuition that essential topological features of a domain are robust to noise, whether arising from sensing, sampling, or approximation. In a barcode, noisy features appear as short bars; in a persistence diagram, as near-diagonal points. To solidify this intuition of robustness, one wants a more specific statement on the *stability* of persistent homology. Can a small change in the input

— whether a sampling of points or a Dowker relation or a perturbation of the metric — have a large impact on how the barcode appears?

There have of late been a plethora of stability theorems in persistent homology, starting with the initial result of Cohen-Steiner et al. [31] and progressing to more general and categorical forms [12, 18, 21, 22, 30]. In every one of these settings, the stability result is given in the context of persistence over a continuous parameter, ϵ, such as one might use in the case of a Vietoris-Rips filtration on a point-cloud. The original stability theorem is further framed in the setting of sublevel sets of a function $h: X \to \mathbb{R}$ and the filtration is by sublevel sets $X_t = \{h \leqslant t\}$. Both the statements of the stability theorems and their proofs are technical; yet the technicalities lie in the difficulties of the continuum parameter. For the sake of clarity and simplicity, we will assume that one has imposed a uniform discretization of the real line with step size a fixed $\epsilon > 0$ (as would often be the case in practice).

Interleaving. At present, the best language for describing the stability of persistent homology and barcode descriptors is the recent notion of *interleaving*. In keeping with the spirit of these lectures, we will present the theory in the context of sequences of vector spaces and linear transformations. Assume that one has a pair of sequences, $V_\bullet$ and $W_\bullet$, of vector spaces and linear transformations. We say that a T-*interleaving* is a pair of degree-T mappings:

$$f_\bullet\colon V_\bullet \to W_{\bullet+T} \qquad g_\bullet\colon W_\bullet \to V_{\bullet+T}, \tag{2.21}$$

such that the diagram commutes:

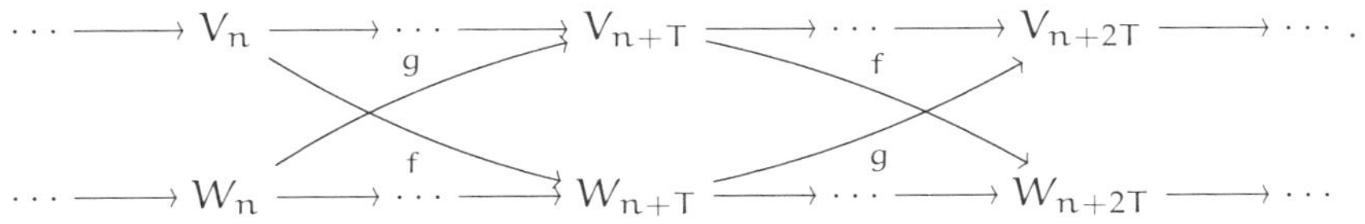

In particular, at each n, the composition $g_{n+T} \circ f_n$ equals the composition of the 2T horizontal maps in $V_\bullet$ starting at n; and, likewise with $f \circ g$ on $W_\bullet$. One defines the *interleaving distance* between two sequences of vector spaces to be the minimal $T \in \mathbb{N}$ such that there exists a T-interleaving.

Exercise 2.22. Verify that the interleaving distance of two sequences is zero if and only if the two sequences are isomorphic.

Exercise 2.23. Assume that $V_\bullet$ is an interval indecomposable of length 5. Describe the set of all $W_\bullet$ that are within interleaving distance one of $V_\bullet$.

Note that the conclusions of Exercises 2.22-2.23 are absolutely dependent on the discrete nature of the problem. In the case of a continuous parameter, the interleaving distance is *not* a metric on persistence complexes, but is rather a *pseudometric*, as one can have the infimum of discretized interleaving distances become zero without an isomorphism in the limit.

Application: TDA

Topological Data Analysis, or TDA, is the currently popular nomenclature for the set of techniques surrounding persistence, persistent homology, and the extraction of significant topological features from data. The typical input to such a problem is a point cloud $\mathcal{Q}$ in a Euclidean space, though any finite metric space will work the same. Given such a data set, assumed to be a noisy sampling of some domain of interest, one wants to characterize that domain. Such questions are not new: linear regression assumes an affine space and returns a best fit; a variety of locally-linear or nonlinear methods look for nonlinear embeddings of Euclidean spaces.

Topological data analysis looks for global structure — homology classes — in a manner that is to some degree decoupled from rigid geometric considerations. This, too, is not entirely novel. Witness clustering algorithms take a point cloud and return a partition that is meant to approximate *connected components*. Of course, this reminds one of H_0, and the use of a Vietoris-Rips complex makes this precise: *single linkage clustering* is precisely the computation of $H_0(\mathsf{VR}_\epsilon(\mathcal{Q}))$ for a choice of $\epsilon > 0$. Which choice is best? The lesson of persistence is to take all ϵ and build the homology barcode. Notice however, that the barcode returns only the dimension of H_0 — the number of clusters — and to more carefully specify the clusters, one needs an appropriate basis. There are many other clustering schemes with interesting functorial interpretations [26,27].

The ubiquity and utility of clustering is clear. What is less clear is the prevalence and practicality of higher-dimensional persistent homology classes in "organic" data sets. Using again a Vietoris-Rips filtration of simplicial complexes on a point cloud $\mathcal{Q}$ allows the computation of homology barcodes in gradings larger than zero. To what extent are they prevalent? The grading of homology is reminiscent of the grading of polynomials in Taylor expansions. Though Taylor expansions are undoubtedly useful, it is acknowledged that the lowest-order terms (zeroth and first especially) are most easily seen and used. Something like this holds in TDA, where one most readily sees clusters (H_0) and simple loops (H_1) in data. The following is a brief list of applications known to the author. The literature on TDA has blown-up of late to a degree that makes it impossible to give an exhaustive account of applications. The following are chosen as illustrative of the basic principles of persistent homology.

Medical imaging data: Some of the earliest and most natural applications of TDA were to image analysis [2,13,25,79]. One recent study by Benditch et al. looks at the structure of arteries in human brains [14]. These are highly convoluted pathways, with lots of branching and features at multiple scales, but which vary in dramatic and unpredictable ways from patient to patient. The topology of arterial structures are globally trivial — sampling the arterial structure though standard imaging techniques yields a family of trees (acyclic graphs). Nevertheless, since the geometry is measurable, one can filter these trees by sweeping a plane across the three-dimensional ambient domain, and look at the persistent H_0. Results show statistically significant correlations between the vector of lengths of the top 100 bars in the persistent H_0 barcode and features such as patient age and sex. For example, older brains tend to have shorter *longest bars* in the H_0 barcode. The significance of the correlation is very strong and outperforms methods derived from graph theory and phylogenetic-based tree-space geometry methods. It is interesting that it is not the "longest bar" that matters so much as the ensemble of longest bars in this barcode. Work in progress includes using H_0 barcode statistics to characterize global structure of graph-like geometries, including examinations of insect wing patterns, tree leaf vein networks, and more. Other exciting examples of persistent H_0 to medical settings feature an analysis of breast cancer by Nicolau et al. [73].

Distinguishing illness from health and recovey: Where does persistent homology beyond H_0 come into applications? A recent excellent paper of Torres et al. uses genetic data of individuals with illnesses to plot a time series of points in a *disease space* of traits [88]. Several examples are given of studies on human and mouse subjects tracking the advancement and recovery from disease (including, in the mice, malaria). Genetic data as a function of time and for many patients gives a point cloud in an abstract space for which geometry is not very relevant (for example, axes are of differing and incomparable units). What is interesting about this study is the incorporation of data from subjects that extends from the onset of illness, through its progress, and including a full recovery phase back to health. Of interest is the question of recovery — does recovery from illness follow the path of illness in reverse? Does one recover to the same state of health, or is there a monodromy? The study of Torres et al. shows a single clear long-bar in the H_1-barcode in disease space, indicating that all instances of the illness are *homologous,* as are all instances of recovery, but that *"illness and recovery are homologically distinct events"* [88]. In contrast to prior studies that performed a linear regression on a pair of variables and concluded a linear relationship between these variables (with a suggestion of a causal relationship), the full-recovery data set with its loopy phenomenon of recovery suggests skepticism: indeed, a careful projection of a generator for the H_1 barcode into this plane recovers the loop.

Robot path planning: A very different set of applications arises in robot motion planning, in which an autonomous agent needs to navigate in a domain $X \subset \mathbb{R}^n$ (either physical or perhaps a configuration space of the robot) from an initial state to a goal state in X. In the now-familiar case of self-driving vehicles, autonomous drones, or other agents with sensing capabilities, the navigable ("obstacle-free") subset of X is relatively uncertain, and known only as a probabilistic model. Bhattacharya et al. consider such a probability density $\rho\colon X \to [0, \infty)$ and use a combination of persistent homology and graph search-based algorithms in order to compute a set of best likely paths-to-goal [15]. The interesting aspects of this application are the following. (1) The persistence parameter is the probability, used to filter the density function on X. This is one natural instance of a continuous as opposed to a discrete persistence parameter. (2) The H_1 homology barcode is used, but on a slight modification of X obtained by abstractly identifying the initial and goal states (by an outside edge if one wants to be explicit), so that a path from initial to goal corresponds precisely to a 1-cycle that intersects this formal edge. (3) Once again, the problem of "where to threshhold" the density is avoided by the use of a barcode; the largest bars in the H_1 barcode correspond to the path classes most likely to be available and robust to perturbations in the sensing. For path classes with short bars, a slight update to the system might invalidate this path.

Localization and mapping: Computing persistent homology is also useful as a means of building topological maps of an unknown environment, also of relevance to problems in robotics, sensing, and localization. Imagine a lost traveler wandering through winding, convoluted streets in an unfamiliar city with unreadable street signs. Such a traveler might use various landmarks to build up an internal map: *"From the jewelry store, walk toward the cafe with the red sign, then look for the tall church steeple."* This can be accomplished without reference to coordinates or odometry. For the general setting, assume a domain $\mathcal{D}$ filled with landmarks identifiable by observers that register landmarks via local sensing/visibility. A collection of observations are taken, with each observation recording only those landmarks "visible" from the observation point. Both the landmarks and observations are each a discrete set, with no geometric or coordinate data appended. The sensing data is given in the form of an unordered sequence of pairs of observation-landmark identities encoding who-sees-what. From this abstract data, one has a Dowker relation from which one can build a pair of (dual, homotopic) Dowker complexes that serve as approximations to the domain topology. In [54], two means of inferring a topological map from persistent homology are given. (1) If observers record in the sensing relation a visibility strength (as in, say, the strengths of signals to all nearby wireless SSIDs), then filtering the Dowker complexes on this (as we did in the neuroscience applications of the previous lecture) gives meaning to long bars as significant map features. (2) In the binary (hidden/seen) sensing case, the existence of non-unique landmarks (*"Ah, look! a Starbucks! I know exactly where I am now..."*) confounds the topology, but filtering according to witness weight can eliminate spurious simplices: see [54] for details and [38] for an experimental implementation.

Protein compressibility: One of the first uses of persistent H_2 barcodes has appeared recently in the work of Gameiro et al. [48] in the context of characterizing compressibility in certain families of protein chains. Compressibility is a particular characterization of a protein's *softness* and is key to the interface of structure and function in proteins, the determination of which is a core problem in biology. The experimental measurement of protein compressibility is highly nontrivial, and involves fine measurement of ultrasonic wave velocities from pressure waves in solutions/solvents with the protein. Hollow cavities within the protein's geometric conformation are contributing factors, both size and number. Gameiro et al. propose a *topological compressibility* that is argued to measure the relative contributions of these features, but with minimal experimental measurement, using nothing more as input than the standard molecular datasets that record atom locations as a point cloud, together with a van der Waals radius about each atom. What is interesting in this case is that one does not have the standard Vietoris-Rips filtered complex, but rather a filtered complex obtained by starting with the van der Waals radii (which vary from atom to atom) and then adding to these radii the filtration parameter $\epsilon > 0$. The proposed topological compressibility is a ratio of the number of persistent H_2 intervals divided by the number of persistent H_1 intervals (where the intervals are restricted to certain parameter ranges). This ratio is meant to serve as proxy to the experimental measurement of cavities and tunnels in the protein's structure. Comparisons with experimental data suggest, with some exceptions, a tight linear correlation between the expensive experimentally-measured compressibility and the (relatively inexpensive) topological compressibility.

These varied examples are merely summaries: see the cited references for more details. The applications of persistent homology to data are still quite recent, and by the time of publication of these notes, there will have been a string of novel applications, ranging from materials science to social networks and more.

Lecture 3: Compression and Computation

We now have in hand the basic tools for topological data analysis: complexes, homology, and persistence. We are beginning to develop the theories and perspectives into which these tools fit, as a higher guide to how to approach qualitative phenomena in data. We have not yet dealt with issues of computation and effective implementation. Our path to doing so will take us deeper into sequences, alternate homology theories, cohomology, and Morse theory.

Sequential Manipulation

Not surprisingly, these lectures take the perspective that the desiderata for homological data analysis include a calculus for complexes and sequences of complexes. We have seen hints of this in Lecture 2; now, we proceed to introduce a bit more of the rich structure that characterizes (the near-trivial linear-algebraic version of) homological algebra. Instead of focusing on spaces or simplicial complexes per se, we focus on algebraic complexes; this motivates our examination of certain types of algebraic sequences.

Exact Complexes Our first new tool is inspired by the question: among all complexes, which are simplest? Simplicial complexes might suggest that the simplest sort of complex is that of a single simplex, which has homology vanishing in all gradings except zero. However, there are simpler complexes still. We say that an algebraic complex $\mathcal{C} = (C_\bullet, \partial)$ is *exact* if its homology completely vanishes, $H_\bullet(\mathcal{C}) = 0$. This is often written termwise as:

$$\ker \partial_k = \operatorname{im} \partial_{k+1} \quad \forall k. \tag{3.1}$$

Exercise 3.2. What can you say about the barcode of an exact complex? (This means the barcode of the complex, not its [null] homology).

The following simple examples of exact complexes help build intuition:

- Two vector spaces are isomorphic, $V \cong W$, iff there is an exact complex of the form:
$$0 \longrightarrow V \longrightarrow W \longrightarrow 0 \,.$$
- The 1st Isomorphism Theorem says that for a linear transformation φ of V, the following sequence is exact:
$$0 \longrightarrow \ker \varphi \longrightarrow V \xrightarrow{\varphi} \operatorname{im} \varphi \longrightarrow 0 \,.$$
Such a 5-term complex framed by zeroes is called a *short exact complex*. In any such short exact complex, the second map is injective; the penultimate, surjective.
- More generally, the kernel and cokernel of any linear transformation $\varphi \colon V \to W$ fit into an exact complex:
$$0 \longrightarrow \ker \varphi \longrightarrow V \xrightarrow{\varphi} W \longrightarrow \operatorname{coker} \varphi \longrightarrow 0 \,.$$

Exercise 3.3. Consider $C = C^\infty(\mathbb{R}^3)$, the vector space of differentiable functions and $\mathcal{X} = \mathcal{X}(\mathbb{R}^3)$, the vector space of C^∞ vector fields on $\mathbb{R}^3$. Show that these fit together into an exact complex,

$$0 \longrightarrow \mathbb{R} \longrightarrow C \xrightarrow{\nabla} \mathcal{X} \xrightarrow{\nabla\times} \mathcal{X} \xrightarrow{\nabla\cdot} C \longrightarrow 0 \,, \tag{3.4}$$

where ∇ is the gradient differential operator from vector calculus, and the initial $\mathbb{R}$ term in the complex represents the constant functions on $\mathbb{R}^3$. This one exact complex compactly encodes many of the relations of vector calculus.

Mayer-Vietoris Complex There are a number of exact complexes that are used in homology, the full exposition of which would take us far afield. Let us focus on one particular example as a means of seeing how exact complexes assist in computational issues. The following is presented in the context of simplicial complexes. Let $X = A \cup B$ be a union of two simplicial complexes with intersection $A \cap B$. The following exact complex is the *Mayer-Vietoris* complex:

$$\longrightarrow H_n(A \cap B) \xrightarrow{H(\phi)} H_n(A) \oplus H_n(B) \xrightarrow{H(\psi)} H_n(X) \xrightarrow{\delta} H_{n-1}(A \cap B) \longrightarrow .$$

The linear transformations between the homologies are where all the interesting details lie. These consist of: (1) $H(\psi)$, which adds the homology classes from A and B to give a homology class in X; (2) $H(\phi)$, which reinterprets a homology class in $A \cap B$ to be a homology class of A and a (orientation reversed!) homology class in B respectively; and (3) δ, which decomposes a cycle in X into a sum of chains in A and B, then takes the boundary of one of these chains in $A \cap B$. This unmotivated construction has a clean explanation, to be addressed soon.

For the moment, focus on what the Mayer-Vietoris complex means. This complex captures the additivity of homology. When, for example, $A \cap B$ is empty, then every third term of the complex vanishes — the $H_\bullet(A \cap B)$ terms. Because every-third-term-zero implies that the complementary pairs of incident terms are isomorphisms, this quickly yields that homology of a disjoint union is additive, using $\oplus$. When the intersection is nonempty, the Mayer-Vietoris complex details exactly how the homology of the intersection impacts the homology of the union: it is, precisely, an inclusion-exclusion principle.

Exercise 3.5. Assume the following: for any $k \geqslant 0$, (1) $H_n(\mathbb{D}^k) = 0$ for all $n > 0$; (2) the 1-sphere $\mathbb{S}^1$ has $H_1(\mathbb{S}^1) \cong \mathbb{F}$ and $H_n(\mathbb{S}^1) = 0$ for all $n > 1$. The computation of $H_\bullet(\mathbb{S}^k)$ can be carried out via Mayer-Vietoris as follows. Let A and B be upper and lower hemispheres of $\mathbb{S}^k$, each homeomorphic to $\mathbb{D}^k$ and intersecting at an equatorial $\mathbb{S}^{k-1}$. Write out the Mayer-Vietoris complex in this case: what can you observe? As $H_\bullet(\mathbb{D}^k) \cong 0$ for $k > 0$, one obtains by exactness that $\delta\colon H_n(\mathbb{S}^k) \cong H_{n-1}(\mathbb{S}^{k-1})$ for all n and all k. Thus, starting from a knowledge of $H_\bullet \mathbb{S}^1$, show that $H_n(\mathbb{S}^k) \cong 0$ for $k > 0$ unless $n = k$, where it has dimension equal to one.

Sequences of Sequences One of the themes of these lectures is the utility of composed abstraction: if spaces are useful, so should be spaces of spaces. Later, we will argue that an idea of "homologies of homologies" is sensible and useful (in the guise of the *sheaf theory* of Lecture 4). In this lecture, we argue that sequences of sequences and complexes of complexes are useful. We begin with an elucidation of what is behind the Mayer-Vietoris complex: what are the maps, and how does it arise?

Consider the following complex of algebraic complexes,

$$0 \longrightarrow C_\bullet(A \cap B) \xrightarrow{\phi_\bullet} C_\bullet(A) \oplus C_\bullet(B) \xrightarrow{\psi_\bullet} C_\bullet(A+B) \longrightarrow 0, \tag{3.6}$$

with chain maps $\phi_\bullet\colon c \mapsto (c, -c)$, and $\psi_\bullet\colon (a, b) \mapsto a + b$. The term on the right, $C_\bullet(A + B)$, consists of those chains which can be expressed as a sum of chains on A and chains on B. In cellular homology with A, B subcomplexes, $C_\bullet(A + B) \cong C_\bullet(X)$.

Exercise 3.7. Show that this complex-of-complexes is exact by construction.

In general, any such *short* exact complex of complexes can be converted to a *long exact complex* on homologies using a method from homological algebra called the *Snake Lemma* [50, 58]. Specifically, given:

$$0 \longrightarrow A_\bullet \xrightarrow{i_\bullet} B_\bullet \xrightarrow{j_\bullet} C_\bullet \longrightarrow 0, \tag{3.8}$$

there is an induced exact complex of homologies

$$\longrightarrow H_n(\mathcal{A}) \xrightarrow{H(i)} H_n(\mathcal{B}) \xrightarrow{H(j)} H_n(\mathcal{C}) \xrightarrow{\delta} H_{n-1}(\mathcal{A}) \xrightarrow{H(i)} . \tag{3.9}$$

Moreover, the long exact complex is *natural*: a commutative diagram of short exact complexes and chain maps

$$\begin{array}{ccccccccc} 0 & \longrightarrow & A_\bullet & \longrightarrow & B_\bullet & \longrightarrow & C_\bullet & \longrightarrow & 0\,. \\ & & \downarrow{\scriptstyle f_\bullet} & & \downarrow{\scriptstyle g_\bullet} & & \downarrow{\scriptstyle h_\bullet} & & \\ 0 & \longrightarrow & \tilde{A}_\bullet & \longrightarrow & \tilde{B}_\bullet & \longrightarrow & \tilde{C}_\bullet & \longrightarrow & 0 \end{array} \tag{3.10}$$

induces a commutative diagram of long exact complexes

$$\begin{array}{cccccccccc} \longrightarrow & H_n(\mathcal{A}) & \longrightarrow & H_n(\mathcal{B}) & \longrightarrow & H_n(\mathcal{C}) & \xrightarrow{\delta} & H_{n-1}(\mathcal{A}) & \longrightarrow & . \\ & \downarrow{\scriptstyle H(f)} & & \downarrow{\scriptstyle H(g)} & & \downarrow{\scriptstyle H(h)} & & \downarrow{\scriptstyle H(f)} & & \\ \longrightarrow & H_n(\tilde{\mathcal{A}}) & \longrightarrow & H_n(\tilde{\mathcal{B}}) & \longrightarrow & H_n(\tilde{\mathcal{C}}) & \xrightarrow{\delta} & H_{n-1}(\tilde{\mathcal{A}}) & \longrightarrow & \end{array} \tag{3.11}$$

The induced *connecting homomorphism* $\delta\colon H_n(\mathcal{C}) \to H_{n-1}(\mathcal{A})$ comes from the boundary map in $\mathcal{C}$ as follows:

(1) Fix $[\gamma] \in H_n(\mathcal{C})$; thus, $\gamma \in C_n$.
(2) By exactness, $\gamma = j(\beta)$ for some $\beta \in B_n$.
(3) By commutativity, $j(\partial\beta) = \partial(j\beta) = \partial\gamma = 0$.
(4) By exactness, $\partial\beta = i\alpha$ for some $\alpha \in A_{n-1}$.
(5) Set $\delta[\gamma] = [\alpha] \in H_{n-1}(\mathcal{A})$.

Exercise 3.12. This is a tedious but necessary exercise for anyone interested in homological algebra: (1) show that $\delta[\gamma]$ is well-defined and independent of all choices; (2) show that the resulting long complex is exact. Work at *ad tedium*: for help, see any textbook on algebraic topology, [58] recommended.

Euler Characteristic, Redux Complexes solve the mystery of the topological invariance of the Euler characteristic. Recall that we can define the Euler characteristic of a (finite, finite-dimensional) complex $\mathcal{C}$ as in Equation (1.13). The alternating sum is a binary exactness. A short exact complex of vector spaces $0 \to A \to B \to C \to 0$ has $\chi = 0$, since $C \cong B/A$. By applying this to individual rows of a short exact complex of (finite, finite-dimensional) chain complexes, we can lift once again to talk about the Euler characteristic of a (finite-enough) complex of complexes:

$$0 \longrightarrow A_\bullet \longrightarrow B_\bullet \longrightarrow C_\bullet \longrightarrow 0\,. \tag{3.13}$$

One sees that χ of this complex also vanishes: $\chi(A_\bullet) - \chi(B_\bullet) + \chi(C_\bullet) = 0$.

The following lemma is the homological version of the Rank-Nullity Theorem from linear algebra:

Lemma 3.14. *The Euler characteristic of a chain complex* $C_\bullet$ *and its homology* $H_\bullet$ *are identical, when both are defined.*

Proof. From the definitions of homology and chain complexes, one has two short exact complexes of chain complexes:

$$
\begin{array}{c}
0 \longrightarrow B_\bullet \longrightarrow Z_\bullet \longrightarrow H_\bullet \longrightarrow 0\,. \\
0 \longrightarrow Z_\bullet \longrightarrow C_\bullet \longrightarrow B_{\bullet-1} \longrightarrow 0
\end{array}
\tag{3.15}
$$

Here, $B_{\bullet-1}$ is the shifted boundary complex whose k^{th} term is B_{k-1}. By exactness, the Euler characteristic of each of these two complexes is zero; thus, so is the Euler characteristic of their concatenation.

$$0 \longrightarrow B_\bullet \longrightarrow Z_\bullet \longrightarrow H_\bullet \longrightarrow 0 \longrightarrow Z_\bullet \longrightarrow C_\bullet \longrightarrow B_{\bullet-1} \longrightarrow 0\,.$$

Count the $+/-$ signs: the Z terms cancel and, since $\chi(B_{\bullet-1}) = -\chi(B_\bullet)$, the B terms cancel. This leaves two terms and we conclude that $\chi(H_\bullet) - \chi(C_\bullet) = 0$. □

Euler characteristic thus inherits its topological invariance from that of homology. Where does the invariance of homology come from? Something more complicated still?

Homology Theories

Invariance of homology is best discerned from a singularly uncomputable variant that requires a quick deep dive into the plethora of homologies available. We begin with a reminder: homology is an algebraic compression scheme — a way of collapsing a complex to the simplest form that respects its global features. The notion of homology makes sense for any chain complex. Thus far, our only means of generating a complex from a space X has been via some finite auxiliary structure on X, such as a simplicial, cubical, or cellular decomposition. There are other types of structures a space may carry, and, with them, other complexes. In the same way that the homology of a simplicial complex is independent of the simplicial decomposition, the various homologies associated to a space under different auspices tend to be isomorphic.

Reduced homology Our first alternate theory is not really a different type of homology at all; merely a slight change in the chain complex meant to make contractible (or rather acyclic) spaces fit more exactly into homology theory. Recall that a contractible cell complex — such as a single simplex — has homology $H_k = 0$ for all $k > 0$, with H_0 being one-dimensional, recording the fact that the cell complex is connected. For certain results in algebraic topology, it would be convenient to have the homology of a contractible space vanish completely. This can be engineered by an *augmented complex* in a manner that is applicable to any $\mathbb{N}$-graded complex. Assume for simplicity that $\mathcal{C} = (C_k, \partial)$ is a $\mathbb{N}$-graded complex of vector spaces over a field $\mathbb{F}$. The *reduced complex* is the following augmentation:

$$\cdots \xrightarrow{\partial} C_3 \xrightarrow{\partial} C_2 \xrightarrow{\partial} C_1 \xrightarrow{\partial} C_0 \xrightarrow{\epsilon} \mathbb{F} \longrightarrow 0\,, \tag{3.16}$$

where the aumentation map $\epsilon\colon C_0 \to \mathbb{F}$ sends a vector in C_0 to the sum of its components (having fixed a basis for C_0). The resulting homology of this complex is called the *reduced homology* and is denoted $\tilde{H}_\bullet$.

Exercise 3.17. Show that the reduced complex is in fact a complex: i.e., that $\epsilon\partial = 0$. How does the reduced homology of a complex differ from the "ordinary" homology? What is the dependence on the choice of augmentation map ϵ? Show that the augmented complex of a contractible simplicial complex is exact.

Čech Homology One simple structure associated to a topological space is an open cover — a collection $\mathcal{U}$ of open sets $\{U_\alpha\}$ in X. The *Čech complex* of $\mathcal{U}$ is the complex $\mathcal{C}(\mathcal{U})$ with basis for $C_k(\mathcal{U})$ being all (unordered) sets of $k+1$ distinct elements of $\mathcal{U}$ with nonempty intersection. (The usual complexities arise for coefficients not in $\mathbb{F}_2$, as one needs to order the elements of $\mathcal{U}$ up to even permutations.) The boundary maps $\partial\colon C_\bullet(\mathcal{U}) \to C_{\bullet-1}(\mathcal{U})$ act on a basis element by forgetting one of the terms in the set, yielding *face maps*.

For a finite collection $\mathcal{U}$ of sets, the Čech complex is identical to the simplicial chain complex of the nerve $\mathcal{N}(\mathcal{U})$. With the further assumption of contractible sets and intersections, the resulting homology is, by the Nerve Lemma, identical to $X = \cup_\alpha U_\alpha$. However, even in the non-finite case, the result still holds if the analogous contractibility assumptions hold. In short, if all the basis elements for the Čech complex are nullhomologous, then the *Čech homology*, $H_\bullet(\mathcal{C}(\mathcal{U}))$ is isomorphic to $H_\bullet(X)$. Of course, the Čech complex and its homology are still well-defined even if the local simplicity assumptions are violated. This Čech homology has in the past been used in the context of a sequence of covers, with limiting phenomena of most interest for complex fractal-like spaces [60]. This is perhaps one of the earliest incarnations of persistent homology.

Singular Homology If you find it risky to think of the Čech homology of a cover of non-finite size, then the next homology theory will seem obscenely prodigal. Given a topological space X, the *singular chain complex* is the complex $\mathcal{C}^{\text{sing}}$ whose k-chains have as basis elements all maps $\sigma\colon \Delta^k \to X$, where Δ^k is the Platonic k-simplex. Note: there are no restrictions on the maps σ other than continuity: images in X may appear crushed or crumpled. The boundary maps are the obvious restrictions of σ to the $(k-1)$-dimensional faces of Δ^k, taking a linear combination with orientations if the field $\mathbb{F}$ demands. The resulting singular chain complex is wildly uncountable, unless X should happen to be a trivial space.

Exercise 3.18. Do the one explicit computation possible in singular homology: show that for X a finite disjoint union of points, the singular homology $H_n^{\text{sing}}(X)$ vanishes except when $n = 0$, in which case it has β_0 equal to the number of points in X. Note: you cannot assume, as in cellular homology, that the higher-dimensional chains C_n vanish for $n > 0$.

There is little hope in computing the resulting *singular homology*, save for the fact that this homology is, blessedly, an efficient compression.

Theorem 3.19. *For a cell complex, singular and cellular homology are isomorphic.*

The proof of this is an induction argument based on the n-skeleton of X. The previous exercise establishes the isomorphism on the level of H_0. To induct to higher-dimensional skeleta requires a few steps just outside the bounds of these lectures: see [51, 58] for details.

Homotopy Invariance Why pass to the uncomputable singular theory? There is so much *room* in $\mathcal{C}^{\text{sing}}$ that it is easy to deform continuously and prove the core result.

Theorem 3.20. *Homology [singular] is a homotopy invariant of spaces.*

When combined with Theorem 3.19, we obtain a truly useful, computable result. The proof of Theorem 3.20 does not focus on spaces at all, but rather, in the spirit of these lectures, pulls back the notion of homotopy to complexes. Recall that $f, g\colon X \to Y$ are homotopic if there is a map $F\colon X \times [0,1] \to Y$ which restricts to f on $X \times \{0\}$ and to g on $X \times \{1\}$. A *chain homotopy* between chain maps $\varphi_\bullet, \psi_\bullet\colon \mathcal{C} \to \mathcal{C}'$ is a graded linear transformation $F\colon \mathcal{C} \to \mathcal{C}'$ sending n-chains to $(n+1)$-chains so that $\partial F - F\partial = \varphi_\bullet - \psi_\bullet$:

$$
\begin{array}{ccccccccc}
\cdots & \longrightarrow & C_{n+1} & \xrightarrow{\partial} & C_n & \xrightarrow{\partial} & C_{n-1} & \xrightarrow{\partial} & \cdots \\
 & \swarrow F & \psi_\bullet \big\downarrow \big\downarrow \varphi_\bullet & \swarrow F & \psi_\bullet \big\downarrow \big\downarrow \varphi_\bullet & \swarrow F & \psi_\bullet \big\downarrow \big\downarrow \varphi_\bullet & \swarrow F & \\
\cdots & \longrightarrow & C'_{n+1} & \xrightarrow[\partial]{} & C'_n & \xrightarrow[\partial]{} & C'_{n-1} & \xrightarrow[\partial]{} & \cdots
\end{array} \tag{3.21}
$$

One calls F a map of *degree* $+1$, indicating the upshift in the grading.[5] Note the morphological resemblance to homotopy of maps: a chain homotopy maps each n-chain to a $n+1$-chain, the algebraic analogue of a 1-parameter family. The difference between the ends of the homotopy, $\partial F - F\partial$, gives the difference between the chain maps.

Exercise 3.22. Show that two chain homotopic maps induce the same homomorphisms on homology. Start by considering $[\alpha] \in H_\bullet(\mathcal{C})$, assuming $\varphi_\bullet$ and $\psi_\bullet$ are chain homotopic maps from $\mathcal{C}$ to $\mathcal{C}'$.

The proof of Theorem 3.20 follows from constructing an explicit chain homotopy [58].

Morse Homology All the homology theories we have looked at so far have used simplices or cells as basis elements of chains and dimension as the grading. There is a wonderful homology theory that breaks this pattern in a creative and, eventually, useful manner. Let M be a smooth, finite-dimensional Riemannian

---

[5] The overuse of the term *degree* in graphs, maps of spheres, and chain complexes is unfortunate.

manifold. There is a homology theory based on a dynamical system on M. One chooses a function $h\colon M \to \mathbb{R}$ and considers the (negative) gradient flow of h on M — the smooth dynamical system given by $dx/dt = -\nabla h$.

The dynamics of this vector field are simple: solutions either are fixed points (critical points of h) or flow *downhill* from one fixed point to another. Let $\mathsf{Cr}(h)$ denote the set of critical points, and assume for the sake of simplicity that all such critical points are *nondegenerate* – the second derivative (or *Hessian*) is nondegenerate (has nonzero determinant) at these points. These nondegenerate critical points are the basis elements of a *Morse complex*. What is the grading?

Nondegenerate critical points have a natural grading – the number of negative eigenvalues of the Hessian of h at p. This is called the *Morse index*, $\mu(p)$, of $p \in \mathsf{Cr}(h)$ and has the more topological interpretation as the dimension of the set of points that converge to p in negative time. The Morse index measures how unstable a critical point is: minima have the lowest Morse index; maxima the highest. Balancing a three-legged stool on k legs leads to an index $\mu = 3 - k$ equilibrium.

One obtains the *Morse complex*, $\mathcal{C}^h = (MC_\bullet, \partial)$, with MC_k the vector space with basis $\{p \in \mathsf{Cr}(h)\,;\, \mu(p) = k\}$. The boundary maps encode the global flow of the gradient field: ∂_k counts (modulo 2 in the case of $\mathbb{F}_2$ coefficients) the number of *connecting orbits* – flowlines from a critical point with $\mu = k$ to a critical point with $\mu = k-1$. One hopes (or assumes) that this number is well-defined. The difficult business is to demonstrate that $\partial^2 = 0$: this involves careful analysis of the connecting orbits, as in, e.g., [8,84]. The use of $\mathbb{F}_2$ coefficients is highly recommended. The ensuing *Morse homology*, $MH_\bullet(h)$, captures information about M.

Theorem 3.23 (Morse Homology Theorem). *Fix a compact manifold* M *and a Morse function*$h\colon M \to \mathbb{R}$. *Then* $MH_\bullet(h;\mathbb{F}_2) \cong H_\bullet(M;\mathbb{F}_2)$, *independent of* h.

Exercise 3.24. Compute the Morse homology of a 2-sphere, $\mathbb{S}^2$, outfitted with a Morse function having two maxima and two minima. How many saddle points must it have?

Our perspective is that Morse homology is a *precompression* of the complex onto its *critical* elements, as measured by h.

Discrete Morse Theory As given, Morse homology would seem to be greatly disconnected from the data-centric applications and perspectives of this chapter — it uses smooth manifolds, smooth functions, smooth flows, and nondegenerate critical points, all within the delimited purview of smooth analysis. However, as with most of algebraic topology, the smooth theory is the continuous limit of a correlative discrete theory. In ordinary homology, the discrete [that is, simplicial] theory came first, followed by the limiting case of the singular homology theory. In the case of Morse theory, the smooth version came long before the following discretization, which first seems to have appeared in a mature form in the work of Forman [45,46]; see also the recent book of Kozlov [65].

Consider for concreteness a simplicial or cell complex X. The critical ingredient for Morse theory is *not* the Morse function but rather its gradient flow. A *discrete vector field* is a pairing V which partitions the cells of X (graded by dimension) into pairs $V_\alpha = (\sigma_\alpha \lessdot \tau_\alpha)$ where σ_α is a codimension-1 face of τ_α. All leftover cells of X not paired by V are the *critical cells* of V, $\mathrm{Cr}(V)$. A *discrete flowline* is a sequence (V_i) of distinct paired cells with codimension-1 faces, arranged so that

$$\overbrace{\sigma_1 \lessdot \tau_1}^{V_1} \gtrdot \overbrace{\sigma_2 \lessdot \tau_2}^{V_2} \gtrdot \cdots \gtrdot \overbrace{\sigma_N \lessdot \tau_N}^{V_N} . \tag{3.25}$$

A flowline is *periodic* if $\tau_N \gtrdot \sigma_1$ for $N > 1$. A *discrete gradient field* is a discrete vector field devoid of periodic flowlines.

The best approach is to lift everything to algebraic actions on the chain complex $\mathcal{C} = (C^{\mathrm{cell}}_\bullet, \partial)$ associated to the cell complex X. By linearity, the vector field V induces a chain map $V\colon C_k \to C_{k+1}$ induced by the pairs $\sigma \lessdot \tau$ – one visualizes an arrow from the face σ to the cell τ. As with classical Morse homology, $\mathbb{F}_2$ coefficients are simplest.

To every discrete gradient field we can associate a discrete Morse complex, $\mathcal{C}^V = (MC_\bullet, \tilde{\partial})$ with MC_k defined to bethe vector space with basis the critical cells $\{\sigma \in \mathrm{Cr}(V)\,;\, \dim(\sigma) = k\}$. Note that dimension plays the role of Morse index.

Exercise 3.26. Place several discrete gradient fields on a discretization of a circle and examine the critical cells. What do you notice about the number and dimension of critical cells? Does this make sense in light of the Euler characteristic of a circle?

The boundary maps $\tilde{\partial}_k$ count (modulo 2 in the case of $\mathbb{F}_2$ coefficients; with a complicated induced orientation else) the number of discrete flowlines from a critical simplex of dimension k to a critical simplex of dimension $k-1$. Specifically, given τ a critical k-simplex and σ a critical $(k-1)$-simplex, the contribution of $\tilde{\partial}_k(\tau)$ to σ is the number of gradient paths from a face of τ to a coface of σ. In the case that $\sigma \lessdot \tau$, then this number is 1, ensuring that the trivial V for which all cells are critical yields $\mathcal{C}^V$ the usual cellular chain complex. It is not too hard to show that $\tilde{\partial}^2 = 0$ and that, therefore, the homology $MH_\bullet(V) = H_\bullet(\mathcal{C}^V)$ is well-defined. As usual, the difficulty lies in getting orientations right for $\mathbb{Z}$ coefficients.

Theorem 3.27 ([45]). *For any discrete gradient field* V, $MH_\bullet(V) \cong H^{\mathrm{cell}}_\bullet(X)$.

Discrete Morse theory shows that the classical constraints – manifolds, smooth dynamics, nondegenerate critical points – are not necessary. This point is worthy of emphasis: the classical notion of a critical point (maximum, minimum, saddle) is distilled away from its analytic and dynamical origins until only the algebraic spirit remains.

Applications of discrete Morse theory are numerous and expansive, including to combinatorics [65], mesh simplification [67], image processing [79], configuration spaces of graphs [43, 44], and, most strikingly, efficient computation of homology of cell complexes [69]. This will be our focus for applications to computation at the end of this lecture.

Application: Algorithms

Advances in applications of homological invariants have been and will remain inextricably linked to advances in computational methods for such invariants. Recent history has shown that potential applications are impotent when divorced from computational advances, and computation of unmotivated quantities is futile. The reader who is interested in applying these methods to data is no doubt interested in knowing the best and easiest available software. Though this is not the right venue for a discussion of cutting-edge software, there are a number of existing software libraries/packages for computing homology and persistent homology of simplicial or cubical complexes, some of which are exciting and deep. As of the time of this writing, the most extensive and current benchmarking comparing available software packages can be found in the preprint of Otter et al. [75]. We remark on and summarize a few of the issues involved with computing [persistent] homology, in order to segue into how the theoretical content of this lecture impacts how software can be written.

Time complexity: Homology is known to be output-sensitive, meaning that the complexity of computing homology is a function of how large the homology is, as opposed to how large the complex is. What this means in practice is that the homology of a simple complex is simple to compute. The time-complexity of computing homology is, by output-sensitivity, difficult to specify tightly. The standard algorithm to compute $H_\bullet(X)$ for X a simplicial complex is to compute the *Smith normal form* of the graded boundary map $\partial : C_\bullet \to C_\bullet$, where we concatenate the various gradings into one large vector space. This graded boundary map is, by definition, nilpotent: it has a block structure with zero blocks on the block-diagonal (since $\partial_k : C_k \to C_{k-1}$) and is nonzero on the superdiagonal blocks. The algorithm for computing Smith normal form is really a slight variant of the ubiquitous Gaussian elimination, with reduction to the normal form via elementary row and column operations. For field coefficients in $\mathbb{F}_2$ this reduction is easily seen to be of time-complexity $O(n^3)$ in the size of the matrix, with an expected run time of $O(n^2)$. This is not encouraging, given the typical sizes seen in applications. Fortunately, compression preprocessing methods exist, as we will detail.

Memory and inputs: Time-complexity is not the only obstruction; holding a complex in memory is nontrivial, as is the problem of *inputting* a complex. A typical simplicial complex is specified by fixing the simplices as basis and then specifying the boundary matrices. For very large complexes, this is prohibitive and unnecessary, as the boundary matrices are typically sparse. There are a number of ways to reduce the input cost, including inputting (1) a distance matrix spelling out an explicit metric between points in a point cloud, using a persistent

Dowker complex (see Exercise 2.10) to build the filtered complex; (2) using voxels in a lattice as means of coordinatizing top-dimensional cubes in a cubical complex, and specifying the complex as a list of voxels; and (3) using the Vietoris-Rips complex of a network of nodes and edges, the specification of which requires only a quadratic number of bits of data as a function of nodes.

Exercise 3.28. To get an idea of how the size of a complex leads to an inefficient complexity bound, consider a single simplex, Δ^n, and cube, I^n, each of dimension n. How many total simplices/cubes are in each? Include all faces of all dimensions. Computing the [nearly trivial] homology of such a simple object requires, in principle, computing the Smith normal form of a graded boundary matrix of what net size?

Morse theory & Compression: One fruitful approach for addressing the computation of homology is to consider alternate intermediate compression schemes. If instead of applying Smith Normal Form directly to a graded boundary operator, one modifies the complex first to obtain a smaller chain-homotopic complex, then the resulting complexity bounds may collapse with a dramatic decrease in size of the input. There have been many proposals for reduction and coreduction of chain complexes that preserve homology: see [62] for examples. One clear and successful compression scheme comes from discrete Morse theory. If one puts a discrete gradient field on a cell complex, then the resulting Morse complex is smaller and potentially much smaller, being generated only by critical cells. The process of defining and constructing an associated discrete Morse complex is roughly linear in the size of the cell complex [69] and thus gives an efficient approach to homology computation. This has been implemented in the popular software package Perseus (see [75]).

Modernizing Morse Theory: Morse homology, especially the discrete version, has not yet been fully unfolded. There are several revolutionary approaches to Morse theory that incorporate tools outside the bounds of these lectures. Nevertheless, it is the opinion of this author that we are just realizing the full picture of the centrality of Morse theory in Mathematics and in homological algebra in particular. Two recent developments are worth pointing out as breakthroughs in conceptual frameworks with potentially large impact. The first, in the papers by Nanda et al. [71,72], gives a categorical reworking of discrete Morse that relaxes the notion of a discrete vector field to allow for any acyclic pairing of cells and faces without restriction on the dimension of the face. It furthermore shows how to reconstruct the topology of the original complex (up to homotopy type, not homology type) using only data about critical cells and the critical discrete flowlines. Though the tools used are formidable (2-categories and localization), the results are equally strong.

Matroid Morse Theory: The second contribution on the cusp of impact comes in the thesis of Henselman [59] which proposes *matroid theory* as the missing link between Morse theory and homological algebra. *Matroids* are classical structures in the intersection of combinatorial topology and linear algebra and have no end of interesting applications in optimization theory. Henselman recasts discrete Morse theory and persistent homology both in terms of matroids, then exploits matroid-theoretic principles (rank, modularity, minimal bases) in order to generate efficient algorithms for computing persistent homology and barcodes. This work has already led to an initial software package Eirene[6] that, as of the time of this writing, has computed persistent H_k for $0 \leqslant k \leqslant 7$ of a filtered 8-dimensional simplicial complex obtained as the Vietoris-Rips complex of a random sampling of 50 points in dimension 20 with a total of $3.3E + 11$ simplices on a PC laptop with i7 quadcore and 16meg RAM in 11.1 seconds with peak memory use of 1.8GB. This computation compares very favorably with the fastest-available software, Ripser[7], on a cluster of machines, as recorded in the survey of Otter et al. [75]: Ripser computes the persistent homology of this complex in 349 seconds with a peak memory load of 24.7GB. This portends much more to come, both at the level of conceptual understanding and computational capability.

This prompts the theme of our next lecture, that in order to prepare for increased applicability, one must ascend and enfold tools and perspectives of increasing generality and power.

Lecture 4: Higher Order

Having developed the basics of topological data analysis, we focus now on the theories and principles to which these point.

Cohomology and Duality

One of the first broad generalizations of all we have described in these lectures is the theory of *cohomology*, an algebraic dual of homology. There are many ways to approach cohomology — dual spaces, Morse theory, differential forms, and configuration spaces all provide useful perspectives in this subject. These lectures will take the low-tech approach most suitable for a first-pass. We have previously considered a general chain complex $\mathcal{C}$ to be a graded sequence of vector spaces $C_\bullet$ with linear transformations $\partial_k \colon C_k \to C_{k-1}$ satisfying $\partial^2 = \partial_{k-1}\partial_k = 0$. These chain complexes are typically graded over the naturals $\mathbb{N}$, and any such complex compresses to its homology, $H_\bullet(\mathcal{C})$, preserving homological features and forgetting all extraneous data.

In a chain complex, the boundary maps descend in the grading. If that grading is tied to dimension or local complexity of an assembly substructure, then the boundary maps encode how more-complex objects are related, attached, or projected to their less-complex components. Though this is a natural data structure in many contexts, there are instances in which one knows instead how objects are

[6] Available at gregoryhenselman.org/eirene.

[7] Available at https://github.com/Ripser/ripser.

related to larger superstructures rather than smaller substructures. This prompts the investigation of *cochain complexes*. For purposes of these lectures, a cochain complex is a sequence $\mathcal{C} = (C^\bullet, d^\bullet)$ of vector spaces and linear transformations which increment the grading ($d^k \colon C^k \to C^{k+1}$) and satisfy the complex condition ($d^{k+1}d^k = 0$). One uses subscripts for chain complexes and superscripts for cochain complexes. The *cohomology* of the cochain complex is the complex $H^\bullet(\mathcal{C}) = \ker d/\operatorname{im} d$ consisting of *cocycles* equivalent up to *coboundaries*.

The simplest example of a cochain complex comes from dualizing a chain complex. Given a chain complex (C_k, ∂, k) of $\mathbb{F}$-vector spaces, define $C^k = C_k^\vee$, the vector space of linear functionals $C_k \to \mathbb{F}$. The coboundary d^k is then the adjoint (*de facto*, transpose) of the boundary ∂_{k+1}, so that

$$d \circ d = \partial^\vee \circ \partial^\vee = (\partial \circ \partial)^\vee = 0^\vee = 0. \tag{4.1}$$

In the case of a simplicial complex, the standard simplicial cochain complex is precisely such a dual to the simplicial chain complex. The coboundary operator d is explicit: the coboundary of a functional on k-simplices, $f \in C^k$, acts as $(df)(\tau) = f(\partial\tau)$. For σ a k-simplex, d implicates the *cofaces* – those $(k+1)$-simplices τ having σ as a face.

Dualizing chain complexes in this manner leads to a variety of cohomology theories mirroring the many homology theories of the previous section: simplicial, cellular, singular, Morse, Čech, and other cohomology theories follow.

Exercise 4.2. Fix a triangulated disc $\mathbb{D}^2$ and consider cochains using $\mathbb{F}_2$ coefficients. What do 1-cocycles look like? Show that any such 1-cocycle is the coboundary of a 0-cochain which labels vertices with 0 and 1 *on the left* and *on the right* of the 1-cocycle, so to speak: this is what a trivial class in $H^1(\mathbb{D}^2)$ looks like. Now fix a circle S^1 discretized as a finite graph and construct examples of 1-cocycles that are (1) coboundaries; and (2) nonvanishing in H^1. What is the difference between the trivial and nontrivial cocycles on a circle?

The previous exercise foreshadows the initially depressing truth: nothing new is gained by computing cohomology, in the sense that $H^n(X)$ and $H_n(X)$ have the same dimension for each n. Recall, however, that there is more to co/homology than just the Betti numbers. Functoriality is key, and there is a fundamental difference in how homology and cohomology transform.

Exercise 4.3. Fix $f \colon X \to Y$ a simplicial map of simplicial complexes, and consider the simplicial cochain complexes $\mathcal{C}(X)$ and $\mathcal{C}(Y)$. We recall that the induced chain map $f_\bullet$ yields a well-defined induced homomorphism on homology $H(f) \colon H_\bullet(\mathcal{C}(X)) \to H_\bullet(\mathcal{C}(Y))$. Using what you know about adjoints, show that the induced homomorphism on cohomology is also well-defined but *reverses direction:* $H^\bullet(f) \colon H^\bullet(Y) \to H^\bullet(X)$. This allows one to lift cohomology cocycles from the codomain to the domain.

Alexander Duality There are numerous means by which duality expresses itself in the form of cohomology. One of the most useful and ubiquitous of these is known as *Alexander duality*, which relates the homology and cohomology of a subset of a sphere $\mathbb{S}^n$ (or, with a puncture, $\mathbb{R}^n$) and its complement. The following is a particularly simple form of that duality theorem.

Theorem 4.4 (Alexander Duality). *Let $A \subset \mathbb{S}^n$ be compact, nonempty, proper, and locally-contractible. There is an isomorphism*

$$\text{(4.5)} \qquad AD\colon \tilde{H}_k(\mathbb{S}^n - A) \xrightarrow{\cong} \tilde{H}^{n-k-1}(A).$$

Note that the reduced theory is used for both homology and cohomology.

Cohomology and Calculus Most students initially view cohomology as more obtuse than homology; however, there are certain instances in which cohomology is the most natural operation. Perhaps the most familiar such setting comes from calculus. As seen in Exercise 3.3 from Lecture 2, the familiar constructs of vector calculus on $\mathbb{R}^3$ fit into an exact complex. This exactness reflects the fact that $\mathbb{R}^3$ is topologically trivial [contractible]. Later, in Exercise 4.2, you looked at simplicial 1-cocycles and hopefully noticed that whether or not they are null in H^1 depends on whether or not these cochains are simplicial *gradients* of 0-chains on the vertex set. These exercises together hint at the strong relationship between cohomology and calculus.

The use of gradient, curl, and divergence for vector calculus is, however, an unfortunate vestige of the philosophy of calculus-for-physics as opposed to a more modern calculus-for-data sensibility. A slight modern update sets the stage better for cohomology. For $U \subset \mathbb{R}^n$ an open set, let $\Omega^k(U)$ denote the differentiable k-form fields on U (a smooth choice of multilinear antisymmetric functionals on ordered k-tuples of tangent vectors at each point). For example, Ω^0 consists of smooth functionals, Ω^1 consists of 1-form fields, viewable (in a Euclidean setting) as duals to vector fields, Ω^n consists of signed densities on U times the volume form, and $\Omega^{k>n}(U) = 0$. There is a natural extension of differentiation (familiar from implicit differentiation in calculus class) that gives a coboundary map $d : \Omega^k \to \Omega^{k+1}$, yielding the *deRham complex*,

$$\text{(4.6)} \qquad 0 \longrightarrow \Omega^0(U) \xrightarrow{d} \Omega^1(U) \xrightarrow{d} \Omega^2(U) \xrightarrow{d} \cdots \xrightarrow{d} \Omega^n(U) \longrightarrow 0\,.$$

As one would hope, $d^2 = 0$, in this case due to the fact that mixed partial derivatives commute: you worked this out explicitly in Exercise 3.3. The resulting cohomology of this complex, the *deRham cohomology* $H^\bullet(U)$, is isomorphic to the singular cohomology of U using $\mathbb{R}$ coefficients.

This overlap between calculus and cohomology is neither coincidental nor concluded with this brief example. A slightly deeper foray leads to an examination of the Laplacian operator (on a manifold with some geometric structure). The well-known *Hodge decomposition theorem* then gives, among other things, an isomorphism between the cohomology of the manifold and the *harmonic* differential forms (those in the kernel of the Laplacian). For more information on these connections, see [19].

What is especially satisfying is that the calculus approach to cohomology and the deRham theory feeds back to the simplicial: one can export the Laplacian and the Hodge decomposition theorem to the cellular world (see [51, Ch. 6]). This, then, impacts data-centric problems of ranking and more over networks.

Cohomology and Ranking Cohomology arises in a surprising number of different contexts. One natural example that follows easily from the calculus-based perspective on cohomology lives in certain Escherian optical illusions, such as impossible tribars, eternally cyclic waterfalls, or neverending stairs. When one looks at an Escher staircase, the drawn perspective is locally realizable – one can construct a local perspective function. – but a global extension cannot be defined. Thus, an Escherlike loop is really a non-zero class in H^1 (as first pointed out by Penrose [78]).

This is not disconnected from issues of data. Consider the problem of ranking. One simple example that evokes nontrivial 1-cocycles is the popular game of *Rock, Paper, Scissors*, for which there are local but not global ranking functions. A local gradient of *rock-beats-scissors* does not extend to a global gradient. Perhaps this is why customers are asked to conduct rankings (*e.g.*, *Netflix* movie rankings or *Amazon* book rankings) as a 0-cochain (*"how many stars?"*), and not as a 1-cochain (*"which-of-these-two-is-better?"*): nontrivial H^1 is, in this setting, undesirable. The *Condorcet paradox* – that locally consistent comparative rankings can lead to global inconsistencies – is an appearance of H^1 in ranking theory.

There are less frivolous examples of precisely this type of application, leveraging the language of gradients and curls to realize cocycle obstructions to perfect rankings in systems. The paper of Jiang et al. [61] interprets the simplicial cochain complex of the clique/flag complex of a network in terms of rankings. For example, the ($\mathbb{R}$-valued) 0-cochains are interpreted as numerical *score* functions on the nodes of the network; the 1-cochains (supported on edges) are interpreted as pairwise preference rankings (with oriented edges and positive/negative values determining which is preferred over the other); and the higher-dimensional cochains represent more sophisticated local orderings of nodes in a clique [simplex]. They then resort to the calculus-based language of grad, curl, and div to build up the cochain complex and infer from its cohomology information about existence and nonexistence of compatible ranking schemes over the network. Their use of the Laplacian and the Hodge decomposition theorem permits projection of noisy or inconsistent ranking schemes onto the nearest consistent ranking.

There are more sophisticated variants of these ideas, with applications passing beyond finding consistent rankings or orderings. Recent work of Gao et al. [49] gives a cohomological and Hodge-theoretic approach to synchronization problems over networks based on pairwise nodal data in the presence of noise. Singer and collaborators [85, 86] have published several works on cryo electron microscopy that is, in essence, a cohomological approach to finding consistent solutions to pairwise-compared data over a network. The larger lesson to be inferred from these types of results is that networks often support data above and beyond what is captured by the network topology alone (nodes, edges). This data blends with the algebra and topology of the system using the language of cohomology. It is this perspective of data that lives above a network that propels our next set of tools.

Cellular Sheaves

One of the most natural uses for cohomology comes in the form of a yet-more-abstract theory that is the stated end of these lectures: sheaf cohomology. Our perspective is that a sheaf is an algebraic data structure tethered to a space (generally) or simplicial complex (in particular). In keeping with the computational and linear-algebraic focus of this series, we will couch everything in the language of linear algebra. The more general approach [20, 64, 83] is much more general.

Fix X a simplicial (or regular cell) complex with $\trianglelefteq$ denoting the face relation: $\sigma\trianglelefteq\tau$ if and only if $\sigma \subset \overline{\tau}$. A *cellular sheaf* over X, $\mathcal{F}$, is generated by (1) an assignment to each simplex σ of X a *stalk*, a vector space $\mathcal{F}(\sigma)$; and (2) to each face pair $\sigma\trianglelefteq\tau$ a *restriction map*, a linear transformation $\mathcal{F}(\sigma\trianglelefteq\tau)\colon \mathcal{F}(\sigma) \to \mathcal{F}(\tau)$. This data must respect that manner in which the simplicial complex is assembled, meaning that faces of faces satisfy the composition rule:

$$\rho \trianglelefteq \sigma \trianglelefteq \tau \;\Rightarrow\; \mathcal{F}(\rho\trianglelefteq\tau) = \mathcal{F}(\sigma\trianglelefteq\tau) \circ \mathcal{F}(\rho\trianglelefteq\sigma). \tag{4.7}$$

The *trivial* face $\tau\trianglelefteq\tau$ by default has the identity isomorphism $\mathcal{F}(\tau\trianglelefteq\tau) = \mathsf{Id}$ as its restriction map. Again, if one thinks of the stalks as the data over the individual simplices, then, in the same manner that the simplicial complex is glued up by face maps, the sheaf is assembled by the system of linear transformations.

One simple example of a sheaf on a cell complex X is that of the *constant sheaf*, $\mathbb{F}_X$, taking values in vector spaces over a field $\mathbb{F}$. This sheaf assigns $\mathbb{F}$ to every cell and the identity map $\mathsf{Id}\colon \mathbb{F} \to \mathbb{F}$ to every face $\sigma\trianglelefteq\tau$. In contrast, the *skyscraper sheaf* over a single cell σ of X is the sheaf $\mathbb{F}_\sigma$ that assigns $\mathbb{F}$ to σ and 0 to all other cells and face maps.

Exercise 4.8. Consider the following version of a random rank-1 sheaf over a simplicial complex X. Assign the field $\mathbb{F}$ to every simplex. To each face map $\sigma\trianglelefteq\tau$ assign either Id or 0 according to some (your favorite) random process. Does this always give you a sheaf? How does this depend on X? What is the minimal set of assumptions you would need to make on either X or the random assignment in order to guarantee that what you get is in fact a sheaf?

One thinks of the values of the sheaf over cells as being data and the restriction maps as something like local constraints or relationships between data. It's very worthwhile to think of a sheaf as programmable – one has a great deal of freedom in encoding local relationships. For example, consider the simple linear recurrence $u_{n+1} = A_n u_n$, where $u_n \in \mathbb{R}^k$ is a vector of states and A_n is a k-by-k real matrix. Such a discrete-time dynamical system can be represented as a sheaf $\mathcal{F}$ of states over the time-line $\mathbb{R}$ with the cell structure on $\mathbb{R}$ having $\mathbb{Z}$ as vertices, where $\mathcal{F}$ has constant stalks $\mathbb{R}^k$. One programs the dynamics of the recurrence relation as follows: $\mathcal{F}(\{n\} \trianglelefteq (n, n+1))$ is the map $u \mapsto A_n u$ and $\mathcal{F}(\{n+1\} \trianglelefteq (n, n+1))$ is the identity. Compatibility of local solutions over the sheaf is, precisely, the condition for being a global solution to the dynamics.

Local and Global Sections One says that the sheaf is *generated* by its values on individual simplices of X: this stalk $\mathcal{F}(\tau)$ over a cell τ is also called the *local sections* of $\mathcal{F}$ on τ: one writes $s_\tau \in \mathcal{F}(\tau)$ for a local section over τ. Though the sheaf is generated by local sections, there is more to a sheaf than its generating data, just as there is more to a vector space than its basis. The restriction maps of a sheaf encode how local sections can be continued into larger sections. One glues together local sections by means of the restriction maps. The value of the sheaf $\mathcal{F}$ on all of X is defined to be those collections of local sections that *continue* according to the restriction maps on faces. The *global sections* of $\mathcal{F}$ on X are defined as:

$$\mathcal{F}(X) = \{(s_\tau)_{\tau \in X} : s_\sigma = \mathcal{F}(\rho \trianglelefteq \sigma)(s_\rho) \ \forall \rho \trianglelefteq \sigma\} \subset \prod_\tau \mathcal{F}(\tau) . \tag{4.9}$$

Exercise 4.10. Show that in the example of a sheaf for the recurrence relation $u_{n+1} = A_n u_n$, the global solutions to this dynamical system are classified by the global sections of the sheaf.

The observed fact that the value of the sheaf over all of X retains the same sort of structure as the type of data over the vertices — say, a vector space over a field $\mathbb{F}$ — is a hint that this space of global solutions is really a type of homological data. In fact, it is cohomological in nature, and, like zero-dimensional cohomology, it is measure of *connected components* of the sheaf.

Cellular Sheaf Cohomology

In the simple setting of a compact cell complex X, it is easy to define a cochain complex based on a sheaf $\mathcal{F}$ on X. Let $C^n(X; \mathcal{F})$ be the product of $\mathcal{F}(\sigma)$ over all n-cells σ of X. These cochains are connected by coboundary maps as follows:

$$0 \longrightarrow \prod_{\dim \sigma = 0} \mathcal{F}(\sigma) \xrightarrow{d} \prod_{\dim \sigma = 1} \mathcal{F}(\sigma) \xrightarrow{d} \prod_{\dim \sigma = 2} \mathcal{F}(\sigma) \xrightarrow{d} \cdots , \tag{4.11}$$

where the coboundary map d is defined on sections over cells using the sheaf restriction maps

$$d(s_\sigma) = \sum_{\sigma \trianglelefteq \tau} [\sigma \colon \tau] \, \mathcal{F}(\sigma \trianglelefteq \tau) s_\sigma , \tag{4.12}$$

where, for a regular cell complex, $[\sigma \colon \tau]$ is either zero or ± 1 depending on the orientation of the simplices involved (beginners may start with all vector spaces using binary coefficients so that $-1 = 1$). Note that $d\colon C^n(X;\mathcal{F}) \to C^{n+1}(X;\mathcal{F})$, since $[\sigma\colon \tau] = 0$ unless σ is a codimension-1 face of τ. This gives a cochain complex: in the computation of d^2, the incidence numbers factor from the restriction maps, and the computation from cellular co/homology suffices to yield 0. The resulting *cellular sheaf cohomology* is denoted $H^\bullet(X;\mathcal{F})$.

This idea of *global compatibility* of sets of local data in a sheaf yield, through the language of cohomology, global qualitative features of the data structure. We have seen several examples of the utility of classifying various types of holes or large-scale qualitative features of a space or complex. Imagine what one can do with a measure of topological features of a data structure *over* a space.

Exercise 4.13. The cohomology of the constant sheaf $\mathbb{F}_X$ on a compact cell complex X is, clearly, $H^\bullet_{\text{cell}}(X;\mathbb{F})$, the usual cellular cohomology of X with coefficients in $\mathbb{F}$. Why the need for compactness? Consider the following cell complex: $X = \mathbb{R}$, decomposed into two vertices and three edges. What happens when you follow all the above steps for the cochain complex of $\mathbb{F}_X$? Show that this problem is solved if you include in the cochain complex only contributions from compact cells.

Exercise 4.14. For a closed subcomplex $A \subset X$, define the constant sheaf over A as, roughly speaking, the constant sheaf on A (as its own complex) with all other cells and face maps in X having data zero. Argue that $H^\bullet(X;\mathbb{F}_A) \cong H^\bullet(A;\mathbb{F})$. Conclude that it is possible to have a contractible base space X with nontrivial sheaf cohomology.

The elements of linear algebra recur throughout topology, including sheaf cohomology. Consider the following sheaf $\mathcal{F}$ over the closed interval with two vertices, a and b, and one edge e. The stalks are given as $\mathcal{F}(a) = \mathbb{R}^m$, $\mathcal{F}(b) = 0$, and $\mathcal{F}(e) = \mathbb{R}^n$. The restriction maps are $\mathcal{F}(b \trianglelefteq e) = 0$ and $\mathcal{F}(a \trianglelefteq e) = A$, where A is a linear transformation. Then, by definition, the sheaf cohomology is $H^0 \cong \ker A$ and $H^1 \cong \operatorname{coker} A$.

Cellular sheaf cohomology taking values in vector spaces is really a characterization of solutions to complex networks of linear equations. If one modifies $\mathcal{F}(b) = \mathbb{R}^p$ with $\mathcal{F}(b \trianglelefteq e) = B$ another linear transformation, then the cochain complex takes the form

$$0 \longrightarrow \mathbb{R}^m \times \mathbb{R}^p \xrightarrow{[A|-B]} \mathbb{R}^n \longrightarrow 0 \longrightarrow \cdots, \tag{4.15}$$

where $d = [A|-B] : \mathbb{R}^{m+p} \to \mathbb{R}^n$ is augmentation of A by $-B$. The zeroth sheaf cohomology H^0 is precisely the set of solutions to the equation $Ax = By$, for $x \in \mathbb{R}^m$ and $y \in \mathbb{R}^p$. These are the global sections over the closed edge. The first sheaf cohomology measures the degree to which $Ax - By$ does not span $\mathbb{R}^n$. All higher sheaf cohomology groups vanish.

Exercise 4.16. Prove that sheaf cohomology of a cell complex in grading zero classifies global sections: $H^0(X;\mathcal{F}) = \mathcal{F}(X)$.

Cosheaves Sheaves are meant for cohomology: the direction of the restriction maps insures this. Is there a way to talk about sheaf homology? If one works in the cellular case, this is a simple process. As we have seen that the only real difference between the cohomology of a cochain complex and the homology of a chain complex is whether the grading ascends or descends, a simple matter of arrow reversal on a sheaf should take care of things. It does. A *cosheaf* $\hat{\mathcal{F}}$ of vector spaces on a simplicial complex assigns (1) to each simplex σ a *costalk*, a vector space $\hat{\mathcal{F}}(\sigma)$; and (2) to each face $\sigma \trianglelefteq \tau$ of τ a *corestriction map*, a linear transformation $\hat{\mathcal{F}}(\sigma \trianglelefteq \tau)\colon \hat{\mathcal{F}}(\tau) \to \hat{\mathcal{F}}(\sigma)$ that reverses the direction of the sheaf maps. Of course, the cosheaf must respect the composition rule:

$$\rho \trianglelefteq \sigma \trianglelefteq \tau \ \Rightarrow\ \hat{\mathcal{F}}(\rho \trianglelefteq \tau) = \hat{\mathcal{F}}(\rho \trianglelefteq \sigma) \circ \hat{\mathcal{F}}(\sigma \trianglelefteq \tau), \tag{4.17}$$

and the identity rule that $\hat{\mathcal{F}}(\tau \trianglelefteq \tau) = \mathsf{Id}$.

In the cellular context, there are very few differences between sheaves and cosheaves — the use of one over another is a matter of convenience, in terms of which direction makes the most sense. This is by no means true in the more subtle setting of sheaves and cosheaves over open sets in a continuous domain.

Splines and Béziers. Cosheaves and sheaves alike arise in the study of splines, Bézier curves, and other piecewise-assembled structures. For example, a single segment of a planar Bézier curve is specified by the locations of two endpoints, along with additional *control points*, each of which may be interpreted as a *handle* specifying tangency data of the resulting curve at each endpoint. The reader who has used any modern drawing software will understand the control that these handles give over the resulting smooth curve. Most programs use a cubic Bézier curve in the plane – the image of the unit closed interval by a cubic polynomial. In these programs, the specification of the endpoints and the endpoint handles (tangent vectors) completely determines the interior curve segment uniquely.

This can be viewed from the perspective of a cosheaf $\hat{\mathcal{F}}$ over the closed interval $I = [0,1]$. The costalk over the interior $(0,1)$ is the space of all cubic polynomials from $[0,1] \to \mathbb{R}^2$, which is isomorphic to $\mathbb{R}^4 \oplus \mathbb{R}^4$ (one cubic polynomial for each of the x and y coordinates). If one sets the costalks at the endpoints of $[0,1]$ to be $\mathbb{R}^2$, the physical locations of the endpoints, then the obvious corestriction maps to the endpoint costalks are nothing more than evaluation at 0 and 1 respectively. The corresponding cosheaf chain complex is:

$$\cdots \longrightarrow 0 \longrightarrow \mathbb{R}^4 \oplus \mathbb{R}^4 \xrightarrow{\ \partial\ } \mathbb{R}^2 \oplus \mathbb{R}^2 \longrightarrow 0. \tag{4.18}$$

Here, the boundary operator ∂ computes how far the cubic polynomial (edge costalk) 'misses' the specified endpoints (vertex costalks).

Exercise 4.19. Show that for this simple cosheaf, $H_0 = 0$ and $H_1 \cong \mathbb{R}^2 \oplus \mathbb{R}^2$. Interpret this as demonstrating that there are four degrees of freedom available for a cubic planar Bézier curve with fixed endpoints: these degrees of freedom are captured precisely by the pair of handles, each of which is specified by a (planar) tangent vector. Repeat this exercise for a 2-segment cubic planar Bézier curve. How many control points are needed and with what degrees of freedom are they needed in order to match the H_1 of the cosheaf?

Note the interesting duality: the global solutions with boundary condition are characterized by the top-dimensional homology of the cosheaf, instead of the zero-dimensional cohomology of a sheaf. This simple example extends greatly, as shown originally by Billera (using cosheaves, without that terminology [16]) and Yuzvinsky (using sheaves [90]). By Billera's work, the (vector) space of splines over a triangulated Euclidean domain is isomorphic to the top-dimensional homology of a particular cosheaf over the domain. This matches what you see in the simpler example of a Bézier curve over a line segment.

Splines and Béziers are a nice set of examples of cosheaves that have natural higher-dimensional generalizations — Bézier surfaces and surface splines are used in design and modelling of surfaces ranging from architectural structures to vehicle surfaces, ship hulls, and the like. Other examples of sheaves over higher-dimensional spaces arise in the broad generalization of the Euler characteristic to the *Euler calculus*, a topological integral calculus of recent interest in topological signal processing applications [9, 10, 80–82].

Towards Generalizing Barcodes One of the benefits of a more general, sophisticated language is the ability to reinterpret previous results in a new light with new avenues for exploration appearing naturally. Let's wrap up our brief survey of sheaves and cosheaves by revisiting the basics of persistent homology, following the thesis of Curry [33]. Recall the presentation of persistent homology and barcodes in Lecture 2 that relied crucially on the Structure Theorem for linear sequences of finite-dimensional vector spaces (Theorem 2.15).

There are a few ways one might want to expand this story. We have hinted on a few occasions at the desirability of a continuous line as a parameter: our story of sequences of vector spaces and linear transformations is bound to the discrete setting. Intuitively, one could take a limit of finer discretizations and hope to obtain a convergence with the appropriate assumptions on variability. Questions of stability and interleaving (recall Exercise 2.23) then arise: see [18, 21, 66]

Another natural question is: what about non-linear sequences? What if instead of a single parameter, there are two or more parameters that one wants to vary? Is it possible to classify higher-dimensional sequences and derive barcodes here? Unfortunately, the situation is much more complex than in the simple, linear setting. There are fundamental algebraic reasons for why such a classification is not directly possible. These obstructions originate from representation theory and *quivers*: see [24, 28, 76]. The good news is that quiver theory implies the existence of barcodes for linear sequences of vector spaces where the directions of the maps do not have to be uniform, as per the *zigzag persistence* of Carlsson and de Silva [24]. The bad news is that quiver theory implies that a well-defined barcode cannot exist as presently conceived for any sequence that is not a *Dynkin diagram* (meaning, in particular, that higher-dimensional persistence has no simple classification).

Nevertheless, one intuits that sheaves and cosheaves should have some bearing on persistence and barcodes. Consider the classical scenario, in which one has a sequence of finite-dimensional vector spaces V_i and linear transformations $\varphi_i \colon V_i \to V_{i+1}$. Consider the following sheaf $\mathcal{F}$ over a $\mathbb{Z}$-discretized $\mathbb{R}$. To each vertex $\{i\}$ is assigned V_i. To each edge $(i, i+1)$ is assigned V_{i+1} with an identity isomorphism from the right vertex stalk to the edge and φ_i as the left-vertex restriction map to the edge data. Note the similarity of this sheaf to that of the recurrence relation earlier in this lecture. As in the case of the recurrence relation, H^0 detects global solutions: something similar happens for intervals in the barcode.

Exercise 4.20. Recall that persistent homology of a persistence complex is really a homology that is attached to an interval in the parameter line. Given the sheaf $\mathcal{F}$ associated to $\mathbb{R}$ as above, and an interval I subcomplex of $\mathbb{R}$, let $\mathcal{F}_I$ be the restriction of the sheaf to the interval, following Exercise 4.14. Prove that the number of bars in the barcode over I is the dimension of $H^0(I; \mathcal{F}_I)$. Can you argue what changes in $\mathcal{F}$ could be made to preserve this result in the case of a sequence of linear transformations φ_i that do not all "go the same direction"? Can you adapt this construction to a collection of vector spaces and linear transformations over an arbitrary *poset* (partially-ordered set)? Be careful, there are some complications.

There are limits to what basic sheaves and cosheaves can do, as cohomology does not come with the descriptiveness plus uniqueness that the representation theoretic approach gives. Nevertheless, there are certain settings in which barcodes for persistent homology are completely captured by sheaves and cosheaves (see the thesis of Curry for the case of level-set persistence [33]), with more characterizations to come [34].

Homological Data, Redux We summarize by updating and expanding the principles that we outlined earlier in the lecture series into a more refined language:

(1) Algebraic co/chain complexes are a good model for converting a space built from local pieces into a linear-algebraic structure.
(2) Co/homology is an optimal compression scheme to collapse inessential structure and retain qualitative features.
(3) The classification of linear algebraic sequences yields barcodes as a decomposition of sequences of co/homologies, capturing the evolution of qualitative features.
(4) Exact sequences permit inference from partial sequential co/homological data to more global characterization.

(5) A variety of different co/homology theories exist, adapted to different types of structures on a space or complex, with functoriality being the tool for relating (and, usually, equating) these different theories.
(6) Sheaves and cosheaves are algebraic data structures over spaces that can be programmed to encode local constraints.
(7) The concomitant sheaf cohomologies [& cosheaf homologies] compress these data structures down to their qualitative core, integrating local data into global features.
(8) Classifications of sheaves and cosheaves recapitulates the classification of co/chain complexes into barcodes, but presage a much broader and more applicable theory in the making.

Application: Sensing and Evasion

Most of the examples of cellular sheaves given thus far have been simplistic, for pedagogical purposes. This lecture ends with an example of what one can do with a more interesting sheaf to solve a nontrivial inference problem. This example is chosen to put together as many pieces as possible of the things we have learned, including homology theories, persistence, cohomology, sheaves, computation, and more. It is, as a result, a bit complicated, and this survey will be highly abbreviated: see [52] for full details.

Consider the following type of *evasion game*, in which an evader tries to hide from a pursuer, and *capture* is determined by being "seen" or "sensed". For concreteness, the game takes place in a Euclidean space $\mathbb{R}^n$ and time progresses over the reals. At each time $t \in \mathbb{R}$, the observer sees or senses a *coverage region* $C_t \subset \mathbb{R}^n$ that is assumed to (1) be connected; and (2) include the region outside a fixed ball (to preclude the evader from running away off to infinity). The *evasion problem* is this: given the coverage regions over time, is it possible for there to be an *evasion path*: a continuous map $e\colon \{t\} \mapsto (\mathbb{R}^n - C_t)$ on all of the timeline $t \in \mathbb{R}$. Such a map is a *section* of the projection $p\colon (\mathbb{R}^n \times \mathbb{R}) - C \to \mathbb{R}$ from the complement of the full coverage region $C \subset \mathbb{R}^n \times \mathbb{R}$ to the timeline.

What makes this problem difficult (i.e., interesting) is that the geometry and topology of the complement of the coverage region, where the evader can hide, is not known: were this known, graph-theoretic methods would handily assist in finding a section or determining nonexistence. Furthermore, the coverage region C is not known geometrically, but rather topologically, with unknown embedding. The thesis of Adams [3] gives examples of two different time-dependent coverage regions, C and C', whose fibers are topologically the same (homotopic) for each t, but which differ in the existence of evasion path. The core difficulty is that, though C and C' each admit "tunnels" in their complements stretching over the entire timeline, one of them has a tunnel that snakes backwards along the time axis: topologically, legal; physically, illegal.

Work of Adams and Carlsson [3] gives a complete solution to the existence of an evasion path in the case of a planar ($n = 2$) system with additional genericity conditions and some geometric assumptions. Recently, a complete solution in all dimensions was given [52] using sheaves and sheaf cohomology. One begins with a closed coverage region $C \subset \mathbb{R}^n \times \mathbb{R}$ whose complement is uniformly bounded over $\mathbb{R}$. For the sake of exposition, assume that the time axis is given a discretization into (ordered) vertices v_i and edges $e_i = (v_i, v_{i+1})$ such that the coverage domains C_t are topologically equivalent over each edge (this is not strictly necessary). There are a few simple sheaves over the discretized timeline relevant to the evasion problem.

First, consider for each time $t \in \mathbb{R}$, the coverage domain $C_t \subset \mathbb{R}^n$. How many different ways are there for an evader to hide from C_t? This is regulated by the number of connected components of the complement, classified by $H_0(\mathbb{R}^n - C_t)$. Since we do not have access to C_t directly (remember – its embedding in $\mathbb{R}^n$ is unknown to the pursuer), we must try to compute this H_0 based only on the topology of C_t. That this can be done is an obvious but wonderful corollary of *Alexander duality*, which relates the homology and cohomology of complementary subsets of $\mathbb{R}^n$. Here, Alexander duality implies that $H_0(\mathbb{R}^n - C_t) \cong H^{n-1}(C_t)$: this, then, is something we can measure, and motivates using the Leray cellular sheaf $\mathcal{H}$ of $n-1$ dimensional cohomology of the coverage regions over the time axis. Specifically, for each edge e_i define $\mathcal{H}(e_i) = H^{n-1}(C_{(v_i, v_{i+1})})$ to be the cohomology of the region over the open edge. For the vertices, use the star: $\mathcal{H}(v_i) = H^{n-1}(C_{(v_{i-1}, v_{i+1})})$.

Exercise 4.21. Why are the stalks over the vertices defined in this way? Show that this gives a well-defined cellular sheaf using as restriction maps the induced homomorphisms on cohomology. *Hint:* which way do the induced maps in cohomology go?

The intuition is that global sections of this sheaf over the time axis, $H^0(\mathbb{R}; \mathcal{H})$, would classify the different complementary "tunnels" through the coverage set that an evader could use to escape detection. Unfortunately, this is incorrect, for reasons pointed out by Adams and Carlsson [3] (using the language of zigzags). The culprit is the commutative nature of homology and cohomology — one cannot discern tunnels which illegally twirl backwards in time. To solve this problem, one *could* try to keep track of some sort of directedness or orientation. Thanks to the assumption that C_t is connected for all time, there is a global orientation class on $\mathbb{R}^n$ that can be used to assign a ± 1 to basis elements of $H^{n-1}(C_t)$ based on whether the complementary tunnel is participating in a time-orientation-preserving evasion path on the time interval $(-\infty, t)$.

However, to incorporate this orientation data into the sheaf requires breaking the bounds of working with vector spaces. As detailed in [52], one may use sheaves that take values in *semigroups*. In this particular case, the semigroups are *positive cones* within vector spaces, where a *cone* $K \subset V$ in a vector space V is a subset closed under vector addition and closed under multiplication by $\mathbb{R}^+$, the [strictly] positive reals. A cone is *positive* if $K \cap -K = \emptyset$. With work, one can formulate co/homology theories and sheaves to take values in cones: for details, see [52]. The story proceeds: within the sheaf $\mathcal{H}$ of $n-1$ dimensional cohomology of C on $\mathbb{R}$, there is (via abuse of terminology) a "subsheaf" $^+\mathcal{H}$ of positive cones, meaning that the stalks of $^+\mathcal{H}$ are positive cones within the stalks of $\mathcal{H}$, encoding all the *positive* cohomology classes that can participate in a legal (time-orientation-respecting) evasion path. It is this sheaf of positive cones that classifies evasion paths.

Theorem 4.22. *For $n > 1$ and $C = \{C_t\} \subset \mathbb{R}^n \times \mathbb{R}$ closed and with bounded complement consisting of connected fibers C_t for all t, there is an evasion path over $\mathbb{R}$ if and only if $H^0(\mathbb{R}; {}^+\mathcal{H})$ is nonempty.*

Note that the theorem statement says *nonempty* instead of *nonzero*, since sheaf takes values in positive cones, which are $\mathbb{R}^+$-cones and thus do not contain zero. The most interesting part of the story is how one can compute this H^0. This is where the technicalities begin to weigh heavily, as one cannot use the classical definition of H^0 in terms of kernels and images. The congenial commutative world of vector spaces requires significant care when passing to the nonabelian setting. One defines H^0 for sheaves of cones using constructs from category theory (*limits*, specifically). Computation of such objects requires a great deal more thought and care than the simpler linear-algebraic notions of these lectures. That is by no means a defect; indeed, it is a harbinger. Increase in resolution requires an increase in algebra.

Conclusion: Beyond Linear Algebra

These lecture notes have approached topological data analysis from the perspective of homological algebra. If the reader takes from these notes the singular idea that linear algebra can be enriched to cover not merely linear transformations, but also sequences of linear transformations that form complexes, then these lectures will have served their intended purpose. The reader for whom this seems to open a new world will be delighted indeed to learn that the vector space version of homological algebra is almost too pedestrian to matter to mathematicians: as hinted at in the evader inference example above, the story begins in earnest when one works with rings, modules, and more interesting categories still. Nevertheless, for applications to data, vector spaces and linear transformations are a safe place to start.

For additional material that aligns with the author's view on topology and homological algebra, the books [50, 51] are recommended. It is noted that the perspective of these lecture notes is idiosyncratic. For a broader view, the interested reader is encouraged to consult the growing literature on topological data analysis. The book by Edelsbrunner and Harer [40] is a gentle introduction, with emphases on both theory and algorithms. The book on computational homology by Kaczynsky, Mischaikow, and Mrozek [62] has even more algorithmic material, mostly in the setting of cubical complexes. Both books suffer from the short shelf life of algorithms as compared to theories. Newer titles on the theory of persistent homology are in the process of appearing: that of Oudot [76] is one of perhaps several soon to come. For introductions to topology specifically geared towards the data sciences, there does not seem to be an ideal book; rather, a selection of survey articles such as that of Carlsson [23] is appropriate.

The open directions for inquiry are perhaps too many to identify properly – the subject of topological data analysis is in its infancy. One can say with certainty that there is a slow spread of these topological methods and perspectives to new application domains. This will continue, and it is not yet clear to this author whether neuroscience, genetics, signal processing, materials science, or some other domain will be the locus of inquiry that benefits most from homological methods, so rapid has been the advances in all these areas. Dual to this spread in applications is the antipodal expansion into Mathematics, as ideas from homological data engage with impact contemporary Mathematics. The unique demands of data have already prompted explorations into mathematical structures (e.g., interleaving distance in sheaf theory and persistence in matroid theory) which otherwise would seem unmotivated and be unexplored. It is to be hoped that the simple applications of representation theory to homological data analysis will inspire deeper explorations into the mathematical tools.

There is at the moment a frenzy of activity surrounding various notions of stability associated to persistent homology, sheaves, and related structures concerning representation theory. It is likely to take some time to sort things out into their clearest form. In general, one expects that applications of deeper ideas from algebraic topology exist and will percolate through applied mathematics into application domains. Perhaps the point of most optimism and uncertainty lies in the intersection with probabilistic and stochastic methods. The rest of this volume makes ample use of such tools; their absence in these notes is noticeable. Topology and probability are neither antithetical nor natural partners; expectation of progress is warranted from some excellent results on the topology of Gaussian random fields [6] and recent work on the homology of random complexes [63]. Much of the material from these lectures appears ripe for a merger with modern probabilistic methods. Courage and optimism — two of the cardinal mathematical virtues — are needed for this.

References

[1] A. Abrams and R. Ghrist. State complexes for metamorphic systems. *Intl. J. Robotics Research*, 23(7,8):809–824, 2004. ←278

[2] H. Adams and G. Carlsson. On the nonlinear statistics of range image patches. *SIAM J. Imaging Sci.*, 2(1):110–117, 2009. MR2486524 ←294

[3] H. Adams and G. Carlsson. Evasion paths in mobile sensor networks. *International Journal of Robotics Research*, 34:90–104, 2014. ←318, 319
[4] R. J. Adler. *The Geometry of Random Fields*. Society for Industrial and Applied Mathematics, 1981. MR3396215 ←280
[5] R. J. Adler, O. Bobrowski, M. S. Borman, E. Subag, and S. Weinberger. Persistent homology for random fields and complexes. In *Borrowing Strength: Theory Powering Applications*, pages 124–143. IMS Collections, 2010. MR2798515 ←285
[6] R. J. Adler and J. E. Taylor. *Random Fields and Geometry*. Springer Monographs in Mathematics. Springer, New York, 2007. MR2319516 ←280, 321
[7] R. H. Atkin. *Combinatorial Connectivities in Social Systems*. Springer Basel AG, 1977. ←278
[8] A. Banyaga and D. Hurtubise. *Morse Homology*. Springer, 2004. ←304
[9] Y. Baryshnikov and R. Ghrist. Target enumeration via Euler characteristic integrals. *SIAM J. Appl. Math.*, 70(3):825–844, 2009. MR2538627 ←280, 316
[10] Y. Baryshnikov and R. Ghrist. Euler integration over definable functions. *Proc. Natl. Acad. Sci. USA*, 107(21):9525–9530, 2010. MR2653583 ←280, 316
[11] Y. Baryshnikov, R. Ghrist, and D. Lipsky. Inversion of Euler integral transforms with applications to sensor data. *Inverse Problems*, 27(12), 2011. MR2854317 ←280
[12] U. Bauer and M. Lesnick. Induced matchings and the algebraic stability of persistence barcodes. *Discrete Comput. Geom.*, 6(2):162–191, 2015. MR3333456 ←293
[13] P. Bendich, H. Edelsbrunner, and M. Kerber. Computing robustness and persistence for images. *IEEE Trans. Visual and Comput. Graphics*, pages 1251–1260, 2010. ←294
[14] P. Bendich, J. Marron, E. Miller, A. Pielcoh, and S. Skwerer. Persistent homology analysis of brain artery trees. *Ann. Appl. Stat.*, 10(1):198–218, 2016. MR3480493 ←294
[15] S. Bhattacharya, R. Ghrist, and V. Kumar. Persistent homology for path planning in uncertain environments. *IEEE Trans. on Robotics*, 31(3):578–590, 2015. ←295
[16] L. J. Billera. Homology of smooth splines: generic triangulations and a conjecture of Strang. *Trans. Amer. Math. Soc.*, 310(1):325–340, 1988. MR965757 ←316
[17] L. J. Billera, S. P. Holmes, and K. Vogtmann. Geometry of the space of phylogenetic trees. *Adv. in Appl. Math.*, 27(4):733–767, 2001. MR1867931 ←278
[18] M. Botnan and M. Lesnick. Algebraic stability of zigzag persistence modules. arXiv:160400655v2. ←293, 316
[19] R. Bott and L. Tu. *Differential Forms in Algebraic Topology*. Springer, 1982. MR658304 ←311
[20] G. Bredon. *Sheaf Theory*. Springer, 1997. MR1481706 ←312
[21] P. Bubenik, V. de Silva, and J. Scott. Metrics for generalized persistence modules. *Found. Comput. Math.*, 15(6):1501–1531, 2015. MR3413628 ←293, 316
[22] P. Bubenik and J. A. Scott. Categorification of persistent homology. *Discrete Comput. Geom.*, 51(3):600–627, 2014. MR3201246 ←293
[23] G. Carlsson. The shape of data. In *Foundations of computational mathematics, Budapest 2011*, volume 403 of *London Math. Soc. Lecture Note Ser.*, pages 16–44. Cambridge Univ. Press, Cambridge, 2013. MR3137632 ←321
[24] G. Carlsson and V. de Silva. Zigzag persistence. *Found. Comput. Math.*, 10(4):367–405, 2010. MR2657946 ←316
[25] G. Carlsson, T. Ishkhanov, V. de Silva, and A. Zomorodian. On the local behavior of spaces of natural images. *Intl. J. Computer Vision*, 76(1):1–12, Jan. 2008. MR3715451 ←294
[26] G. Carlsson and F. Mémoli. Characterization, stability and convergence of hierarchical clustering methods. *J. Mach. Learn. Res.*, 11:1425–1470, Aug. 2010. MR2645457 ←294
[27] G. Carlsson and F. Mémoli. Classifying clustering schemes. *Found. Comput. Math.*, 13(2):221–252, 2013. MR3032681 ←294
[28] G. Carlsson and A. Zomorodian. The theory of multidimensional persistence. *Discrete Comput. Geom.*, 42(1):71–93, 2009. MR2506738 ←316
[29] S. Carson, V. Ruta, L. Abbott, and R. Axel. Random convergence of olfactory inputs in the *drosophila* mushroom body. *Nature*, 497(7447):113–117, 2013. ←285
[30] F. Chazal, V. de Silva, M. Glisse, and S. Oudot. *The Structure and Stability of Persistence Modules*. Springer Briefs in Mathematics, 2016. MR3524869 ←293
[31] D. Cohen-Steiner, H. Edelsbrunner, and J. Harer. Stability of persistence diagrams. *Discrete Comput. Geom.*, 37(1):103–120, 2007. MR2279866 ←293

[32] A. Collins, A. Zomorodian, G. Carlsson, and L. Guibas. A barcode shape descriptor for curve point cloud data. In M. Alexa and S. Rusinkiewicz, editors, *Eurographics Symposium on Point-Based Graphics*, ETH, Zürich, Switzerland, 2004. ←292

[33] J. Curry. *Sheaves, Cosheaves and Applications*. PhD thesis, University of Pennsylvania, 2014. MR3259939 ←316, 317

[34] J. Curry and A. Patel. Classification of constructible cosheaves. arXiv:1603.01587. ←317

[35] C. Curto and V. Itskov. Cell groups reveal structure of stimulus space. *PLoS Comput. Biol.*, 4(10):e1000205, 13, 2008. MR2457124 ←285

[36] M. d'Amico, P. Frosini, and C. Landi. Optimal matching between reduced size functions. Technical Report 35, DISMI, Univ. degli Studi di Modena e Reggio Emilia, Italy, 2003. ←292

[37] V. de Silva and G. Carlsson. Topological estimation using witness complexes. In M. Alexa and S. Rusinkiewicz, editors, *Eurographics Symposium on Point-based Graphics*, 2004. ←278

[38] J. Derenick, A. Speranzon, and R. Ghrist. Homological sensing for mobile robot localization. In *Proc. Intl. Conf. Robotics & Aut.*, 2012. ←296

[39] C. Dowker. Homology groups of relations. *Annals of Mathematics*, pages 84–95, 1952. MR0048030 ←278

[40] H. Edelsbrunner and J. Harer. *Computational Topology: an Introduction*. American Mathematical Society, Providence, RI, 2010. MR2572029 ←275, 321

[41] H. Edelsbrunner, D. Letscher, and A. Zomorodian. Topological persistence and simplification. *Discrete Comput. Geom.*, 28:511–533, 2002. MR1949898 ←292

[42] M. Farber. *Invitation to Topological Robotics*. Zurich Lectures in Advanced Mathematics. European Mathematical Society (EMS), Zürich, 2008. MR2455573 ←280

[43] D. Farley and L. Sabalka. On the cohomology rings of tree braid groups. *J. Pure Appl. Algebra*, 212(1):53–71, 2008. MR2355034 ←305

[44] D. Farley and L. Sabalka. Presentations of graph braid groups. *Forum Math.*, 24(4):827–859, 2012. MR2949126 ←305

[45] R. Forman. Morse theory for cell complexes. *Adv. Math.*, 134(1):90–145, 1998. MR1612391 ←304, 305

[46] R. Forman. A user's guide to discrete Morse theory. *Sém. Lothar. Combin.*, 48, 2002. MR1939695 ←304

[47] Ś. R. Gal. Euler characteristic of the configuration space of a complex. *Colloq. Math.*, 89(1):61–67, 2001. MR1853415 ←280

[48] M. Gameiro, Y. Hiraoka, S. Izumi, M. Kramar, K. Mischaikow, and V. Nanda. Topological measurement of protein compressibility via persistent diagrams. *Japan J. Industrial & Applied Mathematics*, 32(1):1–17, Oct 2014. MR3318898 ←296

[49] T. Gao, J. Brodzki, and S. Mukherjee. The geometry of synchronization problems and learning group actions. arXiv:1610.09051. ←311

[50] S. I. Gelfand and Y. I. Manin. *Methods of Homological Algebra*. Springer Monographs in Mathematics. Springer-Verlag, Berlin, second edition, 2003. MR1950475 ←299, 320

[51] R. Ghrist. *Elementary Applied Topology*. Createspace, 1.0 edition, 2014. ←275, 278, 282, 303, 311, 320

[52] R. Ghrist and S. Krishnan. Positive Alexander duality for pursuit and evasion. To appear, *SIAM J. Appl. Alg. Geom.* MR3763757 ←318, 319, 320

[53] R. Ghrist and S. M. Lavalle. Nonpositive curvature and Pareto optimal coordination of robots. *SIAM J. Control Optim.*, 45(5):1697–1713, 2006. MR2272162 ←278

[54] R. Ghrist, D. Lipsky, J. Derenick, and A. Speranzon. Topological landmark-based navigation and mapping. ←Preprint, 2012. 278, 296

[55] R. Ghrist and V. Peterson. The geometry and topology of reconfiguration. *Adv. in Appl. Math.*, 38(3):302–323, 2007. MR2301699 ←278

[56] C. Giusti, E. Pastalkova, C. Curto, and V. Itskov. Clique topology reveal intrinsic structure in neural connections. *Proc. Natl. Acad. Sci. USA*, 112(44):13455–13460, 2015. MR3429279 ←284, 285

[57] L. J. Guibas and S. Y. Oudot. Reconstruction using witness complexes. In *Proc. 18th ACM-SIAM Sympos. on Discrete Algorithms*, pages 1076–1085, 2007. MR2485259 ←278

[58] A. Hatcher. *Algebraic Topology*. Cambridge University Press, 2002. MR1867354 ←281, 282, 299, 300, 303

[59] G. Henselman and R. Ghrist. Matroid filtrations and computational persistent homology. arXiv:1606.00199. ←307

[60] J. Hocking and G. Young. *Topology*. Dover Press, 1988. MR1016814 ←302
[61] X. Jiang, L.-H. Lim, Y. Yao, and Y. Ye. Statistical ranking and combinatorial Hodge theory. *Math. Program.*, 127(1, Ser. B):203–244, 2011. MR2776715 ←311
[62] T. Kaczynski, K. Mischaikow, and M. Mrozek. *Computational Homology*, volume 157 of *Applied Mathematical Sciences*. Springer-Verlag, New York, 2004. MR2028588 ←275, 278, 307, 321
[63] M. Kahle. Topology of random clique complexes. *Discrete Comput. Geom.*, 45(3):553–573, 2011. MR2770552 ←285, 321
[64] M. Kashiwara and P. Schapira. *Categories and Sheaves*, volume 332 of *Grundlehren der Mathematischen Wissenschaften*. Springer-Verlag, 2006. MR2182076 ←312
[65] D. Kozlov. *Combinatorial Algebraic Topology*, volume 21 of *Algorithms and Computation in Mathematics*. Springer, 2008. MR2361455 ←304, 305
[66] M. Lesnick. The theory of the interleaving distance on multidimensional persistence modules. *J. Found. Comp. Math.*, 15(3):613–650, 2015. MR3348168 ←316
[67] T. Lewiner, H. Lopes, and G. Tavares. Applications of Forman's discrete Morse theory to topology visualization and mesh compression. *IEEE Trans. Visualization & Comput. Graphics*, 10(5):499–508, 2004. ←305
[68] M. Minsky and S. Papert. *Perceptrons: An Introduction to Computational Geometry*. MIT Press, 1987. ←280
[69] K. Mischaikow and V. Nanda. Morse theory for filtrations and efficient computation of persistent homology. *Discrete Comput. Geom.*, 50(2):330–353, 2013. MR3090522 ←306, 307
[70] J. Munkres. *Topology*. Prentice Hall, 2000. MR3728284 ←275
[71] V. Nanda. Discrete Morse theory and localization. arXiv:1510.01907. ←307
[72] V. Nanda, D. Tamaki, and K. Tanaka. Discrete Morse theory and classifying spaces. arXiv:1612.08429vi. ←307
[73] M. Nicolau, A. J. Levine, and G. Carlsson. Topology based data analysis identifies a subgroup of breast cancers with a unique mutational profile and excellent survival. *Proc. Natl. Acad. Sci. USA*, 108(17):7265–7270, 2011. ←295
[74] J. O'Keefe and J. Dostrovsky. The hippocampes as a spatial map. *Brain Research*, 34(1):171–175, 1971. ←285
[75] N. Otter, M. Porter, U. Tillmann, P. Grindrod, and H. Harrington. A roadmap for the computation of persistent homology. *EPJ Data Science* 6(17), 2017. MR3670641 ←306, 307, 308
[76] S. Oudot. *Persistence Theory: From Quiver Representations to Data Analysis*. American Mathematical Society, 2015. MR3408277 ←275, 316, 321
[77] L. Pachter and B. Sturmfels. The mathematics of phylogenomics. *SIAM Rev.*, 49(1):3–31, 2007. MR2302545 ←278
[78] R. Penrose. La cohomologie des figures impossibles. *Structural Topology*, 17:11–16, 1991. MR1140400 ←311
[79] V. Robins, P. Wood, and A. Sheppard. Theory and algorithms for constructing discrete Morse complexes from grayscale digital images. *IEEE Transactions on Pattern Analysis and Machine Intelligence*, 33(8):1646–1658, 2011. ←294, 305
[80] M. Robinson. *Topological Signal Processing*. Springer, Heidelberg, 2014. MR3157249 ←316
[81] P. Schapira. Operations on constructible functions. *J. Pure Appl. Algebra*, 72(1):83–93, 1991. MR1115569 ←316
[82] P. Schapira. Tomography of constructible functions. In *Applied Algebra, Algebraic Algorithms and Error-Correcting Codes*, pages 427–435. Springer, 1995. MR1448182 ←316
[83] J. Schürmann. *Topology of Singular Spaces and Constructible Sheaves*, volume 63 of *Mathematics Institute of the Polish Academy of Sciences. Mathematical Monographs (New Series)*. Birkhäuser Verlag, Basel, 2003. MR2031639 ←312
[84] M. Schwarz. *Morse Homology*, volume 111 of *Progress in Mathematics*. Birkhäuser Verlag, Basel, 1993. MR1239174 ←304
[85] Y. Shkolnisky and A. Singer. Viewing direction estimation in cryo-EM using synchronization. *SIAM J. Imaging Sci.*, 5(3):1088–1110, 2012. MR3022188 ←311
[86] A. Singer. Angular synchronization by eigenvectors and semidefinite programming. *Appl. Comput. Harmonic Anal.*, 30(1):20–36, 2011. MR2737931 ←311
[87] A. Sizemore, C. Giusti, and D. Bassett. Classification of weighted networks through mesoscale homological features. *J. Complex Networks*, 2016. MR3801686 ←285

[88] B. Torres, J. Oliviera, A. Tate, P. Rath, K. Cumnock, and D. Schneider. Tracking resilience to infections by mapping disease space. *PLOS Biology*, 2016. ←295

[89] A. Wilkerson, H. Chintakunta, H. Krim, T. Moore, and A. Swami. A distributed collapse of a network's dimensionality. In *Proceedings of Global Conference on Signal and Information Processing (GlobalSIP)*. IEEE, 2013. ←278

[90] S. Yuzvinsky. Modules of splines on polyhedral complexes. *Math. Z.*, 210(2):245–254, 1992. MR1166523 ←316

[91] A. Zomorodian. *Topology for Computing*. Cambridge Univ Press, 2005. MR2111929 ←292

Departments of Mathematics and Electrical & Systems Engineering, University of Pennsylvania
Email address: ghrist@math.upenn.edu

25 **Michael W. Mahoney, John C. Duchi, and Anna C. Gilbert, Editors,** The Mathematics of Data, 2018

24 **Roman Bezrukavnikov, Alexander Braverman, and Zhiwei Yun, Editors,** Geometry of Moduli Spaces and Representation Theory, 2017

23 **Mark J. Bowick, David Kinderlehrer, Govind Menon, and Charles Radin, Editors,** Mathematics and Materials, 2017

22 **Hubert L. Bray, Greg Galloway, Rafe Mazzeo, and Natasa Sesum, Editors,** Geometric Analysis, 2016

21 **Mladen Bestvina, Michah Sageev, and Karen Vogtmann, Editors,** Geometric Group Theory, 2014

20 **Benson Farb, Richard Hain, and Eduard Looijenga, Editors,** Moduli Spaces of Riemann Surfaces, 2013

19 **Hongkai Zhao, Editor,** Mathematics in Image Processing, 2013

18 **Cristian Popescu, Karl Rubin, and Alice Silverberg, Editors,** Arithmetic of L-functions, 2011

17 **Jeffery McNeal and Mircea Mustaţă, Editors,** Analytic and Algebraic Geometry, 2010

16 **Scott Sheffield and Thomas Spencer, Editors,** Statistical Mechanics, 2009

15 **Tomasz S. Mrowka and Peter S. Ozsváth, Editors,** Low Dimensional Topology, 2009

14 **Mark A. Lewis, Mark A. J. Chaplain, James P. Keener, and Philip K. Maini, Editors,** Mathematical Biology, 2009

13 **Ezra Miller, Victor Reiner, and Bernd Sturmfels, Editors,** Geometric Combinatorics, 2007

12 **Peter Sarnak and Freydoon Shahidi, Editors,** Automorphic Forms and Applications, 2007

11 **Daniel S. Freed, David R. Morrison, and Isadore Singer, Editors,** Quantum Field Theory, Supersymmetry, and Enumerative Geometry, 2006

10 **Steven Rudich and Avi Wigderson, Editors,** Computational Complexity Theory, 2004

9 **Brian Conrad and Karl Rubin, Editors,** Arithmetic Algebraic Geometry, 2001

8 **Jeffrey Adams and David Vogan, Editors,** Representation Theory of Lie Groups, 2000

7 **Yakov Eliashberg and Lisa Traynor, Editors,** Symplectic Geometry and Topology, 1999

6 **Elton P. Hsu and S. R. S. Varadhan, Editors,** Probability Theory and Applications, 1999

5 **Luis Caffarelli and Weinan E, Editors,** Hyperbolic Equations and Frequency Interactions, 1999

4 **Robert Friedman and John W. Morgan, Editors,** Gauge Theory and the Topology of Four-Manifolds, 1998

3 **János Kollár, Editor,** Complex Algebraic Geometry, 1997

2 **Robert Hardt and Michael Wolf, Editors,** Nonlinear partial differential equations in differential geometry, 1996

1 **Daniel S. Freed and Karen K. Uhlenbeck, Editors,** Geometry and Quantum Field Theory, 1995